Runs and Patterns in Probability

Mathematics and Its Applications

Volume 283

Runs and Patterns in Probability: Selected Papers

edited by

Anant P. Godbole
Michigan Technological University,
Houghton, Michigan, U.S.A.

and

Stavros G. Papastavridis
University of Athens,
Athens, Greece

KLUWER ACADEMIC PUBLISHERS
DORDRECHT / BOSTON / LONDON

A C.I.P. Catalogue record for this book is available from the Library of Congress.

ISBN-13: 978-1-4613-3637-2 e-ISBN-13: 978-1-4613-3635-8
DOI: 10.1007/978-1-4613-3635-8

Published by Kluwer Academic Publishers,
P.O. Box 17, 3300 AA Dordrecht, The Netherlands.

Kluwer Academic Publishers incorporates
the publishing programmes of
D. Reidel, Martinus Nijhoff, Dr W. Junk and MTP Press.

Sold and distributed in the U.S.A. and Canada
by Kluwer Academic Publishers,
101 Philip Drive, Norwell, MA 02061, U.S.A.

In all other countries, sold and distributed
by Kluwer Academic Publishers Group,
P.O. Box 322, 3300 AH Dordrecht, The Netherlands.

Printed on acid-free paper

To my nieces Julia-Constantina and Natassa
SGP

To Kalpana, Ramona and Maya
APG

Contents

PART 2. RELIABILITY THEORY AND STATISTICS

PART 3. LIMIT THEOREMS AND APPROXIMATIONS

Contributing Authors

Marilyn A. Agin, School of Statistics, University of Minnesota, Minneapolis, Minnesota 55455, U.S.A.

Hae-Il Ahn, Department of Industrial and Manufacturing Systems Engineering, Iowa State University, Ames, Iowa 50011, U.S.A.

Demetris L. Antzoulakos, Department of Mathematics, University of Patras, 261 10 Patras, GREECE

Dragan Banjević, Department of Statistics, University of Toronto, 100 St. George Street, Toronto, Ontario, CANADA M5S 1A1

Charalambos A. Charalambides, Department of Mathematics, Section of Statistics and Operations Research, University of Athens, Panepistemiopolis, 157 84 Athens, GREECE

Rong-Jaye Chen, Department of Computer Science and Information Engineering, National Chiao-Tung University, Hsinchu 300, TAIWAN

Ourania Chryssaphinou, Department of Mathematics, Section of Statistics and Operations Research, University of Athens, Panepistemiopolis, 157 84 Athens, GREECE

Wenchang Chu, Università degli Studi di Roma "LA SAPIENZA", Dipartimento di Matematica "GUIDO CASTELNUOVO", Piazzale Aldo Moro, 2, I-00185 Roma, ITALIA

Antonia Földes, Department of Mathematics, College of Staten Island, CUNY, 2800 Victory Boulevard, Staten Island, New York 10314, U.S.A.

Anant P. Godbole, Department of Mathematical Sciences, Michigan Technological University, Houghton, Michigan 49931, U.S.A.

William S. Griffith, Department of Statistics, University of Kentucky, Lexington, Kentucky 40506, U.S.A.

Frank K. Hwang, AT&T Bell Laboratories, 600 Mountain Avenue, Murray Hill, New Jersey 07059, U.S.A.

Sanjay Khunger, Software Engineering Division, Laboratory Equipment Corporation, St. Joseph, Michigan 49085, U.S.A.

Markos Koutras, Department of Mathematics, Section of Statistics and Operations Research, University of Athens, Panepistemiopolis, 157 84 Athens, GREECE

Way Kuo, Department of Industrial Engineering, Texas A&M University, College Station, Texas 77843, U.S.A.

Frosso S. Makri, Department of Mathematics, University of Patras, 261 10 Patras, GREECE

Tamás F. Móri, Department of Probability Theory and Statistics, Eötvös Loránd University, Múzeum krt 6-8, Budapest H-1088, HUNGARY

Peter C. O'Brien, Biostatistics Division, Mayo Clinic, Rochester, Minnesota 55905, U.S.A.

Esther Ososanya, Department of Electrical Engineering, Tennessee Technological University, Cookeville, Tennessee 38505, U.S.A.

G. K. Papadopoulos, Department of Mathematics, University of Athens, Panepistemiopolis, 157 84 Athens, GREECE

Stavros G. Papastavridis, Department of Mathematics, University of Athens, Panepistemiopolis, 157 84 Athens, GREECE

Andreas N. Philippou, Department of Mathematics, University of Patras, 261 10 Patras, GREECE

Arthur O. Pittenger, Office of the Provost, University of Maryland Baltimore County, Baltimore, Maryland 21228, U.S.A.

Eugene F. Schuster, Department of Mathematical Sciences, University of Texas, El Paso, Texas 79968, U.S.A.

Kenneth A. Suman, Department of Mathematics and Statistics, Winona State University, Winona, Minnesota 55987, U.S.A.

Theothosios Tsapelas, Department of Mathematics, University of Athens, Panepistemiopolis, 157 84 Athens, GREECE

Jer-Shyan Wu, Department of Computer Science and Information Engineering, National Chiao-Tung University, Hsinchu 300, TAIWAN

Hao Zhang, Department of Statistics, The Wharton School, University of Pennsylvania, Philadelphia, Pennsylvania 19104, U.S.A.

Preface

The Probability Theory of Patterns and Runs has had a long and distinguished history, starting with the work of de Moivre in the 18th century and that of von Mises in the early 1920's, and continuing with the renewal-theoretic results in Feller's classic text *An Introduction to Probability Theory and its Applications, Volume 1.* It is worthwhile to note, in particular, that de Moivre, in the third edition of *The Doctrine of Chances* (1756, reprinted by Chelsea in 1967, pp. 254-259), provides the generating function for the waiting time for the appearance of k consecutive successes. During the 1940's, statisticians such as Mood, Wolfowitz, David and Mosteller studied the distribution theory, both exact and asymptotic, of run-related statistics, thereby laying the foundation for several exact run tests. In the last two decades or so, the theory has seen an impressive re-emergence, primarily due to important developments in Molecular Biology, but also due to related research thrusts in Reliability Theory, Distribution Theory, Combinatorics, and Statistics.

In November 1990, one of the editors of this volume (APG) organized a Special Session entitled *The Probability Theory of Patterns and Runs* at the 861st Meeting of the American Mathematical Society (AMS) in Denton, Texas. The idea of preparing a volume of papers reflecting the unique nature of the invited talks given at that Session grew out of discussions between the editors. We hasten to add, however, that the present project should not be considered to be a "Proceedings" of the above Session; just six of the twenty papers in this volume have any bearing on the work presented at Denton. On the contrary, we made it our mission to significantly enhance the breadth as well as the depth of the contributions, by soliciting papers from leading researchers from across the world, many of whom were unable to attend the AMS Meeting.

The probabilistic analysis of patterns and runs plays a critical role in fields as diverse as network reliability, the distribution theory of kth-order random variables, combinatorial probability, nucleotide sequence analysis, and randomness testing. The twenty papers in this volume have been written by a collection of research workers who are experts in these (and other) fields. The probabilistic and combinatorial methods employed by them include the Stein-Chen technique of Poisson approximation, "generatingfunctionology", Markov chain methods, the "independence principle", exponential probability inequalities, renewal theory, exchangeability, and hypergeometric evaluation.

The volume has been divided into three sections, entitled *Combinatorial Probability*, *Reliability Theory and Statistics*, and *Limit Theorems and Approximations.* Each of the contributions in this volume contains original unpublished research; several of the papers are, however, expository in nature - and can be used as starting points for new investigations. For this reason, we feel that the volume will be particularly useful to persons who wish to enter this exciting area of research, be they graduate students or established investigators.

ANANT P. GODBOLE and STAVROS G. PAPASTAVRIDIS
Houghton and Athens
February 5, 1994

Acknowledgements

This project has benefited from the assistance and encouragement of several people. We would like to thank Brad Burch, Tricia Frantti, Wei Guo, George Hansen, Duhong Liao, Chad Lukkarila, Gomathi Sadhasivan, Steven Sy, Siddharth Takkella, Rebecca Townsend and Qiang Wang for their excellent job in typesetting and/or proofreading various parts of the volume. The secretarial assistance of Marge Shannette was invaluable.

The Editors served as referees of several of the papers in the volume. The majority of articles in this collection were, however, refereed by Ch. Charalambides, O. Chryssaphinou, A. Dallas, Ch. Damianou, S. Graham, S. Kounias, M. Koutras, A. Kyriakoussis *and* by various anonymous referees. We extend our thanks to all of these people for their valuable service.

John Martindale, our editor at Kluwer, deserves a special thanks for his encouragement and patience.

Our work was supported in part by National Science Foundation Grants DMS-9100829 and DMS-9200409.

EXPRESSIONS IN TERMS OF BINOMIAL COEFFICIENTS FOR SOME MULTIVARIATE DISTRIBUTIONS OF ORDER K

Demetris L. Antzoulakos and Andreas N. Philippou

University of Patras

Useful binomial expressions are obtained for several multivariate distributions of order k. Two of these expressions continue to be new even when they are specialized to the respective univariate distributions of the same order, while another two generalize to the multivariate case corresponding univariate results.

MULTIVARIATE DISTRIBUTIONS OF ORDER K; NEGATIVE BINOMIAL; POISSON; LOGARITHMIC SERIES; TYPE I; TYPE II; RUNS; BINOMIAL EXPRESSIONS.

1. Introduction

Philippou et al. (1988) introduced and studied multivariate distributions of order k, among which we presently deal with the type I and type II negative binomial and logarithmic series distributions of order k, as well as the Poisson (type I) distribution of the same order. These distributions constitute appropriate multivariate analogs of respective univariate distributions of order k of Philippou (1983) and (1988), Philippou et al. (1983) and Aki et al. (1984), for some of which useful binomial expressions have been obtained by Charalambides (1986), Philippou and Georghiou (1989) and Godbole (1990). See, also, Philippou (1986) and Uppuluri and Patil (1983).

Motivated by and using some ideas of Godbole (1990) and Philippou and Muwafi (1982), we presently derive useful expressions in terms of binomial coefficients for the above mentioned five multivariate distributions of order k (see Definitions 1-5 and Theorems 1-5). As direct corollaries of these theorems, we obtain two new results for univariate distributions of order k (see Corollaries 2 and 5), and we rederive two

A. P. Godbole and S. G. Papastavridis (eds.), Runs and Patterns in Probability, 1–14.

of the main results of Godbole (1990) (see Corollaries 1 and 3).

First we obtain some preliminary results in Section 2. Specifically, under certain assumptions, we derive the probability that in n independent trials with $m+1$ possible outcomes the length of the longest run of a specific outcome is less than k (see Lemma 1 and Lemma 2), and we give precise upper and lower bounds on the sum of the solutions of the diophantine equation $x_1 + 2x_2 + \ldots + kx_k = x$ (see Lemma 3).

In order to avoid unnecessary repetitions, we mention here that in this paper $x_{11}, \ldots, x_{mk}$ are non-negative integers as specified. Moreover, whenever sums and products are taken over i and j, ranging, respectively, from 1 to m and from 1 to k, we shall omit these limits for notational simplicity.

2. Preliminary results

In our first lemma we derive the probability that the length of the longest run of a specific outcome of an experiment with $m+1$ possible outcomes is less than a fixed number. The probability is given in terms of multinomial coefficients and the proof is based on the methodology of Philippou and Muwafi (1982).

Lemma 1. *Consider a sequence of n (≥ 1) independent trials with possible outcomes $S, F_1, \ldots, F_m$, where $q_i = P(F_i)$ ($1 \leq i \leq m$) and $p = P(S)$ ($0 < q_i < 1$ for $1 \leq i \leq m$, $0 < q_1 + \ldots + q_m < 1$ and $p + q_1 + \ldots + q_m = 1$). Let*

(i) Y_i ($1 \leq i \leq m$) be r.v.'s denoting the number of outcomes F_i,

(ii) Z_i ($1 \leq i \leq m$) be r.v.'s denoting the total number of outcomes S which are before outcomes F_i,

(iii) L_n be a r.v. denoting the longest run of outcome S and

(iv) X_r ($1 \leq r \leq n$) be r.v.'s which take the values 1 and 0 according to whether the outcome of the rth trial is S or not.

Then, for $y_i, z_i \geq 0$ ($1 \leq i \leq m$), $\sum_i y_i > 0$, $\sum_i (y_i + z_i) = n$ and $0 \leq k-1 \leq$

$n-1$, we have

$$P(L_n \le k-1, X_n = 0, \underline{Y} = \underline{y}, \underline{Z} = \underline{z}) = \prod_i q_i^{y_i} p^{z_i} \sum_{\substack{\sum_j j x_{ij} = y_i + z_i \\ \sum_j x_{ij} = y_i \\ i = 1, \dots, m}} \binom{x_{11} + \dots + x_{mk}}{x_{11}, \dots, x_{mk}}$$

Proof. A typical element of the event $(L_n \le k-1, X_n = 0, \underline{Y} = \underline{y}, \underline{Z} = \underline{z})$ is an arrangement $\alpha_1\alpha_2 \dots \alpha_{x_{11}+\dots+x_{mk}}$ of the letters $S, F_1, \dots, F_m$, such that x_{ij} of the α's are $e_{ij} = \underbrace{SS\dots S}_{j-1}F_i$ $(1 \le i \le m$ and $1 \le j \le k)$, and

$$\sum_j x_{ij} = y_i, \qquad \sum_j j x_{ij} = y_i + z_i, \qquad i = 1, \dots, m. \tag{2.1}$$

Fix x_{ij} $(1 \le i \le m$ and $1 \le j \le k)$. Then the number of the above arrangements is

$$\binom{x_{11} + \dots + x_{mk}}{x_{11}, \dots, x_{mk}},$$

and each one of them has probability $\prod_i q_i^{y_i} p^{z_i}$.

The lemma then follows, since the non-negative integers $x_{i1}, \dots, x_{ik}$ $(1 \le i \le m)$ may vary subject to (2.1).

Next, using an idea from Godbole (1990), we derive an alternative formula for the above probability in terms of binomial coefficients.

Lemma 2. *Let Y_i, Z_i $(1 \le i \le m), L_n$ $(n \ge 1)$ and X_r $(1 \le r \le n)$ be as in Lemma 1. Then for positive y_i $(1 \le i \le m)$, say $y_{t_l} \ge 1$ where $1 \le l \le s \le m$ and $1 \le t_1 < \dots < t_s \le m$, we have*

$$P(L_n \le k-1, X_n = 0, \underline{Y} = \underline{y}, \underline{Z} = \underline{z}) = \binom{y_{t_1} + \dots + y_{t_s}}{y_{t_1}, \dots, y_{t_s}} \sum_{j_{t_1}=0}^{[z_{t_1}/k]} \cdots \sum_{j_{t_s}=0}^{[z_{t_s}/k]} (-1)^{\sum_{l=1}^{s} j_{t_l}}$$

$$\times \prod_{l=1}^{s} \binom{y_{t_l}}{j_{t_l}} \binom{z_{t_l} - kj_{t_l} + y_{t_l} - 1}{y_{t_l} - 1} q_{t_l}^{y_{t_l}} p^{z_{t_l}}$$

Proof. First, we shall derive an expression for

$$P(L_n \geq k, X_n = 0, \underline{Y} = \underline{y}, \underline{Z} = \underline{z}).$$

Without loss of generality suppose that $y_l \geq 1$ for $1 \leq l \leq s \leq m$ ($y_{s+1} = \ldots = y_m = 0$), and let $\alpha_j^{(l)}$ ($1 \leq l \leq s$ and $1 \leq j \leq y_l$) denote the number of outcomes S which are before the jth outcome F_l. Then

$$P(L_n \geq k, X_n = 0, \underline{Y} = \underline{y}, \underline{Z} = \underline{z}) = P\left(\cup_{l=1}^{s} \cup_{j=1}^{y_l} A_j^{(l)}\right),$$

where

$$A_j^{(l)} = \left(X_n = 0, \underline{Y} = \underline{y}, \underline{Z} = \underline{z} \text{ and } \alpha_j^{(l)} \geq k\right), \quad 1 \leq l \leq s \text{ and } 1 \leq j \leq y_l .$$

Consider $P(\cap_{l=1}^{s} \cap_{j=1}^{j_l} A_{_lu_j}^{(l)})$, where $0 \leq j_l \leq y_l$ ($1 \leq l \leq s$), $\sum_{l=1}^{s} j_l > 0$, $1 \leq {}_lu_j \leq y_l$ ($1 \leq l \leq s$ and $1 \leq j \leq j_l$), and if some $j_t = 0$ ($1 \leq t \leq s$), $P\left(\cap_{l=1}^{s} \cap_{j=1}^{j_l} A_{_lu_j}^{(l)}\right)$ means $P\left(\cap_{\lambda=1}^{s} \cap_{j=1}^{j_\lambda} A_{_\lambda u_j}^{(\lambda)}\right)$, with $\lambda \neq t$. Suppose that there are y_1 urns, one for each $F_1, \ldots, y_s$ urns, one for each F_s. We are to place z_1 balls into y_1 urns so that the j_1 specified urns ${}_1u_1, \ldots, {}_1u_{j_1}$ each contain at least k balls,..., z_s balls into y_s urns so that the j_s specified urns ${}_su_1, \ldots, {}_su_{j_s}$ each contain at least k balls. The correspondence between this occupancy problem and the event $\cap_{l=1}^{s} \cap_{j=1}^{j_l} A_{_lu_j}^{(l)}$ is clear. We start by placing k balls into each specified urn. There are now $z_1 - kj_1$ balls to be placed into y_1 urns, ..., $z_s - kj_s$ balls to be placed into y_s urns without any restriction. This can be done in

$$\prod_{l=1}^{s} \binom{z_l - kj_l + y_l - 1}{y_l - 1}$$

ways. Therefore,

$$P\left(\cap_{l=1}^{s}\cap_{j=1}^{j_l} A_{i u_j}^{(l)}\right)=\binom{y_1+\ldots+y_s}{y_1,\ldots,y_s}\prod_{l=1}^{s}\binom{z_l-kj_l+y_l-1}{y_l-1}q_l^{y_l}p^{z_l},$$

since the urns can be arranged in

$$\binom{y_1+\ldots+y_s}{y_1,\ldots,y_s}$$

ways, and each arrangement has probability $\prod_{l=1}^{s} q_l^{y_l}p^{z_l}$.

Noting that the above probability depends only on the numbers $j_1, j_2, \ldots, j_s$ of specified urns (not on the urns), and using the inclusion-exclusion principle, we have

$$P(L_n \geq k, X_n = 0, \underline{Y} = \underline{y}, \underline{Z} = \underline{z}) = P\left(\cup_{l=1}^{s}\cup_{j=1}^{y_l} A_j^{(l)}\right)$$

$$=\sum_{\substack{j_1=0\\ j_1+\ldots+j_s>0}}^{y_1}\cdots\sum_{j_s=0}^{y_s}(-1)^{1+\Sigma_{l=1}^{s}j_l}\prod_{l=1}^{s}\binom{y_l}{j_l}P\left(\cap_{l=1}^{s}\cap_{j=1}^{j_l}A_{i u_j}^{(l)}\right)$$

$$=\binom{y_1+\ldots+y_s}{y_1,\ldots,y_s}\sum_{\substack{j_1=0\\ j_1+\ldots+j_s>0}}^{y_1}\cdots\sum_{j_s=0}^{y_s}(-1)^{1+\Sigma_{l=1}^{s}j_l}\prod_{l=1}^{s}\binom{y_l}{j_l}\binom{z_l-kj_l+y_l-1}{y_l-1}q_l^{y_l}p^{z_l}.$$

The lemma then follows, since

$$P(L_n \leq k-1, X_n = 0, \underline{Y} = \underline{y}, \underline{Z} = \underline{z})$$

$$= P(X_n = 0, \underline{Y} = \underline{y}, \underline{Z} = \underline{z}) - P(L_n \geq k, X_n = 0, \underline{Y} = \underline{y}, \underline{Z} = \underline{z}),$$

and

$$P(X_n = 0, \underline{Y} = \underline{y}, \underline{Z} = \underline{z}) = \binom{y_1+\ldots+y_s}{y_1,\ldots,y_s}\prod_{l=1}^{s}\binom{z_l+y_l-1}{y_l-1}q_l^{y_l}p^{z_l},$$

and the j_l 's can clearly be restricted to j_l 's satisfying $0 \le j_l \le [z_l/k]$.

Next, we give precise upper and lower bounds on the sum of the solutions of the diophantine equation $x_1+2x_2+\ldots+kx_k=x$, which may also have some independent interest.

Lemma 3. *Let $x_1, \ldots, x_k$ and x be non-negative integers, denote by $[x]$ the greatest integer in x, and assume that $\sum_j jx_j=x$. Then $x - \left[\frac{(k-1)x}{k}\right] \le \sum_j x_j \le x$.*

Proof. Obviously, $\sum_j x_j \le \sum_j jx_j = x$, and equality holds if $x_1 = x$ and $x_2 = \ldots = x_k = 0$, which shows the right hand-side inequality. To show the left hand-side inequality we observe that any non-negative integer x may be represented by $x = k\nu + p$ ($\nu = 0, 1, \ldots$ and $p = 0, 1, \ldots, k-1$), which implies that $x - [(k-1)x/k] = \nu + \min(1,p)$. Assume, if possible, that $\sum_j x_j < \nu + \min(1,p)$, i.e., $\sum_j x_j = \nu + \min(1,p) - r$ $(r \ge 1)$. Then,

$$\sum_j jx_j \le k\sum_j x_j = k\nu + k\ \min(1,p) - kr \le k\nu + k\ \min(1,p) - k$$
$$= \begin{cases} k\nu - k, & \text{if } p = 0 \\ k\nu, & \text{if } 1 \le p \le k-1 \end{cases} < k\nu + p = x,$$

which contradicts with our assumption, and hence $\sum_j x_j \ge \nu + \min(1,p)$. Equality holds if $x_k = \nu$ and $x_1 = \ldots = x_{k-1} = 0$, if $p = 0$, or $x_k = \nu$, $x_p = 1$ and $x_m = 0$ for $1 \le m \ne p \le k-1$, if $1 \le p \le k-1$. This completes the proof.

3. Binomial expressions for some multivariate distributions of order k.

In the present section, we employ the preliminary results of Section 2 to obtain expressions in terms of binomial coefficients for some multivariate distributions of order k of Philippou et al. (1988). We first recall the following:

Definition 1. *A random vector $\underline{X} = (X_1, \ldots, X_m)$ is said to have the multivariate negative binomial distribution of order k, type I, with parameters $r, q_1, \ldots, q_m$, to be denoted by $\overline{MNB}_{k,I}(r; q_1, \ldots, q_m)$, if for $x_i = 0, 1, \ldots$ $(1 \le i \le m)$,*

$$P(\underline{X}=\underline{x}) = p^{kr} \sum_{\substack{\sum_j jx_{ij} = x_i \\ i=1,\ldots,m}} \frac{\Gamma(r+\sum_i\sum_j x_{ij})}{\Gamma(r)\prod_i\prod_j x_{ij}!} \prod_i p^{x_i}(\frac{q_i}{p})^{\Sigma_j x_{ij}},$$

where $r > 0$, $0 < q_i < 1$ *for* $1 \leq i \leq m$, $q_1 + \ldots + q_m < 1$ *and* $p = 1 - q_1 - \ldots - q_m$.

Definition 2. *A random vector* $\underline{X} = (X_1, \ldots, X_m)$ *is said to have the multivariate negative binomial distribution of order* k*, type II, with parameters* $r, q_1, \ldots, q_m$*, to be denoted by* $MNB_{k,II}(r; q_1, \ldots, q_m)$*, if for* $x_i = 0, 1, \ldots$ $(1 \leq i \leq m)$,

$$P(\underline{X}=\underline{x}) = p^{r} \sum_{\substack{\sum_j jx_{ij} = x_i \\ i=1,\ldots,m}} \frac{\Gamma(r+\sum_i\sum_j x_{ij})}{\Gamma(r)\prod_i\prod_j x_{ij}!} \prod_i (\frac{q_i}{k})^{\Sigma_j x_{ij}},$$

where $r > 0$, $0 < q_i < 1$ *for* $1 \leq i \leq m$, $q_1 + \ldots + q_m < 1$ *and* $p = 1 - q_1 - \ldots - q_m$.

Definition 3. *A random vector* $\underline{X} = (X_1, \ldots, X_m)$ *is said to have the multivariate Poisson distribution of order* k*, type I, with parameters* $\lambda_1, \ldots, \lambda_m$*, to be denoted by* $MP_{k,I}(\lambda_1, \ldots, \lambda_m)$*, if for* $x_i = 0, 1, \ldots$ $(1 \leq i \leq m)$,

$$P(\underline{X}=\underline{x}) = \exp(-k\Sigma_i\lambda_i) \sum_{\substack{\sum_j jx_{ij} = x_i \\ i=1,\ldots,m}} \frac{\prod_i \lambda_i^{\Sigma_j x_{ij}}}{\prod_i\prod_j x_{ij}!},$$

where $\lambda_i > 0$ *for* $1 \leq i \leq m$.

Definition 4. *A random vector* $\underline{X} = (X_1, \ldots, X_m)$ *is said to have the multivariate logarithmic series distribution of order* k*, type I, with parameters* $q_1, \ldots, q_m$*, to be denoted by* $MLS_{k,I}(q_1, \ldots, q_m)$*, if for* $x_i = 0, 1, \ldots$ $(1 \leq i \leq m)$ *and* $\sum_i x_i > 0$,

$$P(\underline{X}=\underline{x}) = (-k\log p)^{-1} \sum_{\substack{\sum_j jx_{ij} = x_i \\ i=1,\ldots,m}} \frac{(\sum_i\sum_j x_{ij} - 1)!}{\prod_i\prod_j x_{ij}!} \prod_i p^{x_i}(\frac{q_i}{p})^{\Sigma_j x_{ij}},$$

where $0 < q_i < 1$ *for* $1 \leq i \leq m$, $q_1 + \ldots + q_m < 1$ *and* $p = 1 - q_1 - \ldots - q_m$.

Definition 5. *A random vector* $\underline{X} = (X_1, \ldots, X_m)$ *is said to have the multivariate logarithmic series distribution of order k, type II, with parameters* $q_1, \ldots, q_m$*, to be denoted by* $MLS_{k,II}(q_1, \ldots, q_m)$*, if for* $x_i = 0, 1, \ldots$ $(1 \le i \le m)$ *and* $\sum_i x_i > 0$,

$$P(\underline{X} = \underline{x}) = (-\log p)^{-1} \sum_{\substack{\sum_j j x_{ij} = x_i \\ i = 1, \ldots, m}} \frac{(\sum_i \sum_j x_{ij} - 1)!}{\prod_i \prod_j x_{ij}!} \prod_i \left(\frac{q_i}{k}\right)^{\Sigma_j x_{ij}},$$

where $0 < q_i < 1$ *for* $1 \le i \le m$, $q_1 + \ldots + q_m < 1$ *and* $p = 1 - q_1 - \ldots - q_m$.

We proceed now to establish our main results. We mention here that in the sequel $\sum_{\underline{y}}^{1}$ will mean $\sum_{y_1=c_1}^{x_1} \cdots \sum_{y_m=c_m}^{x_m}$, where $c_i = x_i - [\frac{(k-1)x_i}{k}]$ $(1 \le i \le m)$, and $\sum_{\underline{j}}^{2}$ will mean $\sum_{j_1=0}^{d_1} \cdots \sum_{j_m=0}^{d_m}$, where $d_i = [\frac{(x_i - y_i)}{k}]$ $(1 \le i \le m)$.

Theorem 1. *Let* $\underline{X} = (X_1, \ldots, X_m)$ *be a random vector distributed as* $\overline{MNB}_{k,I}(r; q_1, \ldots, q_m)$ *and set* $P_{\underline{x}} = P(X_1 = x_1, \ldots, X_m = x_m)$*. Then,* $P_{(0,\ldots,0)} = p^{kr}$*, and if* $\sum_i x_i > 0$*, say* $x_{t_l} \ge 1$ *for* $1 \le l \le s \le m$ *where* $1 \le t_1 < \ldots < t_s \le m$*, we have*

$$P_{\underline{x}} = p^{kr} \sum_{\underline{y}^*}^{1} \frac{\Gamma(r + y_{t_1} + \ldots + y_{t_s})}{\Gamma(r) y_{t_1}! \ldots y_{t_s}!} \sum_{\underline{j}^*}^{2} (-1)^{\Sigma_{l=1}^{s} j_{t_l}}$$
$$\times \prod_{l=1}^{s} \binom{y_{t_l}}{j_{t_l}} \binom{x_{t_l} - k j_{t_l} - 1}{y_{t_l} - 1} q_{t_l}^{y_{t_l}} p^{x_{t_l} - y_{t_l}}$$

where $\underline{y}^* = (y_{t_1}, \ldots, y_{t_s})$ *and* $\underline{j}^* = (j_{t_1}, \ldots, j_{t_s})$.

Proof. If $x_1 = \ldots = x_m = 0$, it follows from Definition 1 that $P_{(0,\ldots,0)} = p^{kr}$. Let now $\sum_i x_i > 0$, say $x_{t_l} \ge 1$ for $1 \le l \le s \le m$ where $1 \le t_1 < \ldots < t_s \le m$. Then

$$P_{\underline{x}} = p^{kr} \sum_{\sum_j j x_{ij} = x_i} \frac{\Gamma(r + \sum_i \sum_j x_{ij})}{\Gamma(r) \prod_i \prod_j x_{ij}!} \prod_i p^{x_i} \left(\frac{q_i}{p}\right)^{\Sigma_j x_{ij}} \qquad \text{(by Definition 1)}$$

$$= p^{kr} \sum_{\underline{y}}^{1} \sum_{\substack{\sum_j jx_{ij} = x_i \\ \sum_j x_{ij} = y_i}} \frac{\Gamma(r + \sum_i \sum_j x_{ij})}{\Gamma(r) \prod_i \prod_j x_{ij}!} \prod_i p^{x_i} (\frac{q_i}{p})^{\Sigma_j x_{ij}} \qquad \text{(by Lemma 3)}$$

$$= p^{kr} \sum_{\underline{y}}^{1} \frac{\Gamma(r + y_1 + \ldots + y_m)}{\Gamma(r)(y_1 + \ldots + y_m)!} \sum_{\substack{\sum_j jx_{ij} = x_i \\ \sum_j x_{ij} = y_i}} \binom{x_{11} + \ldots + x_{mk}}{x_{11}, \ldots, x_{mk}} \prod_i p^{x_i - y_i} q_i^{y_i}$$

$$= p^{kr} \sum_{\underline{y}}^{1} \frac{\Gamma(r + y_1 + \ldots + y_m)}{\Gamma(r)(y_1 + \ldots + y_m)!} P(L_{\Sigma_i x_i} \leq k-1, X_{\Sigma_i x_i} = 0, \underline{Y} = \underline{y}, \underline{Z} = \underline{x} - \underline{y})$$

$$= p^{kr} \sum_{\underline{y}^*}^{1} \frac{\Gamma(r + y_{t_1} + \ldots + y^{t_s})}{\Gamma(r) y_{t_1}! \ldots y_{t_s}!} \sum_{\underline{j}^*}^{2} (-1)^{\Sigma_{l=1}^{s} j_{t_l}} \times$$
$$\prod_{l=1}^{s} \binom{y_{t_l}}{j_{t_l}} \binom{x_{t_l} - kj_{t_l} - 1}{y_{t_l} - 1} q_{t_l}^{y_{t_l}} p^{x_{t_l} - y_{t_l}}.$$

where the next-to-last and last equalities follow from Lemma 1 and Lemma 2, respectively. The proof is completed.

Theorem 2. *Let $\underline{X} = (X_1, \ldots, X_m)$ be a random vector distributed as $MNB_{k,II}(r; q_1, \ldots, q_m)$ and set $P_{\underline{x}} = P(X_1 = x_1, \ldots, X_m = x_m)$. Then, $P_{(0,\ldots 0)} = p^r$, and if $\sum_i x_i > 0$, say $x_{t_l} \geq 1$ for $1 \leq l \leq s \leq m$ where $1 \leq t_1 < \ldots t_s \leq m$, we have*

$$P_{\underline{x}} =$$
$$p^r \sum_{\underline{y}^*}^{1} \frac{\Gamma(r + y_{t_1} + \ldots + y_{t_s})}{\Gamma(r) y_{t_1}! \ldots y_{t_s}!} \sum_{\underline{j}^*}^{2} (-1)^{\Sigma_{l=1}^{s} j_{t_l}} \prod_{l=1}^{s} \binom{y_{t_l}}{j_{t_l}} \binom{x_{t_l} - kj_{t_l} - 1}{y_{t_l} - 1} (\frac{q_{t_l}}{k})^{y_{t_l}},$$

where $\underline{y}^* = (y_{t_1}, \ldots, y_{t_s})$ and $\underline{j}^* = (j_{t_1}, \ldots, j_{t_s})$.

Proof. It follows along the same lines as Theorem 1, by noting that

$$P_{\underline{x}} = p^r \sum_{\underline{y}}^{1} \frac{\Gamma(r + y_1 + \ldots + y_m)}{\Gamma(r)(y_1 + \ldots + y_m)!} (\frac{p}{k})^{\Sigma_i y_i} p^{-\Sigma_i x_i}$$

$$\times P(L_{\Sigma_i x_i} \leq k-1, X_{\Sigma_i x_i} = 0, \underline{Y} = \underline{y}, \underline{Z} = \underline{x} - \underline{y}).$$

Theorem 3. *Let* $\underline{X} = (X_1, \ldots, X_m)$ *be a random vector distributed as* $MP_{k,I}(\lambda_1, \ldots, \lambda_m)$ *and set* $P_{\underline{x}} = P(X_1 = x_1, \ldots, X_m = x_m)$. *Then* $P_{(0,\ldots,0)} = \exp(-k\sum_i \lambda_i)$, *and if* $\sum_i x_i > 0$, *say* $x_{t_l} \geq 1$ *for* $1 \leq l \leq s \leq m$ *where* $1 \leq t_1 < \ldots < t_s \leq m$, *we have*

$$P_{\underline{x}} = \exp(-k\Sigma_i\lambda_i) \sum_{\underline{y}^*}^{1} \frac{1}{y_{t_1}! \ldots y_{t_s}!} \sum_{\underline{j}^*}^{2} (-1)^{\Sigma_{l=1}^s j_{t_l}} \prod_{l=1}^{s} \binom{y_{t_l}}{j_{t_l}} \binom{x_{t_l} - kj_{t_l} - 1}{y_{t_l} - 1} \lambda_{t_l}^{y_{t_l}}$$

where $\underline{y}^* = (y_{t_1}, \ldots, y_{t_s})$ *and* $\underline{j}^* = (j_{t_1}, \ldots, j_{t_s})$.

Proof. It follows along the same lines as Theorem 1, by noting that

$$P_{\underline{x}} = e^{-k\Sigma_i\lambda_i} \sum_{\underline{y}}^{1} \frac{1}{(y_1 + \ldots + y_m)!} p^{-\Sigma_i x_i}$$
$$\times P(L_{\Sigma_i x_i} \leq k-1, X_{\Sigma_i x_i} = 0, \underline{Y} = \underline{y}, \underline{Z} = \underline{x} - \underline{y})$$

where $p = 1/(1 + \lambda_1 + \ldots + \lambda_m)$ and $q_i = \lambda_i/(1 + \lambda_1 + \ldots + \lambda_m)$ for $1 \leq i \leq m$.

Theorem 4. *Let* $\underline{X} = (X_1, \ldots, X_m)$ *be a random vector distributed as* $MLS_{k,I}(q_1, \ldots, q_m)$ *and set* $P_{\underline{x}} = P(X_1 = x_1, \ldots, X_m = x_m)$. *Then, for* $x_{t_l} \geq 1$, $1 \leq l \leq s \leq m$ *where* $1 \leq t_1 < \ldots < t_s \leq m$, *we have*

$$P_{\underline{x}} = \left(\frac{-1}{k \log p}\right) \sum_{\underline{y}^*}^{1} \frac{(y_{t_1} + \ldots + y_{t_s} - 1)!}{y_{t_1}! \ldots y_{t_s}!} \sum_{\underline{j}^*}^{2} (-1)^{\Sigma_{l=1}^s j_{t_l}}$$
$$\times \prod_{l=1}^{s} \binom{y_{t_l}}{j_{t_l}} \binom{x_{t_l} - kj_{t_l} - 1}{y_{t_l} - 1} q_{t_l}^{y_{t_l}} p^{x_{t_l} - y_{t_l}},$$

where $\underline{y}^* = (y_{t_1}, \ldots, y_{t_s})$ *and* $\underline{j}^* = (j_{t_1}, \ldots, j_{t_s})$.

Proof. It follows along the same lines as Theorem 1, by noting that

$$P_{\underline{x}} = (-k \log p)^{-1} \sum_{\underline{y}}^{1} \frac{1}{y_1 + \ldots + y_m} P(L_{\Sigma_i x_i} \leq k-1, X_{\Sigma_i x_i} = 0, \underline{Y} = \underline{y}, \underline{Z} = \underline{x} - \underline{y}).$$

Theorem 5. *Let $\underline{X} = (X_1, \ldots, X_m)$ be a random vector distributed as $MLS_{k,II}(q_1, \ldots, q_m)$ and set $P_{\underline{x}} = P(X_1 = x_1, \ldots, X_m = x_m)$. Then, for $x_{t_l} \geq 1$, $1 \leq l \leq s \leq m$ where $1 \leq t_1 < \ldots < t_s \leq m$, we have*

$$P_{\underline{x}} = (-\log p)^{-1} \sum_{\underline{y}^*}^{1} \frac{(y_{t_1} + \ldots + y_{t_s} - 1)!}{y_{t_1}! \ldots y_{t_s}!} \sum_{\underline{j}^*}^{2} (-1)^{\Sigma_{l=1}^{s} j_{t_l}}$$

$$\times \prod_{l=1}^{s} \binom{y_{t_l}}{j_{t_l}} \binom{x_{t_l} - kj_{t_l} - 1}{y_{t_l} - 1} \left(\frac{q_{t_l}}{k}\right)^{y_{t_l}},$$

where $\underline{y}^* = (y_{t_1}, \ldots, y_{t_s})$ and $\underline{j}^* = (j_{t_1}, \ldots, j_{t_s})$.

Proof. It follows along the same lines as Theorem 1, by noting that

$$P_{\underline{x}} = (-\log p)^{-1} \sum_{\underline{y}}^{1} \frac{1}{y_1 + \ldots + y_m} \left(\frac{p}{k}\right)^{\Sigma_i y_i} p^{-\Sigma_i x_i} \times$$

$$\times P(L_{\Sigma_i x_i} \leq k-1, X_{\Sigma_i x_i} = 0, \underline{Y} = \underline{y}, \underline{Z} = \underline{x} - \underline{y}).$$

Theorems 1-5 are more applicable for calculating the probabilities of the multivariate distributions in comparison with their definitions, since the former involve only simple summations while the latter involve summation over an index set determined by the solution of a set of diophantine equations.

For $m = 1$, Theorems 1-5 reduce to the following corollaries, respectively.

Corollary 1. *Let X be a r.v. distributed as $\overline{NB}_{k,I}(r,p)$, and set $P_x = P(X = x)$. Then $P_0 = p^{kr}$, and for $x \geq 1$*

$$P_x = p^{kr} \sum_{y=x-[(k-1)x/k]}^{x} \binom{y+r-1}{y} q^y p^{x-y} \sum_{j=0}^{[(x-y)/k]} (-1)^j \binom{y}{j} \binom{x-kj-1}{y-1}.$$

Corollary 2. *Let X be a r.v. distributed as $NB_{k,II}(r,p)$, and set $P_x = P(X = x)$. Then $P_0 = p^r$, and for $x \geq 1$*

$$P_x = p^r \sum_{y=x-[(k-1)x/k]}^{x} \binom{y+r-1}{y} \left(\frac{q}{k}\right)^y \sum_{j=0}^{[(x-y)/k]} (-1)^j \binom{y}{j} \binom{x-kj-1}{y-1}.$$

Corollary 3. *Let X be a r.v. distributed as $P_{k,I}(\lambda)$, and set $P_x = P(X = x)$. Then $P_0 = \exp(-k\lambda)$, and for $x \geq 1$*

$$P_x = e^{-k\lambda} \sum_{y=x-[(k-1)x/k]}^{x} \frac{\lambda^y}{y!} \sum_{j=0}^{[(x-y)/k]} (-1)^j \binom{y}{j} \binom{x-kj-1}{y-1}.$$

Corollary 4. *Let X be a r.v. distributed as $LS_{k,I}(p)$, and set $P = P(X = x)$. Then for $x \geq 1$*

$$P_x = (-k \log p)^{-1} \sum_{y=x-[(k-1)x/k]}^{x} \frac{1}{y} q^y p^{x-y} \sum_{j=0}^{[(x-y)/k]} (-1)^j \binom{y}{j} \binom{x-kj-1}{y-1}.$$

Corollary 5. *Let X be a r.v. distributed as $LS_{k,II}(p)$, and set $P_x = P(X = x)$. Then for $x \geq 1$*

$$P_x = (-\log p)^{-1} \sum_{y=x-[(k-1)x/k]}^{x} \frac{1}{y} (\frac{q}{k})^y \sum_{j=0}^{[(x-y)/k]} (-1)^j \binom{y}{j} \binom{x-kj-1}{y-1}.$$

Corollaries 2 and 5 are new results, while Corollaries 1, 3 and 4 have been obtained by Charalambides (1986), by means of the truncated exponential Bell polynomials, provided $k(j+1)$ in his expression for $N(n,r,k)$ is corrected to $(k+1)j$. Corollaries 1 and 3 have also been derived independently, by Godbole (1990), and they constitute, along with an analogous result for the binomial distribution of order k, his main results.

Other binomial expressions for the probabilities of $\overline{NB}_{k,I}(r,p)$ and $NB_{k,\mathrm{II}}(r,p)$, which seem to be more applicable than the present ones for small values of r, have been obtained by Philippou and Georghiou (1989), who employed for this purpose convolutions of Fibonacci-type polynomials of order k. See, also, Corollary 3.2 of Philippou (1986) and Proposition 2.1 of Uppuluri and Patil (1983).

We note in ending that useful recurrences for calculating the probabilities of $\overline{MNB}_{k,I}\ (r; q_1, \ldots, q_m)$ and $\mathrm{MNB}_{k,\mathrm{II}}(r; q_1, \ldots, q_m)$ have been derived by Philippou and Antzoulakos (1991).

References

AKI, S., KUBOKI, H. AND HIRANO. K. (1984) On discrete distributions of order k. *Ann. Inst. Statist. Math.* **36**, 431-440.

CHARALAMBIDES, C. A. (1986) On discrete distributions of order k. *Ann. Inst. Statist. Math.* **38**, 557-568.

GODBOLE, A. P. (1990) Specific formulae for some succes run distributions. *Statist. Probab. Lett.* **10**, 119-124.

PHILIPPOU, A. N. (1983) Poisson and compound Poisson distributions of order k and some of its properties (in Russian, English summary). *Zap. Nauchn. Sem. Leningrad Otdel. Mat. Inst. Steklov. (LOMI)* **130**, 175-180 (in Russian, English summary).

PHILIPPOU, A. N. (1986) Distributions and Fibonacci polynomials of order k, longest runs, and reliability of consecutive-k-out-of-n : F systems. *Fibonacci Numbers and Their Applications: (Patras 1984), (eds. A. N. Philippou, G. E. Bergum and A. F. Horadam)*, 203-227, Reidel, Dordrecht.

PHILIPPOU, A. N. (1988) On multiparameter distributions of order k. *Ann. Inst. Statist. Math.* **40**, 467-475.

PHILIPPOU, A. N. AND ANTZOULAKOS, D. L. (1991) Generalized multivariate Fibonacci polynomials of order k and the multivariate negative binomial distributions of the same order. *Fibonacci Quart.* **29**, 322-328.

PHILIPPOU, A. N. AND GEORGHIOU, C. (1989) Convolutions of Fibonacci-type polynomials of order k and the negative binomial distributions of the same order. *Fibonacci Quart.* **27**, 209-216.

PHILIPPOU, A. N. AND MUWAFI, A. A. (1982) Waiting for the kth consecutive success and the Fibonacci sequence of order k. *Fibonacci Quart.* **20**, 28-32.

PHILIPPOU, A. N., ANTZOULAKOS, D. L. AND TRIPSIANNIS G. A. (1988) Multivariate distributions of order k. *Statist. Probab. Lett.* **7**, 207-216.

PHILIPPOU, A. N., GEORGHIOU, C. AND PHILIPPOU, G. N. (1983) A generalized geometric distribution and some of its properties. *Statist. Probab. Lett.* **1**, 171-175.

UPPULURI, V. R. R. AND PATIL, S. A. (1983) Waiting times and generalized Fibonacci sequences. *Fibonacci Quart.* **21**, 342-349.

SUCCESS RUNS IN A CIRCULAR SEQUENCE OF INDEPENDENT BERNOULLI TRIALS

Ch. A. Charalambides
University of Athens, Greece

The joint probability function of the number $N^c_{n,k}$ of nonoverlapping success runs of length k and the number S_n of successes in a circular sequence of n independent Bernoulli trials, with constant success probability, is combinatorially obtained and its probability generating function is then derived. The marginal probability function of $N^c_{n,k}$ and its probability generating function are deduced. Also, the factorial moments of $N^c_{n,k}$ are calculated. Finally, the conditional probability function of $N^c_{n,k}$ given $S_n = y$ is deduced.

1. Introduction

Consider a sequence of n independent Bernoulli trials, with a constant success probability p, arranged on a line (linear sequence) or on a circle (circular sequence). In the case of a (linear) sequence let $N_{n,k}$ and $M_{n,k}$ be the numbers of nonoverlapping and overlapping success runs of length k respectively, while in the case of a circular sequence let us denote the respective numbers by $N^c_{n,k}$ and $M^c_{n,k}$. Further, in either case, let S_n be the number of successes. The probability generating function of $N_{n,k}$ was given, for the first time, by Feller (1968, page 341). Subsequently, an explicit expression for the probability function $P(N_{n,k} = x)$, in terms of a sum of multiple sums of probabilities with multinomial coefficients, was derived by Philippou and Makri (1986) and independently by Hirano (1986). Recently, a more attractive formula for the same probability function, in terms of a double sum of probabilities with binomial coefficients, was obtained by Godbole (1990) and independently by Papastavridis (1990). An explicit formula for the probability function $P(M_{n,k} = x)$, in terms of a sum of multiple sums of probabilities with multinomial coefficients,

A. P. Godbole and S. G. Papastavridis (eds.), Runs and Patterns in Probability, 15–30.

was derived by Ling (1988). In addition, he obtained recurrence relations of this probability function and its moment generating function. Chryssaphinou et al (1992) derived recurrence relations of the probability functions $P(M_{n,k} = x)$ and $P(M^c_{n,k} = x)$ and their probability generating functions. Philippou and Makri (1993) obtained explicit expressions for the probability functions $P(N^c_{n,k} = x)$ and $P(M^c_{n,k} = x)$ in terms of sums of multiple sums of probabilities with multinomial coefficients. They also derived the means $E(N^c_{n,k})$ and $E(M^c_{n,k})$.

In the present paper, the joint probability function of $N^c_{n,k}$ and S_n is combinatorially obtained and its probability generating function is then derived (Section 2). As corollaries of these expressions, the probability function of $N^c_{n,k}$ and its probability generating function are deduced. It is worth noticing that the probability function of $N^c_{n,k}$ is expressed as a double sum of probabilities with binomial coefficients. Further, the factorial moments of $N^c_{n,k}$ are derived by expanding its moment generating function (Section 3). Finally the conditional probability function of $N^c_{n,k}$ given $S_n = y$ is deduced (Section 4).

2. The Joint Distribution of $N^c_{n,k}$ and S_n

Consider a circular sequence of n independent Bernoulli trials with a constant success probability p and failure probability $q = 1 - p$. Let $N^c_{n,k}$ be the number of nonoverlapping circular success runs of length k and let S_n be the number of successes.

Theorem 2.1. *The probability function $p_n(x,y) = P(N^c_{n,k} = x, S_n = y)$, $x = 0, 1, \ldots, [y/k], y = 0, 1, \ldots, n$ is given by*

$$p_n(x,y) = \begin{cases} \binom{n-y+x-1}{x} N_c(y-kx, n-y, k) p^y q^{n-y}, & x = 0, 1, \ldots, [y/k] \\ & y = 0, 1, \ldots, n-1 \\ p^n, & x = [n/k], y = n, \end{cases} \tag{2.1}$$

where $[s]$ denotes the integral part of s, with $[s] \leq s < [s] + 1$ and

$$\begin{aligned} N_c(y-kx, n-y, k) &= \frac{n}{n-y}\sum_{j=0}^{n-y}(-1)^j\binom{n-y}{j}\binom{n-kx-kj-1}{n-y-1} \\ &= \sum_{j=0}^{n-y}(-1)^j\binom{n-y}{j}\frac{n}{n-kx-kj}\binom{n-kx-kj}{n-y}, \end{aligned} \tag{2.2}$$

is the number of circular permutations of $y-kx$ 1's and $n-y$ 0's, with one of these $n-kx$ symbols marked, containing no run of 1's of length k.

Proof. Recording a success by 1 and a failure by 0, the elements ω of the sample space Ω of a circular sequence of n independent Bernoulli trials are circular permutations $(i_1, i_2, \ldots, i_n)$ of n 1's and 0's, with one of these n symbols marked to indicate that it corresponds to the result of the first trial. The event

$$\{\omega\epsilon\Omega : N^c_{n,k}(\omega) = x, S_n(\omega) = y\}$$

contains the circular permutations $(i_1, i_2, \ldots, i_n)$ of y 1's and $n-y$ 0's, with one of these n symbols marked, containing x nonoverlapping runs of 1's of length k. The number of these permutations can be derived as follows:

Note first that the $n-y$ 0's displaced on a circle form $n-y$ indistinguishable cells, with a cell defined by the space between two zeros. These cells are made distinguishable by labelling them. For $0 \le x \le [y/k]$, $0 \le y \le n-1$, the number of ways x nonoverlapping runs of 1's of length k can be placed in the $n-y$ distinguishable cells is equal to

$$\binom{n-y+x-1}{x},$$

the number of x-combinations of the $n-y$ distinguishable cells with repetitions allowed. For $\max\{0, y-n+[n/k]\} \le x \le [y/k], 0 \le y \le n-1$, whence $0 \le y-kx \le (k-1)(n-y)$, the number of ways the $y-kx$ remaining 1's can be placed in the $n-y$ distinguishable cells with no cell receiving more than $k-1$ 1's is equal to (Riordan (1958), page 104)

$$N(y-kx,n-y,k)=\sum_{j=0}^{n-y}(-1)^j\binom{n-y}{j}\binom{n-kx-kj-1}{n-y} \tag{2.3}$$

Therefore the total number of arrangements (distributions) of the y 1's in the $n-y$ distinguishable cells, yielding x nonoverlapping runs of 1's of length k, is given by

$$\binom{n-y+x-1}{x}N(y-kx,n-y,k).$$

Further, putting a mark on one of the n symbols (1's and 0's), each of these arrangements gives rise to n arrangements (distributions) of the y 1's in the $n-y$ distinguishable cells, with one of the n symbols marked, yielding x nonoverlapping runs of 1's of length k. Removing the labels from the $n-y$ cells, the preceding arrangements fall into classes of $n-y$ like, by rotation, circular permutations of y 1's and $n-y$ 0's, with one of these n symbols marked, containing x nonoverlapping runs of 1's of length k. Thus, the number of these circular permutations is equal to

$$\binom{n-y+x-1}{x}\frac{n}{n-y}N(y-kx,n-y,k).$$

Each of these circular permutations has probability $p^y q^{n-y}$ and hence for $\max\{0,y-n+[n/k]\}\leq x\leq [y/k]$, $0\leq y\leq n-1$,

$$p_n(x,y)=\binom{n-y+x-1}{x}\frac{n}{n-y}N(y-kx,n-y,k)p^y q^{n-y}$$

while for $x=[n/k]$, $y=n$,

$$p_n(x,y)=p^n$$

Finally, introducing the numbers (2.2), which are connected with the numbers (2.3) by

$$N_c(y-kx,n-y,k)=\frac{n}{n-y}N(y-kx,n-y,k)$$

the expression (2.1) follows. Note that, in this expression, the points added by the extension of the range of the integral valued variables x and y from $\max\{0,y-n+$

$[n/k]\} \leq x \leq [y/k], 0 \leq y \leq n$, to $0 \leq x \leq [y/k], 0 \leq y \leq n-1$, and $x = [n/k], y = n$, have probability zero. Indeed, for $0 \leq x < y - n + [n/k]$, $n - [n/k] < y \leq n - 1$, whence $y - kx > (k-1)(n-y)$,

$$N_c(y - kx, n - y, k) = 0$$

and thus for these points

$$p_n(x, y) = 0.$$

Theorem 2.2. *The generating function*

$$g(t, u, z) = \sum_{n=0}^{\infty} g_n(t, u) z^n \tag{2.4}$$

of the sequence of generating functions

$$g_n(t, u) = \sum_{y=0}^{n} \sum_{x=0}^{[y/k]} p_n(x, y) t^x u^y, n = 0, 1, \ldots \tag{2.5}$$

is given by

$$\begin{aligned} g(t, u, z) = {} & z \frac{\partial}{\partial z} \log\left[\frac{(1 - puz)(1 - p^k u^k z^k t)}{1 - z + (1 - u)pz + qp^k u^k z^{k+1} - (1 - puz)p^k u^k z^k t}\right] \\ & + \frac{1 - p^k u^k z^k}{(1 - puz)(1 - p^k u^k z^k t)} \end{aligned} \tag{2.6}$$

Proof. Consider the auxiliary sequence

$$q_n(x, y) = p_n(x, y)/n, x = 0, 1, \ldots, [y/k], y = 0, 1, \ldots, n - 1, n = 1, 2, \ldots$$

and let

$$Q_x(u, z) = \sum_{n=1}^{\infty} \sum_{y=0}^{n-1} q_n(x, y) u^y z^n, x = 0, 1, 2, \ldots.$$

Then

$$\begin{aligned} h_x(u,z) &= \sum_{n=0}^{\infty}\sum_{y=0}^{n} p_n(x,y)u^y z^n \\ &= \sum_{n=1}^{\infty}\sum_{y=0}^{n-1} nq_n(x,y)u^y z^n + \sum_{n=0}^{\infty}(puz)^n \delta_{x,[n/k]} \\ &= z\frac{\partial}{\partial z}\sum_{n=1}^{\infty}\sum_{y=0}^{n-1} q_n(x,y)u^y z^n + (puz)^{kx}\sum_{i=0}^{k-1}(puz)^i \\ &= z\frac{\partial}{\partial z}Q_x(u,z) + \frac{(1-p^k u^k z^k)(p^k u^k z^k)^x}{1-puz}, x = 0,1,2,\ldots. \end{aligned}$$

Since

$$\begin{aligned} q_n(x,y) &= \frac{1}{n-y}\binom{n-y+x-1}{x}p^y q^{n-y}\sum_{j=0}^{n-y}(-1)^j\binom{n-y}{j}\binom{n-kx-kj-1}{n-y-1} \\ &= \frac{1}{x}\binom{n-y+x-1}{n-y}p^y q^{n-y}\sum_{j=0}^{n-y}(-1)^j\binom{n-y}{j}\binom{n-kx-kj-1}{n-y-1}, \end{aligned}$$

it follows that

$$\begin{aligned} Q_0(u,z) &= \sum_{n=1}^{\infty}\sum_{y=0}^{n-1} q_n(0,y)u^y z^n = \sum_{n=1}^{\infty}\sum_{r=1}^{n} q_n(0,n-r)(uz)^{n-r}z^r \\ &= \sum_{r=1}^{\infty}\frac{1}{r}(qz)^r\sum_{j=0}^{r}(-1)^j\binom{r}{j}(puz)^{kj}\sum_{n=r+kj}^{\infty}\binom{n-kj-1}{r-1}(puz)^{n-r-kj} \\ &= \sum_{r=1}^{\infty}\frac{1}{r}(qz)^r(1-p^k u^k z^k)^r(1-puz)^{-r} \\ &= -\log\left(1-\frac{qz-qp^k u^k z^{k+1}}{1-puz}\right) = \log\left(\frac{1-puz}{1-z+(1-u)puz+qp^k u^k z^{k+1}}\right) \end{aligned}$$

and

$$\begin{aligned}
Q_x(u,z) &= \sum_{n=1}^{\infty}\sum_{y=0}^{n-1} q_n(x,y)u^y z^n = \sum_{n=1}^{\infty}\sum_{r=1}^{n} q_n(x,n-r)(uz)^{n-r}z^r \\
&= \frac{1}{x}\sum_{r=1}^{\infty}\binom{x+r-1}{r}(qz)^r\sum_{j=0}^{r}(-1)^j\binom{r}{j}(puz)^{kj} \\
&\times \sum_{n=r+kx+kj}^{\infty}\binom{n-kx-kj-1}{r-1}(puz)^{n-r-kj} \\
&= \frac{1}{x}(puz)^{kx}\sum_{r=1}^{\infty}\binom{x+r-1}{r}(qz)^r(1-p^ku^kz^k)^r(1-puz)^{-r} \\
&= \frac{1}{x}(puz)^{kx}(1-\frac{qz-qp^ku^kz^{k+1}}{1-puz})^{-x}-\frac{1}{x}(puz)^{kx}, x=1,2,\ldots.
\end{aligned}$$

Therefore

$$h_0(u,z) = z\frac{\partial}{\partial z}\log\left(\frac{1-puz}{1-z+(1-u)pz+qp^ku^kz^{k+1}}\right)+\frac{1-p^ku^kz^k}{1-puz},$$

$$\begin{aligned}
h_x(u,z) &= \frac{z}{x}\frac{\partial}{\partial z}(\frac{(1-puz)p^ku^kz^k}{1-z+(1-u)pz+qp^ku^kz^{k+1}})^x-\frac{z}{x}\frac{\partial}{\partial z}(p^ku^kz^k)^x \\
&+\frac{1-p^ku^kz^k}{1-puz}(p^ku^kz^k)^x, x=1,2,\ldots.
\end{aligned}$$

and

$$\begin{aligned}
g(t,u,z) &= \sum_{n=0}^{\infty}g_n(t,u)z^n = \sum_{x=0}^{\infty}h_x(u,z)t^n \\
&= z\frac{\partial}{\partial z}\log\frac{(1-puz)}{1-z+(1-u)pz+qp^ku^kz^{k+1}}+z\frac{\partial}{\partial z}\log(1-p^ku^kz^kt) \\
&- z\frac{\partial}{\partial z}\log(1-\frac{(1-puz)p^ku^kz^kt}{1-z+(1-u)pz+qp^ku^kz^{k+1}})+\frac{1-p^ku^kz^k}{(1-puz)(1-p^ku^kz^kt)} \\
&= z\frac{\partial}{\partial z}\log\left(\frac{(1-puz)(1-p^ku^kz^kt)}{1-z+(1-u)pz+qp^ku^kz^{k+1}-(1-puz)p^ku^kz^kt}\right) \\
&+\frac{1-p^ku^kz^k}{(1-puz)(1-p^ku^kz^kt)}.
\end{aligned}$$

3. The Distribution of $N_{n,k}^c$

The marginal distribution of $N_{n,k}^c$ may be deduced from the joint distribution of $N_{n,k}^c$ and S_n. It is given in the following corollary of Theorem 2.1.

Corollary 3.1. *The probability function $p_n(x) = P(N_{n,k}^c = x)$, $x = 0, 1 \ldots, [n/k]$ is given by*

$$p_n(x) = p^n\Big\{\sum_{r=1}^{n-kx}\binom{r+x-1}{x}N_c(n-r-kx,r,k)(q/p)^r + \delta_{x,[n/k]}\Big\}, \qquad (3.1)$$

$$x = 0, 1, \ldots, [n/k],$$

where $[s]$ denotes the integral part of s, with $[s] \le s < [s]+1$, $\delta_{x,s}$ is the Kronecker delta function and

$$\begin{aligned} N_c(n-r-kx,r,k) &= \frac{n}{r}\sum_{j=0}^{r}(-1)^j\binom{r}{j}\binom{n-kx-kj-1}{r-1} \\ &= \sum_{j=0}^{r}(-1)^j\binom{r}{j}\frac{n}{n-kx-kj}\binom{n-kx-kj}{r} \end{aligned} \qquad (3.2)$$

is the number of circular permutations of $n-r-kx$ 1's and 0's, one of these $n-kx$ symbols marked, containing no run of 1's of length k.

Proof. Summing with respect to y the probability function (2.1), it follows that

$$p_n(x) = \sum_{y=kx}^{n} p_n(x,y) = \sum_{r=1}^{n-kx} p_n(x,n-r) + p_n(x,n)$$

$$= p^n\{\sum_{r=1}^{n-kx}\binom{r+x-1}{x}N_c(n-r-kx,r,k)(q/p)^r + \delta_{x,[n/k]}\}, x = 0, 1, \ldots, [n/k].$$

Remark 3.1 The probability function (3.1) gives, in particular, for $x = 0$, the probability that the length of the longest run in a circular sequence of n independent Bernoulli trials is less than or equal to $k-1$:

$$p_n(0) = \sum_{r=0}^{n} p^{n-r}q^r \sum_{j=0}^{[(n-r)/k]} (-1)^j \binom{r}{j} \frac{n}{n-kj}\binom{n-kj}{j} - p^n + p^n\delta_{0,[n/k]}$$

$$= \sum_{j=0}^{[(n-1)/k]} (-1)^j \frac{n}{n-kj}\binom{n-kj}{j} \sum_{r=j}^{n-kj} \binom{n-kj-j}{r-j} p^{n-r}q^r - p^n + p^n\delta_{0,[n/k]}$$

$$= \sum_{j=0}^{[(n-1)/k]} (-1)^j \frac{n}{n-kj}\binom{n-kj}{j} p^{jk}q^j - p^n + p^n\delta_{0,[n/k]},$$

which, in turn, by interchanging p and q, gives the reliability of a circular consecutive-k-out-of-n: F system:

$$R_c(p,k,n) = \sum_{j=0}^{[(n-1)/k]} (-1)^j \frac{n}{n-kj}\binom{n-kj}{j} p^j q^{jk} - q^n, n \geq k,$$

in the form derived by Charalambides (1991). An equivalent expression of this reliability was given by Lambiris and Papastavridis (1985).

The right tail probabilities of the probability function (3.1):

$$R_n(m) = \sum_{x=m}^{[n/k]} p_n(x) = p^n \sum_{x=m}^{[n/k]-1} \sum_{r=1}^{n-kx} \binom{r+x-1}{x} N_c(n-r-kx, r, k)(q/p)^r + p^n,$$

give the reliability of a circular m-consecutive-k-out-of-n: F system. This expression, apart from a missing factor n, was also obtained by Hwang and Papastavridis (1991).

The probability generating function of $N^c_{n,k}$ may be deduced from the joint probability generating function of $N^c_{n,k}$ and S_n. It is given in the following corollary of Theorem 2.2.

Corollary 3.2. *The generating function*

$$g(t,z) = \sum_{n=0}^{\infty} g_n(t)z^n \tag{3.3}$$

of the sequence of generating functions

$$g_n(t) = \sum_{x=0}^{[n/k]} p_n(x)t^x, n = 0, 1, 2, \ldots \tag{3.4}$$

is given by

$$g(t,z) = z\frac{\partial}{\partial z}\log\left[\frac{(1-pz)(1-p^kz^kt)}{1-z+qp^kz^{k+1}-(1-pz)p^kz^kt}\right] + \frac{1-p^kz^k}{(1-pz)(1-p^kz^kt)} \tag{3.5}$$

Proof. The generating function (3.4), on using (2.5), may be expressed as

$$g_n(t) = \sum_{x=0}^{[n/k]} p_n(x)t^x = \sum_{x=0}^{[n/k]}\sum_{y=kx}^{n} p_n(x,y)t^x = \sum_{y=0}^{n}\sum_{x=0}^{[y/k]} p_n(x,y)t^x = g_n(t,1)$$

and hence the generating function (3.3), on using (2.4), may be expressed as

$$g(t,z) = \sum_{n=0}^{\infty} g_n(t)z^n = \sum_{n=0}^{\infty} g_n(t,1)z^n = g(t,1,z).$$

Therefore from (2.6), with u=1, (3.5) is deduced.

Theorem 3.1. *Let*

$$\mu_{(r)}(n) = \sum_{x=r}^{[n/k]} (x)_r p_n(x), r = 0, 1, \ldots, [n/k], \tag{3.6}$$

be the rth factorial moment of the random variable $N^c_{n,k}$ and

$$f_n(t) = \sum_{r=0}^{[n/k]} \mu_{(r)}(n)t^r/r!, \quad n = 0, 1, 2, \ldots \tag{3.7}$$

Then

(a) the generating function

$$f(t,z) = \sum_{n=0}^{\infty} f_n(t)z^n \tag{3.8}$$

is given by

$$f(t,z) = x\frac{\partial}{\partial z}\log\Big(\frac{(1-pz)(1-p^kz^k)-(1-pz)p^kz^kt}{(1-z)(1-p^kz^k)-(1-pz)p^kz^kt}\Big) + \frac{1-p^kz^k}{(1-pz)(1-p^kz^k-p^kz^kt)} \tag{3.9}$$

and

(b) the rth factorial moment $\mu_{(r)}(n)$ *is given by*

$$\mu_{(r)}(n) = n(r-1)!\sum_{i=0}^{m_r}\sum_{j=r}^{[(n-i)/k]}(-1)^i\binom{r}{i}\binom{n-kj+r-i-1}{r-1}\binom{j-1}{r-1}p^{kj+1} + r!\binom{[n/k]}{r}(1-k\delta_{n,k[n/k]})p^n,\quad r=1,2,\ldots,[n/k],\ n\geq k, \tag{3.10}$$

where $m_r = \min\{r, n-kr\}$.

Proof.

(a) The factorial moment generating function (3.7) is given in terms of the probability generating function (3.4) by

$$f_n(t) = g_n(t+1),\quad n=0,1,2\ldots$$

and therefore the generating function (3.8) is given in terms of generating function (3.3) by the relation

$$f(t,z) = g(t+1,z),$$

which, on using (3.5), implies (3.9).

(b) the generating function (3.9) may be rewritten as

$$f(t,z) = z\frac{d}{dz}\log(\frac{1-pz}{1-z}) - z\frac{\partial}{\partial Z}\log(1 - \frac{(1-pz)p^k z^k t}{(1-z)(1-p^k z^k)})$$
$$+ z\frac{\partial}{\partial Z}\log(1 - \frac{p^k z^k t}{1-p^k z^k}) + \frac{1}{1-pz}(1 - \frac{p^k z^k t}{1-p^k z^k})^{-1}$$
$$= \frac{1}{1-z} - z\frac{\partial}{\partial z}\log[1 - \frac{(1-pz)p^k z^k t}{(1-z)(1-p^k z^k)}] + z\frac{\partial}{\partial Z}\log(1 - \frac{p^k z^k t}{1-p^k z^k})$$
$$- \frac{1}{1-pz} + \frac{1}{1-pz}(1 - \frac{p^k z^k t}{1-p^k z^k})^{-1}$$
$$= \frac{1}{1-z} - z\frac{\partial}{\partial Z}\log(1 - \frac{(1-pz)p^k z^k t}{(1-z)(1-p^k z^k)}) + z\frac{\partial}{\partial Z}\log(1 - \frac{p^k z^k t}{1-p^k z^k})$$
$$+ \frac{1}{1-pz}\frac{p^k z^k t}{1-p^k z^k}(1 - \frac{p^k z^k t}{1-p^k z^k})^{-1}.$$

Let

$$M_r(z) = \sum_{n=0}^{\infty} \mu_{(r)}(n) z^n, \quad r = 0,1,2\ldots \tag{3.11}$$

Then

$$f(t,z) = \sum_{n=0}^{\infty} f_n(t) z^n = \sum_{r=0}^{\infty} M_r(z) t^r / r!.$$

Expanding $f(t,z)$ into powers of t,

$$f(z,t) = \frac{1}{1-z} + z\frac{\partial}{\partial z}\sum_{r=1}^{\infty}\frac{(1-pz)^r p^{kr} z^{kr}}{(1-z)^r(1-p^k z^k)^r}\frac{t^r}{r}$$
$$-z\frac{\partial}{\partial z}\sum_{r=1}^{\infty}\frac{p^{kr} z^{kr}}{(1-p^k z^k)^r}\frac{t^r}{r} + \frac{1}{1-pz}\sum_{r=1}^{\infty}\frac{p^{kr} z^{kr}}{(1-p^k z^k)^r} t^r,$$

it follows that

$$M_r(z) = (r-1)! z\frac{d}{dz}\frac{(1-pz)^r p^{kr} z^{kr}}{(1-z)^r (1-p^k z^k)^k} - (r-1)! z\frac{d}{dz}\frac{p^{kr} z^{kr}}{(1-p^k z^k)^r}$$
$$+ r!\frac{p^{kr} z^{kr}}{(1-pz)(1-p^k z^k)^r}, \quad r = 1,2,\ldots.$$

and

$$M_r(z) = (r-1)!z\frac{d}{dz}\sum_{i=0}^{r}(-1)^i\binom{r}{i}p^iz^i\sum_{s=0}^{\infty}\binom{r+s-1}{s}z^s\sum_{j=r}^{\infty}\binom{j-1}{r-1}p^{kj}z^{kj}$$

$$-(r-1)!z\frac{d}{dz}\sum_{j=r}^{\infty}\binom{j-1}{r-1}p^{kj}z^{kj}+r!\sum_{i=0}^{\infty}p^iu^i\sum_{j=r}^{\infty}\binom{j-1}{r-1}p^{kj}z^{kj}$$

$$= (r-1)!\bullet$$

$$\sum_{n=kr}^{\infty}n\Big\{\sum_{i=0}^{m_r}\sum_{j=r}^{[(n-i)/k]}(-1)^i\binom{r}{i}\binom{n-kj+r-i-1}{r-1}\binom{j-1}{r-1}p^{kj+i}\Big\}z^n$$

$$-r!\sum_{j=r}^{\infty}kj\binom{j-1}{r-1}p^{kj}z^{kj}+r!\sum_{n=kr}^{\infty}\sum_{j=r}^{[n/k]}\binom{j-1}{r-1}p^nz^n$$

$$=(r-1)!\sum_{n=kr}^{\infty}n\Big\{\sum_{i=0}^{m_r}\sum_{j=r}^{[(n-i)/k]}(-1)^i\binom{r}{i}\binom{n-kj+r-1}{r-1}\binom{j-1}{r-1}p^{kj+i}\Big\}z^n$$

$$-r!k\sum_{j=r}^{\infty}\binom{j}{r}p^{kj}z^{kj}+r!\sum_{n=kr}^{\infty}\binom{[n/k]}{r}p^nz^n,$$

Therefore, by virtue of (3.11), the expression (3.10) is deduced.

Corollary 3.3. *The mean* $\mu(n) = E(N^c_{n,k})$ *is given by*

$$\mu(n) = \begin{cases} 0 & 1 \le n \le k-1, \quad k \ge 2 \\ [n/k]p^n + np^kq\frac{1-p^{n-k}}{1-p^k}, & n = k[n/k], \quad k \ge 1 \\ [n/k]p^n + np^kq\frac{1-p^{n-s}}{1-p^k}, & n = k[n/k]+s, \quad 1 \le s \le k-1, \quad k \ge 2 \end{cases} \tag{3.12}$$

Proof. From (3.10) with $r = 1$ it follows that

$$\mu(n) = n\sum_{j=1}^{[n/k]}p^{kj} - n\sum_{j=1}^{[(n-1)/k]}P^{kj+1} + [n/k](1-k\delta_{n,k[n/k]})p^n.$$

Thus, for $n = k[n/k]$

$$\mu(n) = n\sum_{j=1}^{[n/k]} p^{kj} - n\sum_{j=1}^{[n/k]-1} p^{kj+1} - np^n + [n/k]p^n = [n/k]p^n + nq\sum_{j=1}^{[n/k]-1} p^{kj}$$
$$= [n/k]p^n + np^k q\frac{1-p^{n-k}}{1-p^k}$$

and for $n = k[n/k] + s$, $1 \le s \le k-1$,

$$\mu(n) = n\sum_{j=1}^{[n/k]} p^{kj} - n\sum_{j=1}^{[n/k]} p^{kj+1} + [n/k]p^n$$
$$= [n/k]p^n + nq\sum_{j=1}^{[n/k]} p^{kj} = [n/k]p^n + np^k q\frac{1-p^{k[n/k]}}{1-p^k}$$
$$= [n/k]p^n + np^k q\frac{1-p^{n-s}}{1-p^k}$$

4. The Conditional Distribution of $N^c_{n,k}$ Given S_n

The marginal distribution of S_n is the binomial with probability function

$$P(S_n = y) = \binom{n}{y} p^y q^{n-y}, y = 0, 1, 2, \ldots, n.$$

The conditional distribution of $N^c_{n,k}$ given $S_n = y$ is given in the following corollary of Theorem 2.1.

Corollary 4.1. *The conditional probability function*

$$p_n(x|y) = P(N^c_{n,k} = x|S_n = y), \quad x = 0, 1, \ldots [y/k], \quad (y = 0, 1, 2, \ldots n)$$

is given by

$$p_n(x|y) = \binom{n-y+x-1}{x} N_c(y-kx, n-y, k)/\binom{n}{y}, x = 0, 1, \ldots, [y/k] \quad (4.1)$$

for $y = 0, 1, \ldots, n-1$ *and*

$$p_n(x|n) = 1, \quad x = [n/k], \quad (4.2)$$

where $[s]$ *denotes the integral part of* s*, with* $[s] \leq s < [s]+1$*, and the coefficient* $N_c(y-kx, n-y, k)$ *is given by (2.2).*

References

CHARALAMBIDES, Ch. A. (1991). Lucas numbers and polynomials of order k and the length of the longest circular success run. *Fibonacci Quart.* **29**, 290-297.

CHRYSSAPHINOU, O., PAPASTAVRIDIS, S. and TSAPELAS, T. (1992). On the number of overlapping success runs in a sequence of independent Bernoulli trials. In *Applications of Fibonacci Numbers,* Vol. 5, ed. G.E. Bergum, Kluwer Acad. Publishers.

FELLER, W. (1968). *An Introduction to Probability Theory and its Applications,* Vol. 1, 3rd edition, Wiley, New York.

GODBOLE, A. P. (1990). Specific formulae for some success run distributions. *Statist. Probab. Lett.* **10**, 119-124.

HIRANO, K. (1986). Some properties of the distributions of order k. In *Fibonacci Numbers and Their Applications,* ed. A. N. Philippou et al. Reidel, Dordrecht.

HWANG, F. K. and PAPASTAVRIDIS, S. (1991). Binary vectors with exactly k nonoverlapping m-tuples of consecutive ones. *Discrete Applied Math.* **30**, 83-86.

LAMBIRIS, M. and PAPASTAVRIDIS, S. (1985). Exact reliability formulas for linear and circular consecutive-k-out-of-n: F systems. *IEEE Trans. Reliability* **34**, 124-126.

LING, K. D. (1988). On binomial distributions of order k. *Statist. Probab. Lett.* **6**, 247-250.

MAKRI, F. S. and PHILIPPOU, A. N. (1993). Binomial distributions of order k on the circle. In this volume.

PAPASTAVRIDIS, S. (1990). m-consecutive-k-out-of-n : F systems. *IEEE Trans. Reliability* **39**, 386-388.

PHILIPPOU, A. N. amd MAKRI, F. S. (1986). Successes, runs and longest runs. *Statist. Probab. Lett.* **4**, 211-215.

RIORDAN, J. (1958). *An Introduction to Combinatorial Analysis,* Wiley, New York.

INVERSION TECHNIQUES and COMBINATORIAL IDENTITIES:

A Quick Introduction to Hypergeometric Evaluations

CHU Wenchang [1]

Institute of Systems Science, Academia Sinica, Beijing

Based on a pair of reciprocal relations due to Gould and Hsu, a general attack for verifying and discovering binomial identities is described. For ordinary hypergeometric series, the most basic closed formulas are revisited and some new interesting evaluations are established by means of an embedding technique, which may be more accessible to the reader.

1. Introduction

Combinatorial problems arise in many areas of mathematics and theoretical physics, and their solutions often involve the evaluation of combinatorial sums. There are various methods for evaluating such sums scattered in the literature and some procedures are discussed in the books by Riordan [23], Knuth [20] and Egorychev [10].

The impression one gains from treatises on the subject is that, for the most part, such identities cannot be put into a coherent setting. Riodan's [23, page *vii*] statement is perhaps the most pessimistic: *The central part developed is that identities are both inexhaustible and unpredictable, and age-old dream of putting order in this chaos is doomed to failure.*

This situation seems quite unfortunate. Knuth [20] has written that *there are thousands of binomial identities, many excellent mathematicians have expended considerable amount of time in proving them.* Contrary to the prevailing view, Andrews [2] first pointed out, and now it is widely recognized, that hypergeometric series have

[1] Supported by Alexander von Humboldt Foundation

A. P. Godbole and S. G. Papastavridis (eds.), Runs and Patterns in Probability, 31–57.

great relevance to combinatorial sums, i.e., the most of binomial identities are special cases of relatively few hypergeometric identities. Thus a great deal of mathematical ingenuity has been needlessly wasted because many mathematicians have been unable to recognize that a given binomial summation is actually equivalent to one already known, as binomial coefficients are like chameleons, changing their appearance easily. To carry on our discussion, we must introduce the notion of hypergeometric series.

With the shifted factorial

$$(x)_0 = 1, \quad (x)_n = x(x+1)\cdots(x+n-1), \quad (n>0) \tag{1-1}$$

the generalized hypergeometric series is defined by

$${}_{1+p}F_q\left[\begin{matrix} a_0, & a_1, & \cdots, & a_p \\ & b_1, & \cdots, & b_q \end{matrix}; z\right] = \sum_{n=0}^{\infty} \frac{(a_0)_n(a_1)_n\ldots(a_p)_n}{n!(b_1)_n\ldots(b_q)_n} z^n . \tag{1-2}$$

If $z=1$, then we omit it. It has $p+1$ numerator parameters and q denominator parameters, and one variable. Any of these quantities may be real or complex but the b–parameters must not be zero or negative integers, as in that case the series is not defined. If any of the a–parameters is zero or a negative integer, the function reduces to a polynomial (or terminating series).

When $p<q$, the series converges for all values of z. When $p>q$, the series converges only for $z=0$, and is therefore significant only when it terminates. Throughout the paper we shall be concerned with the case when $p=q$. Then the series converges when $|z|<1$, and also when $z=1$ provided that $R(\Sigma b_i - \Sigma a_j) > 0$, and when $z=-1$ provided that $R(1+\Sigma b_i - \Sigma a_j) > 0$.

According to the relations satisfied by the parameters involved in ${}_{1+p}F_p$, the hypergeometric series can be classified as follows: If $\Sigma b = k + \Sigma a$, then the series will be said to be k–balanced, and if $k=1$ further, will be said to be Saalschutzian or balanced. If $1+a_0 = a_1+b_1 = \cdots = a_p+b_p$, the series will be said to be well-poised, and if $a_0/2 = a_1 - 1 = b_1$ further, will be said to be very well-poised. The series will be called nearly-poised if all but one pairs of parameters have the same sum.

To shorten the writing, we shall use the contracted notations (see [26]) for the products of shifted factorials and Gamma functions

$$\left[\begin{matrix} a_1, a_2, \ldots, a_p \\ b_1, b_2, \ldots, b_q \end{matrix}\right]_n = \prod (a_i)_n / \prod (b_j)_n \tag{1-3a}$$

$$\Gamma\left[\begin{matrix} a_1, a_2, \ldots, a_p \\ b_1, b_2, \ldots, b_q \end{matrix}\right] = \prod \Gamma(a_i) / \prod \Gamma(b_j) \tag{1-3b}$$

which will be found to be convenient for understanding our approach.

Instead of ad-hoc tricks, Andrews [2] proposed the unifying principle that can systematically be used to handle a great variety of binomial coefficient summations all at once. And the whole technique consists of transforming the identity expressed in terms of binomial coefficients into a hypergeometric formula in a *canonical* way. Since then, attention has been turned to methods of systematizing the closed forms for hypergeometric series. Along this line, Gosper [12] (cf.[16]) has shown how to find indefinite hypergeometric sums, where they exist, by quite a general procedure. Recently, Wilf and Zeilberger [30] have described a general attack on definite hypergeometrics based on *W-Z pairs*. But these two methods are heavily dependent on computerized certification. One could not imagine how to work on them without a computer. On another side, the theory of hypergeometric series is vast itself, and up to now, it has not been able to revisit the most of related closed formulas in a self contained booklet. Hence a new approach is desirable to treat the known closed formulas of hypergeometric series in a unified way, which could enable non-specialists to get access to such identities. It will be demonstrated that inversion techniques can be used to accomplish this project. The basic tool is a pair of inverse series relations due to Gould and Hsu [15], which could be stated as follows: Let $\{a_i\}$ and $\{b_i\}$ be two sequences of complex numbers such that the polynomials defined by

$$\phi(x;0) = 1, \quad \phi(x;n) = \prod_{k=0}^{n-1}(a_k + xb_k), \quad (n > 0) \tag{1-4}$$

differ from zero for $x, n \in \mathbf{N}_0$ (the set of non-negative integers). Then there hold the inverse relations.

$$f(n) = \sum_{k=0}^{n}(-1)^k \binom{n}{k} \phi(k;n) g(k) \tag{1-4a}$$

$$g(n) = \sum_{k=0}^{n}(-1)^k \binom{n}{k} \frac{a_k + kb_k}{\phi(n;k+1)} f(k) \tag{1-4b}$$

One implication of this pair is that for every identity of the form (1-4a) or (1-4b), there is a companion of the dual form. To prove one is to prove both. For (1-4a–1-4b) involve two arbitrary number sequences, there is a relatively big flexibility in use. Sometimes we call (1-4a–1-4b) the dual pair in short.

The central part of this paper is to show how the embedding technique based on (1-4a–1-4b) can be used to verify and discover combinatorial identities in a unified way. With this elementary and direct approach for terminating hypergeometric identities, the reciprocal pair (1-4a–1-4b) is just like a *black-box*. As described by Greene and Knuth [17], *inverse relations are partially responsible for the proliferation of binomial identities. If one member of an inverse pair can be properly imbedded in a known binomial identity, then the other member of the pair will often provide a new identity.* This enables us to manufacture identities frequently. On another side, if one wants to confirm (prove or disprove) the truth for a combinatorial summation formula in the forms of (1-4a) or (1-4b), then one only needs to check its dual formulation which often reduces to a trival combinatorial fact. We will convince the reader that Gould and Hsu's inverse pair is indeed a powerful tool for terminating hypergeometrics, as major closed formulas (for examples, those displayed in the books by

Bailey [5] and Slater [26]) can be revisited systematically, from the telescoping.

The most striking examples [9] of the insight gained from this new technique are the realization that the Hagen-Rothe identity (cf.[13, 14])

$$\sum_{k=0}^{n} \frac{u+vk}{(a+bk)(c-bk)}\binom{a+bk}{k}\binom{c-bk}{n-k} = \frac{(a+c)u+n(av-bu)}{a(a+c)(c-bn)}\binom{a+c}{n} \tag{1-5a}$$

and the Abel formula (cf.[13, 14])

$$\sum_{k=0}^{n} \frac{u+vk}{(a+bk)(c-bk)}\binom{n}{k}(a+bk)^k(c-bk)^{n-k} = \frac{(a+c)u+n(av-bu)}{a(a+c)(c-bn)}(a+c)^n \tag{1-5b}$$

are trivally implied by the Chu–Vandermonde convolution

$$\sum_{k=0}^{n}\binom{x}{k}\binom{y}{n-k} = \binom{x+y}{n} \tag{1-6a}$$

and the binomial theorem

$$\sum_{k=0}^{n}\binom{n}{k}x^k y^{n-k} = (x+y)^n \tag{1-6b}$$

respectively.

It should be pointed out that neither (1-5a) nor (1-5b) in general can be attributed to the known hypergeometric identities. Even for the special case $b=2$ in (1-5a), the method of *WZ-pairs* could not succeed in verification [29].

As a supplementary means, we shall occasionally use the composition for inversions (or *series arrangement*) to lift up the order of hypergeometric summations. On the surface (or *formally*), this process seems to retrogress, but it often leads to the progression on summations in practice (e.g., *Whipple transformation and Dougall formulae*). To make the paper self-contained, two appendices on the reciprocity-certification between Gould–Hsu pair and the preliminaries for Gamma functions are presented at the end of the paper. As for the development of inversion techniques on strange evaluations [11] of ordinary hypergeometric series and the q-analogue for basic hypergeometric series of the materials treated here, they will appear in the forthcoming papers.

2. Chu–Vandermonde–Gauss and Pfaff–Saalschutz–Sheppard

Chu-Vandermonde Convolutions

Denote by Δ the difference operator with unit increment in x. Induction on n will yield

$$\Delta^n \left\{ \frac{(a)_x}{(b)_x} \right\}_{x=0} = (-1)^n (b-a)_n/(b)_n. \qquad (2-1a)$$

This is equivalent to the Chu-Vandermonde (see e.g.[5]) convolution

$${}_2F_1 \left[\begin{matrix} -n, & a \\ & b \end{matrix} \right] = \frac{(b-a)_n}{(b)_n} \qquad (2-1b)$$

which can be reformulated in terms of binomial coefficients:

$$\sum_{k=0}^{n} \binom{x}{k} \binom{y}{n-k} = \binom{x+y}{n}, \qquad (2-1c)$$

$$\sum_{k=0}^{n} \binom{u+k}{k} \binom{v+n-k}{n-k} = \binom{u+v+n+1}{n}. \qquad (2-1d)$$

For non-negative integers ε and k, we have, by Newton interpolation

$$\frac{(x+\varepsilon)_k}{(x)_k} = \frac{(x+k)_\varepsilon}{(x)_\varepsilon} = \sum_{i=0}^{\varepsilon} (-1)^i \binom{k}{i} (-\varepsilon)_i/(x)_i. \qquad (2-2)$$

Interchanging order of summation and substituting $k = i + j$ on the summation indices, we can proceed as follows:

$$
{}_3F_2\left[\begin{matrix}-n, & x+\varepsilon, & a\\ & x, & b\end{matrix}\right] = \sum_{k=0}^{n}\left[\begin{matrix}-n, & a\\ 1, & b\end{matrix}\right]_k \sum_{i=0}^{k}(-1)^i\binom{k}{i}(-\varepsilon)_i/(x)_i
$$

$$
= \sum_{i=0}^{n}(-1)^i\left[\begin{matrix}-n, & a, & -\varepsilon\\ 1, & b, & x\end{matrix}\right]_i {}_2F_1\left[\begin{matrix}-n+i, & a+i\\ & b+i\end{matrix}\right].
$$

In view of Chu-Vandermonde theorem, we have obtained the symmetric formula:

$$
{}_3F_2\left[\begin{matrix}-n, & x+\varepsilon, & a\\ & x, & b\end{matrix}\right] = \frac{(b-a)_n}{(b)_n}\,{}_3F_2\left[\begin{matrix}-n, & -\varepsilon, & a\\ & 1+a-b-n, & x\end{matrix}\right]. \quad (2-3)
$$

For $\varepsilon = 0$, it reduces to Eq.[2-1b]. For $\varepsilon = 1$ and $\varepsilon = 2$, the corresponding formulas can be displayed as follows:

$$
{}_3F_2\left[\begin{matrix}-n, & x+1, & a\\ & x, & b\end{matrix}\right] = \frac{(b-a)_n}{(b)_n}\{1+\frac{an}{x(1+a-b-n)}\} \quad (2-3a)
$$

$$
{}_3F_2\left[\begin{matrix}-n, & x+2, & a\\ & x, & b\end{matrix}\right] = \frac{(b-a)_n}{(b)_n} \quad (2-3b)
$$

$$
\times\{1+\frac{2an}{x(1+a-b-n)}+\frac{an(a+1)(n-1)}{x(x+1)(1+a-b-n)(2+a-b-n)}\}.
$$

Pfaff–Saalschutz Formulas

A similar method as for (2-3) also works for ε–balanced ${}_3F_2$ series which can be demonstrated as follows:

$$
{}_3F_2\left[\begin{matrix}-n, & a, & b\\ c+\varepsilon, & 1+a+b-c-n\end{matrix}\right] =
$$

$$\sum_{k=0}^{n}\begin{bmatrix} -n, & b \\ 1, 1+a+b-c-n \end{bmatrix}_k {}_2F_1\begin{bmatrix} -k, & c-a+\varepsilon \\ & c+\varepsilon \end{bmatrix}$$

$$=\sum_{i=0}^{n}(-1)^i\begin{bmatrix} -n, c-a+\varepsilon, b \\ 1, 1+a+b-c-n, c+\varepsilon \end{bmatrix}_i {}_2F_1\begin{bmatrix} -n+i, b+i \\ 1+a+b-c-n+i \end{bmatrix}$$

$$=\frac{(c-a)_n}{(c-a-b)_n}\,{}_3F_2\begin{bmatrix} -n, & b, & c-a+\varepsilon \\ & c+\varepsilon, & c-a \end{bmatrix}.$$

The last line can be transformed by (2-3) and thus we have (cf.[25])

$${}_3F_2\begin{bmatrix} -n, & a, & b \\ c+\varepsilon, 1+a+b-c-n \end{bmatrix} =$$

$$\begin{bmatrix} c-a, c-b+\varepsilon \\ c-a-b, c+\varepsilon \end{bmatrix}_n {}_3F_2\begin{bmatrix} -n, & -\varepsilon, & b \\ 1+b-c-n-\varepsilon, c-a \end{bmatrix}.$$

For $\varepsilon = 0$, 1 and 2, it degenerates, respectively, to the Pfaff-Saalschutz formula (see e.g.[5])

$${}_3F_2\begin{bmatrix} -n, & a, & b \\ c, 1+a+b-c-n \end{bmatrix} = \begin{bmatrix} c-a, c-b \\ c, c-a-b \end{bmatrix}_n, \qquad (2-3a)$$

Anderson-Sheppard summation (cf.[1, 25])

$${}_3F_2\begin{bmatrix} -n, & a, & b \\ c+1, 1+a+b-c-n \end{bmatrix} = \begin{bmatrix} c-a, c-b \\ c+1, c-a-b \end{bmatrix}_n \left\{1+\frac{n(c-a-b)}{(c-a)(c-b)}\right\} \qquad (2-3b)$$

and

$${}_3F_2\begin{bmatrix} -n, & a, & b \\ c+2, 1+a+b-c-n \end{bmatrix} = \begin{bmatrix} c-a, c-b+2 \\ c+2, c-a-b \end{bmatrix}_n \qquad (2-3c)$$

$$\times\{1+\frac{2bn}{(c-a)(b-c-n-1)}+\frac{bn(b+1)(n-1)}{(c-a)(c-a+1)(b-c-n)(b-c-n-1)}\}.$$

Gauss Theorem

For n tending to infinity, (2-3a) reduces to the Gauss summation formula (cf.[5, page 2])

$$_2F_1\left[\begin{matrix} a, & b \\ & c \end{matrix}\right]=\Gamma\left[\begin{matrix} c, c-a-b \\ c-a, c-b \end{matrix}\right]. \qquad (2-4)$$

Taking $b, c \to \infty$ and $b/c = z$ in the above, the Stirling formula may be used to yield the binomial theorem

$$_1F_0\left[\begin{matrix} a & \\ - & ;z \end{matrix}\right]=(1-z)^{-a} \qquad (2-5)$$

which is a hypergeometric version of (1-6b).

In the following, we shall demonstrate, by Gould–Hsu inverse pair (1-4a–1-4b), a couple of binomial identities.

First, we have, from (2-3b)

$$_3F_2\left[\begin{matrix} -b, & a-2b, & -n \\ & 1+a-b, & 1-2b-n \end{matrix}\right]=\left[\begin{matrix} a, & 1+a/2, & b \\ 1+a-b, & a/2, & 2b \end{matrix}\right]_n. \qquad (2-6a)$$

This 2-balanced summation can be reformulated as

$$\sum_{k=0}^{n}(-1)^k\binom{n}{k}\frac{-2b}{(-2b-n)_{k+1}}\left[\begin{matrix} a-2b, -b \\ 1+a-b \end{matrix}\right]_k= \qquad \text{(2-6b)}$$

$$\left[\begin{matrix} a, & 1+a/2, & b \\ 1+a-b, & a/2, & 1+2b \end{matrix}\right]_n . \tag{2-7b}$$

Embedding this relation in (1-4b), (1-4a) gives the dual form

$$\sum_{k=0}^{n}(-1)^k\binom{n}{k}(-2b-k)_n\left[\begin{matrix} a, & 1+a/2, & b \\ 1+a-b, & a/2, & 1+2b \end{matrix}\right]_k = \left[\begin{matrix} a-2b, -b \\ 1+a-b \end{matrix}\right]_n . \tag{2-7c}$$

Noting that

$$(x-k)_n = \frac{(x)_n(1-x)_k}{(1-x-n)_k},$$

the above dual summation can be expressed as a nearly-poised hypergeometric identity

$${}_4F_3\left[\begin{matrix} a, & 1+a/2, & b, & -n \\ & a/2, & 1+a-b, & 1+2b-n \end{matrix}\right] = \left[\begin{matrix} a-2b, & -b \\ 1+a-b, & -2b \end{matrix}\right]_n . \tag{2-7d}$$

Next, from (2-3a), we have

$${}_3F_2\left[\begin{matrix} b-a, & 1+a/2, & -n \\ & a/2, & 1+b+n \end{matrix}\right] = \frac{(a)_{2n}(1+b)_n}{(b)_{2n}(1+a)_n}. \tag{2-8a}$$

Then the same procedure as that for (2-6a) will yield a Saalschutzian hypergeometrics (cf.[4])

$${}_4F_3\left[\begin{matrix} -n, & a/2, & (1+a)/2, & b+n \\ & b/2, & (1+b)/2, & 1+a \end{matrix}\right] = \frac{(b-a)_n}{(b)_n}, \tag{2-8b}$$

which corresponds to one special case $b=2$ of (1-5a).

Finally, notice that

$${}_4F_3\left[\begin{matrix} a, & 1+a/2, & b, & -n \\ & a/2, & 1+a-b, & 2+2b-n \end{matrix}\right] =$$

$$
{}_3F_2\left[\begin{matrix} a, & b, & -n \\ & 1+a-b, & 2+2b-n \end{matrix}\right]
$$

$$
-\frac{2bn}{(1+a-b)(2+2b-n)}\,{}_3F_2\left[\begin{matrix} 1+a, & 1+b, & -n+1 \\ & 2+a-b, & 3+2b-n \end{matrix}\right].
$$

From (2-4), one can derive its closed expression as follows:

$$
{}_4F_3\left[\begin{matrix} a, & 1+a/2, & b, & -n \\ & a/2, & 1+a-b, & 2+2b-n \end{matrix}\right] =
$$

$$
\left[\begin{matrix} a-2b-1, -b-1, -b+(1+a)/2 \\ 1+a-b, -2b-1, -b-(1-a)/2 \end{matrix}\right]_n
$$

which is a self-reciprocal identity via (1-4a–1-4b).

3. Kummer–Dixon–Dougall

The surprising fact implied by Gould–Hsu pair is the reciprocity between the terminating Saalschutzian theorem [5, page 9] and the very well-poised Dixon formula [5, page 13].

$$
{}_3F_2\left[\begin{matrix} -n, & a+n, & 1+a-b-c \\ & 1+a-b, & 1+a-c \end{matrix}\right] = \left[\begin{matrix} b, & c \\ 1+a-b, & 1+a-c \end{matrix}\right]_n \tag{3-1a}
$$

$$
{}_5F_4\left[\begin{matrix} a, & 1+a/2, & b, & c, & -n \\ & a/2, & 1+a-b, & 1+a-c, & 1+a+n \end{matrix}\right] \tag{3-1b}
$$

$$
= \left[\begin{matrix} 1+a, 1+a-b-c \\ 1+a-b, 1+a-c \end{matrix}\right]_n
$$

By means of composition based on this inverse pair, we can derive the Whipple transformation as follows:

$$
{}_7F_6\left[\begin{matrix} a, & 1+a/2, & b, & c, & d, & e, & -n \\ & a/2, & 1+a-b, & 1+a-c, & 1+a-d, & 1+a-e, & 1+a+n \end{matrix}\right] =
$$

$$
\sum_{k=0}^{n}\left[\begin{matrix} a, & 1+a/2, & b, & c, & -n \\ 1, & a/2, & 1+a-b, & 1+a-c, & 1+a+n \end{matrix}\right]_k
$$

$$
\times\ {}_3F_2\left[\begin{matrix} -k, & a+k, & 1+a-d-e \\ & 1+a-d, & 1+a-e \end{matrix}\right]
$$

$$
\overset{*}{=} \sum_{i=0}^{n}(-1)^i\frac{(1+a)_{2i}}{i!}\left[\begin{matrix} 1+a-d-e, & b, & c, & -n \\ 1+a-b, 1+a-c, & 1+a-d, 1+a-e, 1+a+n \end{matrix}\right]_i
$$

$$
\times\ {}_5F_4\left[\begin{matrix} a+2i, & 1+\frac{a}{2}+i, & b+i, & c+i, & -n+i \\ & \frac{a}{2}+i, & 1+a-b+i, & 1+a-c+i, & 1+a+n+i \end{matrix}\right],
$$

where $\overset{*}{=}$ follows from the interchange of the summation-order and the substitution on summation variables as in (2-3). Evaluating the inner–sum in the above with Eq.(3-2) leads to

Whipple transformation [28]

$$
{}_7F_6\left[\begin{matrix} a, & 1+a/2, & b, & c, & d, & e, & -n \\ & a/2, & 1+a-b, & 1+a-c, & 1+a-d, & 1+a-e, & 1+a+n \end{matrix}\right] = \tag{3-2}
$$

$$
\left[\begin{matrix} 1+a, 1+a-b-c \\ 1+a-b, 1+a-c \end{matrix}\right]_n {}_4F_3\left[\begin{matrix} -n, & b, & c, & 1+a-d-e \\ & 1+a-d, & 1+a-e, & b+c-a-n \end{matrix}\right].
$$

Dougall formulas

Taking $1+2a+n=b+c+d+e$, then ${}_4F_3$ in (3-2) reduces to a Saalschutzian series which can be evaluated by (2-3a). Thus we have established the Dougall theorem [28]

$$
{}_7F_6\left[\begin{matrix} a, & 1+a/2, & b, & c, & d, & e, & -n \\ & a/2, & 1+a-b, & 1+a-c, & 1+a-d, & 1+a-e, & 1+a+n \end{matrix}\right]
= \left[\begin{matrix} 1+a, 1+a-b-c, 1+a-b-d, 1+a-c-d \\ 1+a-b, 1+a-c, 1+a-d, 1+a-b-c-d \end{matrix}\right]_n,
$$

where $1+2a+n=b+c+d+e$. This is the most general hypergeometric identity which serves as the starting point for deriving binomial identities (cf.[8]).

Taking $2a+n=b+c+d+e$ instead, then ${}_4F_3$ in (3-2) could be evaluated through (2-4) and leads to a 4-balanced summation formula due to Lakin (cf.[3, 21])

$$
\begin{aligned}
{}_7F_6\left[\begin{matrix} a, & 1+a/2, & b, & c, & d, & e, & -n \\ & a/2, & 1+a-b, & 1+a-c, & 1+a-d, & 1+a-e, & 1+a+n \end{matrix}\right]
&= \left\{1+\frac{n\,(2a+n-b-c-d)\,(a-b-c-d)}{(a-b-c)\,(a-b-d)\,(a-c-d)}\right\} \\
&\times \left[\begin{matrix} 1+a, a-b-c, a-b-d, a-c-d \\ 1+a-b, 1+a-c, 1+a-d, a-b-c-d \end{matrix}\right]_n
\end{aligned}
\tag{3-3}
$$

When $n\to\infty$, both the Dougall and Lakin theorems reduce to the well-known non-terminating Dixon formula [5, 28]

$$
{}_5F_4\left[\begin{matrix} a, & 1+a/2, & b, & c, & d \\ & a/2, & 1+a-b, & 1+a-c, & 1+a-d \end{matrix}\right] = \tag{3-4}
$$

$$\Gamma\left[\begin{matrix}1+a-b,1+a-c,1+a-d,1+a-b-c-d\\1+a,1+a-b-c,1+a-b-d,1+a-c-d\end{matrix}\right]_n.$$

Dixon formulas

For $d=a/2$, $(1+a)/2$ and ∞, (3-4) will generate the following non-terminating identities (cf.[5, 26]), respectively.

$$ {}_3F_2\left[\begin{matrix}a, & b, & c\\ & 1+a-b, & 1+a-c\end{matrix}\right]$$

$$\Gamma\left[\begin{matrix}1+\frac{a}{2},1+\frac{a}{2}-b-c,1+a-b,1+a-c\\1+a,1+a-b-c,1+\frac{a}{2}-b,1+\frac{a}{2}-c\end{matrix}\right]. \qquad (3-5a)$$

$$ {}_4F_3\left[\begin{matrix}a, & 1+a/2, & b, & c\\ & a/2, & 1+a-b, & 1+a-c\end{matrix}\right]= \qquad (3-5b)$$

$$\Gamma\left[\begin{matrix}\frac{1}{2}+\frac{a}{2},\frac{1}{2}+\frac{a}{2}-b-c,1+a-b,1+a-c\\1+a,1+a-b-c,\frac{1}{2}+\frac{a}{2}-b,\frac{1}{2}+\frac{a}{2}-c\end{matrix}\right]$$

$$ {}_4F_3\left[\begin{matrix}a, & 1+a/2, & b, & c\\ & a/2, & 1+a-b, & 1+a-c\end{matrix};-1\right]=\Gamma\left[\begin{matrix}1+a-b,1+a-c\\1+a,1+a-b-c\end{matrix}\right]. \qquad (3-5c)$$

Kummer formulae

Kummer theorems (cf.[5, 26, 28]) follow from the limiting case $c\to\infty$ of (3-5a–3-5b);

$$ {}_2F_1\left[\begin{matrix}a, & b\\ & 1+a-b\end{matrix};-1\right]=\Gamma\left[\begin{matrix}1+\frac{a}{2},1+a-b\\1+a,1+\frac{a}{2}-b\end{matrix}\right]_n, \qquad (3-6a)$$

$$ {}_3F_2\left[\begin{matrix} a, & 1+a/2, & b \\ & a/2, & 1+a-b \end{matrix} ;-1\right] = \Gamma\left[\begin{matrix} \frac{1}{2}+\frac{a}{2}, 1+a-b \\ 1+a, \frac{1}{2}+\frac{a}{2}-b \end{matrix}\right]_n . \qquad (3-6b) $$

Some reciprocal pairs

A short list of reciprocal pairs may be obtained from the terminating forms of (3-5a–3-5c) and (3-6a–3-6b) through the Gould–Hsu pair (1-4a–1-4b) as follows:

$$ {}_3F_2\left[\begin{matrix} a, & b, & -n \\ & 1+a-b, & 1+a+n \end{matrix}\right] = \left[\begin{matrix} 1+a, & 1+\frac{a}{2}-b \\ 1+a/2, & 1+a-b \end{matrix}\right]_n \qquad (3-7a) $$

$$ {}_3F_2\left[\begin{matrix} a, & 1+\frac{a}{2}-b, & a+n \\ & 1+a/2, & 1+a-b \end{matrix}\right] = \frac{a}{a+2n}\left[\begin{matrix} b \\ 1+a-b \end{matrix}\right]_n \qquad (3-7b) $$

$$ {}_4F_3\left[\begin{matrix} a, & 1+a/2, & b, & -n \\ & a/2, & 1+a-b, & 1+a+n \end{matrix}\right] = \left[\begin{matrix} 1+a, & \frac{1}{2}+\frac{a}{2}-b \\ 1+a-b, & \frac{1}{2}+\frac{a}{2} \end{matrix}\right]_n \qquad (3-8a) $$

$$ {}_3F_2\left[\begin{matrix} -n, & \frac{1}{2}+\frac{a}{2}-b, & a+n \\ & \frac{1}{2}+\frac{a}{2}, & 1+a-b \end{matrix}\right] = \left[\begin{matrix} b \\ 1+a-b \end{matrix}\right]_n \qquad (3-8b) $$

$$ {}_4F_3\left[\begin{matrix} a, & 1+a/2, & b, & -n \\ & a/2, & 1+a-b, & 1+a+n \end{matrix} ;-1\right] = \left[\begin{matrix} 1+a \\ 1+a-b \end{matrix}\right]_n \qquad (3-9a) $$

$$ {}_2F_1\left[\begin{matrix} -n, & a+n \\ & 1+a-b \end{matrix}\right] = (-1)^n\left[\begin{matrix} b \\ 1+a-b \end{matrix}\right]_n \qquad (3-9b) $$

$$ {}_2F_1\left[\begin{matrix} a, & -n \\ & 1+a+n \end{matrix};-1\right] = \left[\begin{matrix} 1+a \\ 1+a/2 \end{matrix}\right]_n \tag{3-10a} $$

$$ {}_2F_1\left[\begin{matrix} -n, & a+n \\ & 1+a/2 \end{matrix}\right] = (-1)^n \frac{a}{a+2n} \tag{3-10b} $$

$$ {}_3F_2\left[\begin{matrix} a, & 1+a/2, & -n \\ & a/2, & 1+a+n \end{matrix};-1\right] = \left[\begin{matrix} 1+a \\ (1+a)/2 \end{matrix}\right]_n \tag{3-11a} $$

$$ {}_2F_1\left[\begin{matrix} -n, & a+n \\ & (1+a)/2 \end{matrix}\right] = (-1)^n \tag{3-11b} $$

4. Watson–Gauss and Whipple–Bailey

Watson's original formula and Gauss' second theorem

From the Saalschutz formula we have

$$ {}_3F_2\left[\begin{matrix} -n/2, & (1-n)/2, & c+(1-b)/2 \\ & c+1/2, & -n+(3-b)/2 \end{matrix}\right] = \left[\begin{matrix} b-1, & c \\ (b-1)/2, & 2c \end{matrix}\right]_n, \tag{4-1a} $$

which can be reformulated as

$$ \sum_{k=0}^{n} \binom{n}{2k} \left(\frac{b-1}{2}-k\right)_n \left[\begin{matrix} 1/2, c+(1-b)/2 \\ (3-b)/2, c+1/2 \end{matrix}\right]_k = \left[\begin{matrix} (3-b)/2, b-1, & c \\ (b-1)/2, & 2c \end{matrix}\right]_n. \tag{4-1b} $$

Embedding this in (1-4a), the dual relation (1-4b) gives

$$ \sum_{k=0}^{n} (-1)^k \binom{n}{k} \frac{(b+k-1)/2}{(\frac{b-n-1}{2})_{k+1}} \left[\begin{matrix} (3-b)/2, b-1, & c \\ (b-1)/2, & 2c \end{matrix}\right]_k = \tag{4-1c} $$

$$= \begin{cases} \left[\begin{matrix} 1/2, c+(1-b)/2 \\ (3-b)/2, c+1/2 \end{matrix}\right]_m & (n=2m) \\ 0 & (n-odd). \end{cases}$$

Rewriting it in terms of hypergeometrics, we obtain the original Watson's formula [27]:

$${}_3F_2\left[\begin{matrix} -n, & b, & c \\ & (b-n+1)/2, & 2c \end{matrix}\right] = \begin{cases} \left[\begin{matrix} 1/2, c+(1-b)/2 \\ (1-b)/2, c+1/2 \end{matrix}\right]_m & (n=2m) \\ 0 & (n-odd). \end{cases} \quad (4-1d)$$

Again, from the Saalschutz theorem, we have

$${}_3F_2\left[\begin{matrix} -n/2, & -(1+n)/2, & -n+(1-a-b)/2 \\ & -n+(1-a)/2, & -n+(1-b)/2 \end{matrix}\right]$$

$$= 4^{-n}\left[\begin{matrix} a, & b \\ (a+1)/2, & (b+1)/2 \end{matrix}\right]_n. \quad (4-2a)$$

Then the dual pair related to its reversal can be obtained from (1-4a–1-4b) similarly.

$$\sum_{k=0}^{n}(-1)^k\binom{n}{k}\frac{(-1-2k)_n}{1+2k}\left[\begin{matrix} (a+1)/2, (b+1)/2 \\ 1/2, (a+b+1)/2 \end{matrix}\right]_k = \frac{1}{n+1}\left[\begin{matrix} a, & b \\ (a+b+1)/2 \end{matrix}\right]_n \quad (4-2b)$$

$$\sum_{k=0}^{n}(-1)^k\binom{n}{k}\frac{-1}{(-1-2n)_{k+1}}\left[\begin{matrix} a, & b \\ (a+b+1)/2 \end{matrix}\right]_k$$

$$= \frac{1}{1+2n}\left[\begin{matrix} (a+1)/2, (b+1)/2 \\ 1/2, (a+b+1)/2 \end{matrix}\right]_n \quad (4-2c)$$

Among them, the last one could be restated as [6, Eq.(2.2)]

$$ {}_3F_2\left[\begin{matrix} a, & b, & -n \\ (a+b+1)/2, & -2n \end{matrix}\right] = \left[\begin{matrix} (a+1)/2, (b+1)/2 \\ 1/2, (a+b+1)/2 \end{matrix}\right]_n, \tag{4-2d} $$

which is another terminating version of the Watson formula, but not implied by that one.

Taking $n \to \infty$ in (4-2d), we get the Gauss' second summation theorem [5, page 11]

$$ {}_2F_1\left[\begin{matrix} a, & b, \\ (a+b+1)/2 \end{matrix} ;1/2\right] = \Gamma\left[\begin{matrix} 1/2, (a+b+1)/2 \\ (a+1)/2, (b+1)/2 \end{matrix}\right]. \tag{4-3} $$

Whipple and Bailey theorems

Similar to the derivation from Saalschutz theorem to Watson's formula, the reciprocity between Saalschutz formula and the terminating version of the Whipple theorem [6] can be established as follows:

$$ {}_3F_2\left[\begin{matrix} -n, & (1+e-a)/2, & (e+a)/2 \\ & (1+e-n)/2, & 1+(e-n)/2 \end{matrix}\right] = \left[\begin{matrix} a, & 1-a \\ -e, & 1-e \end{matrix}\right]_n \tag{4-4a} $$

$$ {}_3F_2\left[\begin{matrix} -n, & a, & 1-a \\ & e, & 1-2n-e \end{matrix}\right] = \left[\begin{matrix} (e+a)/2, & (1+e-a)/2 \\ e/2, & (1+e)/2 \end{matrix}\right]_n \tag{4-4b} $$

where the last identity [6, Eq.(3.1)] could not be derived from the non-terminating Whipple theorem.

The limiting case of (4-4b) for n tending to infinity reduces to Bailey theorem [5, page 11]

$$ {}_2F_1\left[\begin{matrix} a, & 1-a \\ & e \end{matrix} ;1/2\right] = \Gamma\left[\begin{matrix} e/2, (1+e)/2 \\ (1+e-a)/2, (e+a)/2 \end{matrix}\right]. \tag{4-5} $$

Non-terminating Watson and Whipple formulas

To make the paper self-contained, we reproduce the proofs (cf.[5, page 16]) for the non-terminating versions of

Watson formula

$$ {}_3F_2\left[\begin{matrix} a, & b, & c \\ (a+b+1)/2, & & 2c \end{matrix}\right] = \Gamma\left[\begin{matrix} \frac{1}{2}, \frac{1}{2}+c, \frac{1+a+b}{2}, \frac{1-a-b}{2}+c \\ \frac{1+a}{2}, \frac{1+b}{2}, \frac{1-a}{2}+c, \frac{1-b}{2}+c \end{matrix}\right], \qquad (4-6) $$

and Whipple formula

$$ {}_3F_2\left[\begin{matrix} a, & 1-a, & c \\ & e, & 1+2c-e \end{matrix}\right] $$

$$ = 2^{1-2c}\pi\,\Gamma\left[\begin{matrix} e, & 1+2c-e \\ \frac{a+e}{2}, \frac{1-a+e}{2}, \frac{1+a-e}{2}+c, 1-\frac{a+e}{2}+c \end{matrix}\right]. \qquad (4-7) $$

Rewrite Gauss' theorem in the form

$$ \Gamma\left[\begin{matrix} a+n, e+f-a+n \\ e+n, \quad f+n \end{matrix}\right] = {}_2F_1\left[\begin{matrix} e-a, f-a \\ e+f-a+n \end{matrix}\right]. $$

Then the series-composition can be carried out as follows:

$$ {}_3F_2\left[\begin{matrix} a, & b, & c \\ & e, & f \end{matrix}\right] $$

$$ = \Gamma\left[\begin{matrix} e, \quad f \\ a, e+f-a \end{matrix}\right] \sum_n \left[\begin{matrix} b, \quad c \\ 1, e+f-a \end{matrix}\right]_n {}_2F_1\left[\begin{matrix} e-a, f-a \\ e+f-a+n \end{matrix}\right] $$

$$ = \Gamma\left[\begin{matrix} e, \quad f \\ a, e+f-a \end{matrix}\right] \sum_m \left[\begin{matrix} e-a, f-a \\ 1, e+f-a \end{matrix}\right]_m {}_2F_1\left[\begin{matrix} b, \quad c \\ e+f-a+m \end{matrix}\right], $$

which leads to the Thomae's transformation (where $s=e+f-a-b-c$)

$$
{}_3F_2\left[\begin{matrix}a, & b, & c\\ & e, & f\end{matrix}\right]=\Gamma\left[\begin{matrix}s, & e, & f\\ a, & s+b, & s+c\end{matrix}\right]{}_3F_2\left[\begin{matrix}s, & e-a, & f-a\\ & b+s, & c+s\end{matrix}\right]. \tag{4-8}
$$

Thus (4-6) follows from (4-8) and Dixon's formula, and (4-7) from (4-8) and (4-6) immediately.

In addition, another terminating version of the Whipple theorem

$$
{}_3F_2\left[\begin{matrix}-n, & 1+n, & c\\ & e, & 1+2c-e\end{matrix}\right]=\left[\begin{matrix}(e-n)/2, e-2c\\ -c+(e-n)/2, e\end{matrix}\right]_n \tag{4-9}
$$

fits into the general relation (1-4a). The corresponding reciprocal relations are displayed as follows:

$$
\sum_{k=0}^{n}(-1)^n\binom{n}{k}(1+k)_n\left[\begin{matrix}1, \quad c\\ e, 1+2c-e\end{matrix}\right]_k=\left[\begin{matrix}1, & e-2c, (e-n)/2\\ e, & -c+(e-n)/2\end{matrix}\right]_n \tag{4-10a}
$$

$$
\sum_{k=0}^{n}(-1)^n\binom{n}{k}\frac{1+2k}{(1+n)_{k+1}}\left[\begin{matrix}1, & e-2c, (e-k)/2\\ e, & -c+(e-k)/2\end{matrix}\right]_k=\left[\begin{matrix}1, \quad c\\ e, 1+2c-e\end{matrix}\right]_n \tag{4-10b}
$$

Among them, the last summation is a new strange evaluation. It could be expressed, by splitting the summation with respect to the parity of summation index k, as a hypergeometric identity in terms of two nearly-poised ${}_6F_5$ series

$$
\left[\begin{matrix}2, c\\ e, 1+2c-e\end{matrix}\right]_n =
$$

$$
{}_6F_5\left[\begin{matrix}1, & -n/2, & (1-n)/2, & 1+1/4, & 1-e/2, & -c+(1+e)/2\\ & (3+n)/2, & (2+n)/2, & 1/4, & (1+e)/2, & 1+c-e/2\end{matrix}\right]
$$

$$-\frac{3n(e-1)(e-2c)}{(n+2)e(e-2c-1)}$$

$$\times {}_6F_5\left[\begin{matrix}1, & (1-n)/2, & 1-n/2, & 1+3/4, & (3-e)/2, & -c+(2+e)/2\\ & (4+n)/2, & (3+n)/2, & 3/4, & (2+e)/2, & c+(3-e)/2\end{matrix}\right].$$

5. Some More Evaluations

In this section, some new hypergeometric evaluations will be established through the combination of the Whipple transformation and the Hagen–Rothe identity.

Denote by $_7W_6[a;b,c,d,e,-n]$ the very well-poised series

$${}_7F_6\left[\begin{matrix}a, & 1+a/2, & b, & c, & d, & e, & -n\\ & a/2, & 1+a-b, & 1+a-c, & 1+a-d, & 1+a-e, & 1+a+n\end{matrix}\right].$$

The reversal of balanced formula (2-8b) can be restated as another balanced formula:

$${}_4F_3\left[\begin{matrix}-n, & (1+a)/2, & (2+a)/2, & c+n\\ & (1+c)/2, & (2+c)/2, & 1+a\end{matrix}\right]=\frac{c}{c+2n}\left[\begin{matrix}c-a\\ c\end{matrix}\right]_n. \quad (5-1)$$

Specify the parameters in (3-2) so that the $_4F_3$ could be evaluated by (2-8b) and (5-1). Then we can get the following four very well-poised hypergeometric identities:

$${}_7W_6[u;v/2,(1+v)/2,u-v,1+2u-v+n,-n]= \quad (5\text{-}2)$$

$$\left[\begin{matrix}1+u,1+2u-2v\\ 1+u-v,1+2u-v\end{matrix}\right]_n \quad (5\text{-}3)$$

$${}_7W_6[u;v/2,(1+v)/2,u-v,2u-v+n,-n]= \quad (5\text{-}4)$$

$$\frac{2u-v}{2u-v+2n}\left[\begin{matrix}1+u,2u-2v\\ u-v,2u-v\end{matrix}\right]_n \quad (5\text{-}5)$$

$$_7W_6[u; v/2, (1+v)/2, 1+u-v, 2u-v+n, -n] = \tag{5-6}$$

$$\frac{2u-v}{2u-v+2n}\begin{bmatrix} 1+u, 1+2u-2v \\ u-v, 2u-v \end{bmatrix}_n \tag{5-7}$$

$$_7W_6u; v/2, (1+v)/2, 1+u-v, 1+2u-v+n, -n] = \tag{5-8}$$

$$\begin{bmatrix} 1+u, 2+2u-2v \\ 1+u-v, 1+2u-v \end{bmatrix}_n \tag{5-9}$$

All the formulas displayed above have not appeared previously except for (5-3), which is due to Bailey [4] (cf.[5, page 98]).

APPENDIX A. Gould–Hsu Inversions

Substituting (1-4b) into (1-4a), the reciprocity between (1-4a) and (1-4b) reduces to the orthogonality

$$\sum_{k=m}^{n}(-1)^{k+m}\binom{n-m}{k-m}\frac{\phi(k;n)}{\phi(k;m+1)}(a_{m+1}+mb_{m+1}) = \delta_{m,n} \tag{A1}$$

which follows from the difference operation (cf.[15]).

However, if substituting (1-4a) into (1-4b), the orthogonality induced by Gould–Hsu pair would be

$$\sum_{k=m}^{n}(-1)^{k+m}\binom{n-m}{k-m}\frac{\phi(m;k)}{\phi(n;k+1)}(a_{k+1}+kb_{k+1}) = \delta_{m,n}. \tag{A2}$$

It was claimed in [15] that *it would be more tedious to prove this dual relation.* In fact, (A2) simply follows from the summand-splitting

$$\binom{n-m}{k-m}\frac{\phi(m;k)}{\phi(n;k+1)}(a_{k+1}+kb_{k+1})$$

$$= \binom{n-m-1}{k-m} \frac{\phi(m;k+1)}{\phi(n;k+1)} + \binom{n-m-1}{k-m-1} \frac{\phi(m;k)}{\phi(n;k)}$$

and the diagonal-cancelling.

APPENDIX B. Gamma Functions

Definition

Euler integral

$$\Gamma(z) = \int_0^\infty t^{z-1} e^{-t}\, dt, \qquad R(z) > 0 \tag{B1}$$

Infinite products

$$\Gamma(z) = \lim_{n\to\infty} n!\, n^z/(z)_{n+1} = z^{-1} \prod_{n=1}^{\infty} \{(1+1/n)^z/(1+z/n)\} \tag{B2}$$

$$1/\Gamma(z) = z e^{\gamma z} \prod_{n=1}^{\infty} \{(1+z/n)\, e^{-z/n}\}, \qquad (\gamma - \textit{Euler constant}). \tag{B3}$$

[B2 and B3 follow from $\Gamma(z) = \lim_{n\to\infty} \int_0^n t^{z-1}(1-t/n)^n\, dt$]

Recurrences

$$\Gamma(z+n) = (z)_n \Gamma(z) \tag{B4}$$

$$\Gamma(z-n) = (-1)^n \Gamma(z)/(1-z)_n \tag{B5}$$

Special evaluations

$$\Gamma(n+1) = n! \tag{B6}$$

$$\Gamma(1/2) = \sqrt{\pi} \tag{B7}$$

$$Res_{|z=-n} \Gamma(z) = (-1)^n/n! \tag{B8}$$

[B8 follows from the infinite product-expression]

Multiplicative formulae

$$\Gamma(2z) = \frac{2^{2z-1}}{\sqrt{\pi}}\Gamma(z)\Gamma(z+1/2) \tag{B9}$$

$$\Gamma(mz) = (2\pi)^{(1-m)/2} m^{mz-1/2} \prod_{k=0}^{m-1} \Gamma(z+k/m) \tag{B10}$$

Reciprocity

$$\Gamma(z)\Gamma(1-z) = \pi/\sin \pi z \tag{B11}$$

$$\Gamma(\frac{1}{2}+z)\Gamma(\frac{1}{2}-z) = \pi/\cos \pi z \tag{B12}$$

[B11 and B12 follow from the infinite product $\sin \pi z = \pi z \prod_{n=1}^{\infty}(1 - z^2/n^2)$]

Asymptotics

$$\Gamma(z+n) \sim n!\, n^{z-1}, \qquad (n \to \infty) \tag{B13}$$

[It follows from B2 and plays the same part as the Stirling formula, displayed below, in the asymptotics for hypergeometric series]

$$\Gamma(z+n) \sim \sqrt{2\pi}\, n^{z-1/2} (n/e)^n, \qquad (n \to \infty) \quad [\mathit{Stirling formula}] \tag{B14}$$

References

[1] E. A. Anderson, *Two summation formulae for product sums of binomial coefficients*, Math. Scand.1(1953), 261-262.

[2] G. E. Andrews, *Applications of basic hypergeometric functions*, SIAM Rev.16(1974), 441-484.

[3] R. Askey, *Variants of Clausen's formula for the square of a special* ${}_2F_1$, in *"Number Theorey and Related Topics"*, Tata Institute of Fundamental Research, Bombay, 1988, 1-12.

[4] W. N. Bailey, *Some identities involving generalized hypergeometric serie*, Proc. London Math. Soc.29(1929), 503-516.

[5] W. N. Bailey, *Generalized Hypergeometric Series*, Cambridge University Press, London, 1935.

[6] W. N. Bailey, *On the sum of a terminating* ${}_3F_2$, Quart. J. Math. Oxford (2), 4(1953), 237-240.

[7] W. C. Chu, *A class of bivariate reciprocal formulas and applications to combinatorial identities*, Math. Appl.1:3(1988), 63-66.

[8] W. C. Chu & L. C. Hsu, *The hypergeometric series method for combinatorial sums*, Appl. Math.(A J. Chinese Univ.),5:1(1990), 125-136.

[9] W. C. Chu & L. C. Hsu, *Some new applications of Gould-Hsu inversions*, JCISS.14(1990), 1-4.

[10] G. P. Egorychev, *Integral Representation and Computation of Combinatorial Sums* , Transl. Math. Monographs (AMS), vol.59, 1984.

[11] I. Gessel & D. Stanton, *Strange evaluations of hypergeometric series*, SIAM J. Math. Anal.13(1982), 295-308.

[12] R. W. Gosper, *Decision procedure for indefinite hypergeometric summations*, Proc. Nat. Acad. Sci. USA 75(1978), 40-42.

[13] H. W. Gould, *Some generalizations of Vandermonde's convolution*, Amer. Math. Month.63:1(1956), 84-91.

[14] H. W. Gould, *Combinatorial identities*, Morgantown, West-Virginia, 1972.

[15] H. W. Gould & L. C. Hsu, *Some new inverse series relations*, Duke Math. J.40(1973), 885-891.

[16] R. L. Graham, D. E. Knuth and O. Patashnik, *Concrete Mathematics*, Addison-Wesley, Reading, MA 1989.

[17] D. H. Greene & D. E. Knuth, *Mathematics for the Analysis of Algorithms*, Birkhäuser,1981.

[18] D. M. Jackson, *Algebraic Combinatorics*, Manuscript, Beijing, 1989.

[19] J. Kaucky, *Combinatorial Identities* (Slovak), Veda Bratislava, 1975.

[20] D. E. Knuth, *The Art of Computer Programming: vol.1. Fundamental Algorithms*, Addison-Wesley, Reading, MA 1969.

[21] A. Lakin, *A hypergeometric identity related to Dougall's theorem*, J. London Math. Soc.27(1952), 229-234.

[22] B. W. Minton, *Generalized hypergeometric functions of unit argument*, J. Math. Phys.11(1975), 1175-1176.

[23] J. Riodan, *Combinatorial Identities*, Wiley, New York, 1968.

[24] R. Roy, *Binomial identities and hypergeometric functions*, Amer. Math. Month.94:1(1987), 36-46.

[25] W. F. Sheppard, *Summations of the coefficients of some terminating hypergeometric series*, Proc. London Math. Soc.10(1912), 469-478.

[26] I. J. Slater, *Generalized Hypergeometric Functions*, Cambridge University Press, Cambridge, 1966.

[27] G. N. Watson, *A note on Generalized hypergeometric series*, Proc. Camb. Phil. Soc.23(1925), xiii-xv.

[28] F. J. W. Whipple, *On well-poised series, generalized hypergeometric series having parameters in pairs, each pair with the same sum*, Proc. London Math. Soc.(2), 24(1926), 247-263.

[29] H. S. Wilf, *Computer-generated proofs of 54 binomial coefficient identities*, Preprint, 1990.

[30] H. S. Wilf & D. Zeilberger, *Rational functions certify combinatorial identities*, J. Amer. Math. Soc.3:1(1990), 147-158.

THE TWO COLOR WHITWORTH PROBLEM FOR LONG CYCLES OR LONG LINES

F.K Hwang

AT&T Bell Laboratories

One interpretation of the two-color Whitworth bracelet problem is to count the number of ways of selecting k vertices from an n-cycle such that the selected vertices induce exactly l edges of the cycle. The problem can also be studied by substituting an n-line for the n-cycle. The special case of $l = 0$ has been studied by Kaplansky for both the cycle and the line case as lemmas to solve the famous problème des ménages. Recently, these selection problems have been studied for m cycles or m lines. Although progress has been made on the theoretical side, the results are not easily applicable since the numbers of selections are given as sums of $kl2^m$ terms and depend on the lengths of the cycles or lines. In this paper we show that when the lengths of the cycles or lines are large relative to k, the numbers depend only on the total number of vertices, and simple formulas are available.

1. Introduction

The problem of counting the number of ways $g(n, k, l)$ of selecting k vertices from an n-cycle to contain exactly l edges of the cycle was shown [2] to be equivalent to the two-color Whitworth bracelet problem [1] which gives

$$g(n,k,l) = \begin{cases} 1 & l = k = n \\ \frac{n}{k-l}\binom{k-1}{k-l-1}\binom{n-k-1}{k-l-1} & l < k < n \\ 0 & \text{otherwise} \end{cases} \tag{1}$$

(The special case of $l = 0$ was also obtained by Kaplansky [5]). Hwang [2] has extended the above problem to m cycles. Let n_i denote the size of the ith cycle and let $\sum_{i=1}^{m} n_i = n$. Let $g(\{n_1, \ldots, n_m\}, k, l)$ denote the number of ways of selecting k vertices from the m cycles to contain exactly l edges, and let

A. P. Godbole and S. G. Papastavridis (eds.), Runs and Patterns in Probability, 59–63.

$$G_{\{n_1,\ldots,n_m\},k}(t) = \sum_{l=0}^{k} g(\{n_1\ldots,n_m\},k,l)t^l$$

denote its generating function. Hwang proved that

$$G_{\{n_1,\ldots,n_m\},k}(t) = \sum_{S\subseteq M} (-1)^{|S|} G_{n-2n(S),k-n(S)}(t), \tag{2}$$

where M is the set $\{1,\ldots,m\}$ and $n(S) = \sum_{i\in S} n_i$. However, no explicit formula for $g(\{n_1,\ldots,n_m\},k,l)$ is available.

Let $f(\{n_1,\ldots,n_m\},k,l)$ and $F_{\{n_1,\ldots,n_m\},k}(t)$ denote the corresponding number and generating function when we replace cycles by lines. The knowledge about the line case is even less complete. Kaplansky [5] gave

$$f(n,k,0) = \binom{n+1-k}{k}. \tag{3}$$

Konvalina [8] and Hwang, Korner and Wei [3] entended Kaplansky's result to more than one line. In particular the latter showed

$$f(\{n_1,\ldots,n_m\},k,0) = \sum_{S\subseteq M} (-1)^{n(S)+|S|} \sum_{i\geq 0} \binom{i+m-2}{i} \times f(n-2n(S)-4\mid S\mid -4i, k-n(S)-2\mid S\mid -2i, 0). \tag{4}$$

Liu [9] was the first to obtain results for general l. He proved

$$f(n,k,l) = \binom{k-1}{l}\binom{n-k+1}{k-l}$$

and

$$F_{\{n_1,n_2\},k}(t) = F_{n,k}(t) - \sum_{r=1}^{\bar{n}} (t-1)^r [F_{n-2r+1,k-r}(t) - F_{n-2r,k-r}(t)],$$

where $\bar{n} = \min\{n_1,n_2,\ldots,n_m\}$. Recently, Kirschenhofer and Prodinger [7] found the general solution

$$F_{\{n_1,\ldots,n_m\},k}(t) = \sum_{S\subseteq M}\sum_{i,j\geq 0}\binom{-(m-1)}{i}(t-1)^{n(S)+|S|+i+j}\binom{k-1-n(S)}{j} \times\binom{n-2n(S)-2\mid S\mid-2i-j}{k-n(S)-2\mid S\mid-2i-j} \quad (5)$$

and

$$f(\{n_1,\ldots,n_m\},k,l) = \sum_{S\subseteq M}\sum_{i,j\geq 0}\binom{i+m-2}{i}\binom{n(S)+\mid S\mid+i}{l-j} \times(-1)^{n(S)+|S|-l+j}\binom{k-1-n\mid S\mid}{j}\binom{n-n(S)-2\mid S\mid-2i-k+1}{k-n(S)-2\mid S\mid-2i-j} \quad (6)$$

Note that Eqs. (2), (4) and (5) all involve exponentially (in m) many terms and are hard to evaluate. Thus these formulas are of theoretical interest but of little practical use. In this note we make a simple observation that when the lengths of cycles and lines are large relative to k, $g(\{n_1,\ldots,n_m\},k,l)$ and $f(\{n_1,\ldots,n_m\},k,l)$ have simple formulas.

2. The Main Results

Define $\bar{n} = \min\{n_1,\ldots,n_m\}$.

Theorem 1. *For* $k \leq \bar{n}-1$,

$$g(\{n_1,\ldots,n_m\},k,l) = \begin{cases} 1 & l=k=n \\ \frac{n}{k-l}\binom{k-1}{k-l-1}\binom{n-k-1}{k-l-1} & l<k<n \\ 0 & \text{otherwise} \end{cases}$$

Proof. In Eq.(2) for all nonempty S

$$G_{n-2n(S),k-n(S)}(t) = 0$$

since

$$k-n(S) \leq k-\bar{n} < 0.$$

Thus

$$G_{\{n_1,\ldots,n_m\},k}(t) = G_{n,k}(t).$$

Theorem 1 now follows from eq. (1).

Theorem 2. *For $k \le \bar{n} + 1$,*

$$f(\{n_1,\ldots,n_m\},k,l) = \sum_{i,j\ge 0} \binom{i+m-2}{i}\binom{i}{l-j}(-1)^{-l+j}\binom{k-1}{j}\binom{n-2i-k+1}{k-2i-j}.$$

Proof. In eq. (5) for all nonempty S

$$\binom{n-n(S)-2\mid S\mid -2i-k+1}{k-n(S)-2\mid S\mid -2i-j} = 0$$

since

$$k - n(S) - 2\mid S\mid -2i - j \le k - \bar{n} - 2 < 0.$$

Thus Theorem 2.

Corollary. *For $k \le \bar{n} + 1$*

$$f(\{n_1,\ldots,n_m\},k,0) = \sum_{i\ge 0}\binom{i+m-2}{i}\binom{n+1-k-2i}{k-2i}. \tag{7}$$

Proof. By taking only the $S = \emptyset$ term in eq. (4) and using eq. (3).

3. Conclusion

Although the mathematics is not complicated, it is quite unexpected to find that for $k \le \bar{n} - 1$ in the cycle case and $k \le \bar{n} + 1$ in the line case, the many-cycle and many-line generating functions depend on n and k only, but not on m or the sizes of the cycles and lines. A happy consequence is that simple formulas for counting are available for those cases, so that using these formulas becomes a practical matter.

It is of some interest to note that the righthand-side of eq. (7) is the same expression as shown in [4, 6] to equal

$$\sum_{p\in P}\prod_{i=1}^{n}\binom{n_i+1-k_i}{k_i}$$

where P is the set of all partitions of k into nonnegative integers $k_1, \ldots, k_m$, allowing $k_i > n_i$.

References

[1] D. E. BARTON and F. N. DAVID (1958), Runs in a ring, *Biometrika* **45**, 572-578.

[2] F. K. HWANG (1981), Cycle polynomial, *Proc. Amer. Math. Soc.* **83**, 215-219.

[3] F. K. HWANG, J. KORNER and V. K. WEI (1984), Selecting non-consecutive balls arranged in many lines, *J. Combin. Theory A* **37**, 327-336.

[4] F. K. HWANG and V. K. WEI (1983), A partition identity, *Disc. Math.* **46**, 323-326.

[5] I. KAPLANSKY (1943), Solution of the problème des ménages, *Bull. Amer. Math. Soc.* **49**, 784-785.

[6] P. KIRSCHENHOFER and H. PRODINGER (1984), A short proof for a partition identity of Hwang and Wei, *Disc. Math.* **52**, 305-306.

[7] P. KIRSCHENHOFER and H. PRODINGER (1986), Two selection problems revisited, *J. Combin. Theory A* **42**, 310-316.

[8] J. KONVALINA (1981), On the number of combinations without unit separation, *J. Combin. Theory A* **31**, 101-107.

[9] C. J. LIU (1984), The generating function of Whitworth runs, *Disc. Math.* **51**, 221-228.

BINOMIAL DISTRIBUTIONS OF ORDER K ON THE CIRCLE

Frosso S. Makri and Andreas N. Philippou

University of Patras

Two circular binomial distributions of order k are introduced presently, as the distributions of nonoverlapping and possibly overlapping success runs of length k (≥ 1) in n (≥ 1) Bernoulli trials arranged on a circle, and their means are obtained. They are also related to the usual binomial distributions of order k. As a consequence of the circular distributions, two exact formulas are derived for the reliability of a consecutive-k-out-of-n: F system whose components are ordered circularly and function independently with probability p, and a Theorem of Derman et al. is directly reestablished.

Binomial distributions of order k, circular, type I, type II, success runs, overlapping, nonoverlapping, reliability, consecutive-k-out-of-n: F system.

1. Introduction

Let $N_{n,k}$ denote the number of nonoverlapping success runs of length k (≥ 1) in n (≥ 1) Bernoulli trials. Modifying an argument of Philippou and Muwafi (1982), Philippou and Makri (1986) and independently Hirano (1986) derived the following exact formula for the probability distribution of $N_{n,k}$,

$$(1.1)\qquad P(N_{n,k}=x)=\sum_{i=0}^{k-1}\sum_{x_1,\ldots,x_k}\binom{x_1+\cdots+x_k+x}{x_1,\ldots,x_k,x}p^n\left(\frac{q}{p}\right)^{x_1+\cdots+x_k}$$

$$(x=0,1,\ldots,[n/k])$$

where the inner summation is over all non-negative integers $x_1,\ldots,x_k$ such that $\sum_{j=1}^{k} jx_j=n-i-kx$, and $[x]$ denotes the greatest integer in x.

The probability distribution (1.1) is known as binomial distribution of order k, type I, with parameter vector (n,p), and it is denoted by $B_{k,I}(n,p)$. The

A. P. Godbole and S. G. Papastavridis (eds.), Runs and Patterns in Probability, 65–81.

asymptotic normality of a normalized version of $N_{n,k}$ was first established by von Mises (see Feller (1968, p.324), where a simpler proof is presented). Furthermore, if n and k tend to ∞ so that $nqp^k \to \lambda$, von Mises also proved (cf. Feller (1968, p.341), Problem 26) that $P(N_{n,k} = x) \to e^{-\lambda}\lambda^x/x!$, $x = 0, 1, \ldots$. Several other problems, however, pertaining to $B_{k,I}(n,p)$, were still open and Philippou and Makri (1986) offered them as an objective and challenge for further investigations. One of them, referring to the derivation of the probability generating function of $B_{k,I}(n,p)$, was solved by Aki and Hirano (1988). A second one, referring to the possible weak convergence of $B_{k,I}(n,p)$ to the Poisson distribution of order k as $n \to \infty$, $p \to 0$ and $np \to \lambda$, was answered by Godbole (1990b) in the negative. Godbole (1990a) and independently Papastavridis (1990, Theorem 2) derived the following binomial expression for $P(N_{n,k} = x)$,

$$P(N_{n,k} = x) = \sum_{y=\left[\frac{n-kx}{k}\right]}^{n-kx} q^y p^{n-y} \binom{x+y}{x} \sum_{j=0}^{\left[\frac{n-kx-y}{k}\right]} (-1)^j \binom{y+1}{j} \binom{n-kx-jk}{y}$$

$$(x = 0, 1, \ldots, [n/k]), \tag{1.2}$$

which is alternative to and simpler than (1.1).

Let $M_{n,k}$ denote the number of overlapping success runs of length k in n Bernoulli trials. Following Philippou and Makri (1986), Ling (1988) derived the following exact formula for the probability distribution of $M_{n,k}$,

$$P(M_{n,k} = x) = \sum_{i=0}^{n} \sum_{x_1,\ldots,x_n} \binom{x_1 + \cdots + x_n}{x_1, \ldots, x_n} p^n \left(\frac{q}{p}\right)^{x_1+\ldots+x_n} \tag{1.3}$$

$$(x = 0, 1, \ldots, n-k)$$

where the inner summation is over all non-negative integers $x_1, \ldots, x_n$ such that $\sum_{j=1}^{n} jx_j = n - i$ and $\max\{0, i-k+1\} + \sum_{j=k+1}^{n}(j-k)x_j = x$.

He also derived the mean and variance of $M_{n,k}$ and two recurrences; one for calculating the probabilities $P(M_{n,k} = x)$, the other for calculating the probability generating function of $M_{n,k}$.

The probability distribution (1.3) has been called binomial distribution of order k, type II, with parameter vector (n,p) and it was denoted by $B_{k,II}(n,p)$.

If we set $x=0$ and interchange p with q in formulas (1.1), (or (1.3)) and (1.2), we readily get two exact alternative expressions for $R_L(p;k,n)$, the reliability of a consecutive-k-out-of-n: F system whose components are ordered linearly and function independently with probability p. The first one easily reduces to Theorem 4.6 of Philippou (1986), which provides two simpler exact formulas for $R_L(p;k,n)$ as well as a third one in terms of the Fibonacci-type polynomials of order k.

In the present paper we introduce two circular binomial distributions of order k, say $B^c_{k,I}(n,p)$ and $B^c_{k,II}(n,p)$, as the distributions of nonoverlapping $(N^c_{n,k})$ and possibly overlapping $(M^c_{n,k})$ success runs of length k (≥ 1) in n (≥ 1) Bernoulli trials ordered circularly (see Theorems 2.1 and 2.2). We also derive the means of $N^c_{n,k}$ and $M^c_{n,k}$ and the reliability $R_C(p;k,n)$ of a consecutive-k-out-of-n: F system whose components are ordered circularly and function independently with probability p (see Propositions 2.1 and 2.2 and Corollary 2.1). In Section 3, we relate the circular binomial distributions of order k to the usual binomial distributions of the same order (see Theorems 3.1 and 3.2), and we reestablish a theorem of Derman et al. (1982) as a simple consequence of either one of them.

2. Circular binomial distributions of order k

In this section we introduce two circular binomial distributions of order k as the distributions of $N^c_{n,k}$ and $M^c_{n,k}$, and we derive the reliability $R_C(p;k,n)$. We also obtain the means of $N^c_{n,k}$ and $M^c_{n,k}$.

Theorem 2.1. *Let $N^c_{n,k}$ be a random variable (rv) denoting the number of non-overlapping success runs of length k (≥ 1) in n (≥ 1) Bernoulli trials ordered* *circularly. Then, for* $x = 0, 1, \ldots, [n/k]$,

$$(2.1)\quad P(N_{n,k}^c = x) = qp^{n-1}\sum_{i=1}^{k} i\sum\nolimits_1 \binom{x_1+\cdots+x_k+x}{x_1,\ldots,x_k,x}\left(\frac{q}{p}\right)^{x_1+\cdots+x_k}$$

$$+kqp^{n-1}\sum_{i=1}^{k}\sum\nolimits_1 \frac{x}{x_1+\cdots+x_k+1}\binom{x_1+\cdots+x_k+x}{x_1,\ldots,x_k,x}\left(\frac{q}{p}\right)^{x_1+\cdots+x_k}$$

$$+p^n\delta_{x,[n/k]},$$

where the summation $\sum_1$ *is taken over all non-negative integers* $x_1, \ldots, x_k$ *satisfying the condition* $\sum_{j=1}^{k} jx_j = n - i - kx$, *and* δ *is the Kronecker delta function.*

We shall first establish two preliminary lemmas. The first one gives a formula for $P(N_{n,k}^c = x)$, $x = 0, 1, \ldots, [n/k]$.

Lemma 2.1. *Let* $N_{n,k}^c$, $\sum_1$ *and* δ *be as in Theorem 2.1. Then, for* $x = 0, 1, \ldots, [n/k]$,

$$P(N_{n,k}^c = x) = qp^{n-1}\sum_{r=0}^{x}\sum_{i=1}^{k} i\sum\nolimits_1 \binom{x_1+\cdots+x_k+x-r}{x_1,\ldots,x_k,x-r}\left(\frac{q}{p}\right)^{x_1+\cdots+x_k}$$

$$+qp^{n-1}\zeta(1,x)\zeta(1,k)\sum_{r=0}^{x-1}\sum_{i=1}^{k-1}(k-i)\sum\nolimits_1 \binom{x_1+\cdots+x_k+x-r-1}{x_1,\ldots,x_k,x-r-1}\left(\frac{q}{p}\right)^{x_1+\cdots+x_k}$$

$$+p^n\delta_{x,[n/k]},$$

where $\zeta(.,.)$ *is the zeta function defined by* $\zeta(v,u) = 1$ *if* $u \geq v$ *and* 0 *otherwise.*

Proof. Let 1 denote success and 0 denote failure. We observe that a typical element of the event $(N_{n,k}^c = x)$ is either a circular arrangement

(2.2)

$$\underbrace{\underbrace{11\ldots1}_{k}\ldots\underbrace{11\ldots1}_{k}}_{r}\underbrace{11\ldots1}_{\alpha}0a_1a_2\ldots a_{x_1+\ldots+x_k+x-r}\underbrace{11\ldots1}_{\beta}\qquad (0 \leq x \leq [n/k])$$

such that x_j of the a's are $e_j = \underbrace{11\ldots1}_{j-1}0$ $(1 \leq j \leq k)$, $x - r$ of the a's are $\tilde{e}_k = \underbrace{11\ldots1}_{k}$, and

(2.2)′

$$\sum_{j=1}^{k} jx_j = n-1-kx-(\alpha+\beta),\ 0 \leq r \leq x,\ 0 \leq \alpha, \beta \leq k-1 \text{ and } \alpha+\beta \leq k-1,$$

or a circular arrangement

(2.3)

$$\underbrace{\underbrace{11\ldots1}_{k}\ldots\underbrace{11\ldots1}_{k}}_{r}\underbrace{11\ldots1}_{\alpha}0a_1a_2\ldots a_{x_1+\cdots+x_k+x-r-1}\underbrace{11\ldots1}_{\beta} \qquad (1 \leq x \leq [n/k])$$

such that x_j $(1 \leq j \leq k)$ and $x - r - 1$ now of the a's are as above, and

$$(2.3)'\ \sum_{j=1}^{k} jx_j = n-1-k(x-1)-(\alpha+\beta),\ x \geq 1,\ 0 \leq r \leq x-1,\ 1 \leq \alpha, \beta \leq k-1$$
$$\text{and } k \leq \alpha+\beta \leq 2k-2,$$

or a circular arrangement $\underbrace{11\ldots1}_{n}$ $(x = [n/k])$.

Fix r, x_j $(1 \leq j \leq k)$, α and β. Then the number of the circular arrangements of type (2.2) is

$$\binom{x_1+\cdots+x_k+x-r}{x_1,\ldots,x_k,x-r},$$

and each one of them has probability $qp^{n-1}(q/p)^{x_1+\cdots+x_k}$. Furthermore, the number of the circular arrangements of type (2.3) is

$$\binom{x_1+\cdots+x_k+x-r-1}{x_1,\ldots,x_k,x-r-1},$$

and each one of them has probability $qp^{n-1}(q/p)^{x_1+\cdots+x_k}$. Finally, we observe that $P(\underbrace{11\ldots1}_{n}) = p^n$. But r, x_j $(1 \leq j \leq k)$, α and β may vary subject to (2.2)′ and (2.3)′ for the elements of type (2.2) and type (2.3), respectively. Denote by $\sum_1{}'$ and $\sum_1{}''$ summation over all non-negative integers $x_1, \ldots, x_k$ satisfying

(2.2)′ and (2.3)′, respectively. Then, for $x = 0, 1, \ldots, [n/k]$ and $\sum_1$ as in the theorem, we have

$$P(N_{n,k}^c = x) = \sum_{r=0}^{x}\sum_{\alpha=0}^{k-1}\sum_{\substack{\beta=0 \\ \alpha+\beta\leq k-1}}^{k-1}\sum\nolimits_1' \binom{x_1+\cdots+x_k+x-r}{x_1,\ldots,x_k,x-r} qp^{n-1}\left(\frac{q}{p}\right)^{x_1+\cdots+x_k}$$

$$+\zeta(1,x)\zeta(1,k)\sum_{r=0}^{x-1}\sum_{\alpha=1}^{k-1}\sum_{\substack{\beta=1 \\ k\leq\alpha+\beta\leq 2k-2}}^{k-1}\sum\nolimits_1'' \binom{x_1+\cdots+x_k+x-r-1}{x_1,\ldots,x_k,x-r-1} qp^{n-1}\left(\frac{q}{p}\right)^{x_1+\cdots+x_k}$$

$$+p^n\delta_{x,[n/k]}$$

$$= \sum_{r=0}^{x}\sum_{\alpha+\beta=0}^{k-1}(\alpha+\beta+1)\sum\nolimits_1' \binom{x_1+\cdots+x_k+x-r}{x_1,\ldots,x_k,x-r} qp^{n-1}\left(\frac{q}{p}\right)^{x_1+\cdots+x_k}$$

$$+\zeta(1,x)\zeta(1,k)\sum_{r=0}^{x-1}\sum_{\alpha+\beta=k}^{2k-2}(2k-\alpha-\beta-1) \tag{1}$$

$$\sum\nolimits_1'' \binom{x_1+\cdots+x_k+x-r-1}{x_1,\ldots,x_k,x-r-1} qp^{n-1}\left(\frac{q}{p}\right)^{x_1+\cdots+x_k}$$

$$+p^n\delta_{x,[n/k]}$$

$$= qp^{n-1}\sum_{r=0}^{x}\sum_{i=1}^{k} i\sum\nolimits_1 \binom{x_1+\cdots+x_k+x-r}{x_1,\ldots,x_k,x-r}\left(\frac{q}{p}\right)^{x_1+\cdots+x_k}$$

$$+qp^{n-1}\zeta(1,x)\zeta(1,k)\sum_{r=0}^{x-1}\sum_{i=1}^{k-1}(k-i) \tag{2}$$

$$\sum\nolimits_1 \binom{x_1+\cdots+x_k+x-r-1}{x_1,\ldots,x_k,x-r-1}\left(\frac{q}{p}\right)^{x_1+\cdots+x_k}$$

$$+p^n\delta_{x,[n/k]},$$

which establishes the lemma.

We proceed now to the second lemma, which provides a formula for executing the summation with respect to r appearing in Lemma 2.1.

Lemma 2.2. *Let*

$$\binom{x_1+\cdots+x_k}{x_1,\ldots,x_k} = \frac{(x_1+\cdots+x_k)!}{x_1!\ldots x_k!}$$

denote the multinomial coefficient. Then, for $x = 0, 1, \ldots,$ we have

$$\sum_{r=0}^{x} \binom{x_1 + \cdots + x_k + r}{x_1, \ldots, x_k, r} = \frac{x+1}{x_1 + \cdots + x_k + 1} \binom{x_1 + \cdots + x_k + x + 1}{x_1, \ldots, x_k, x+1}.$$

Proof. We observe that

$$\begin{aligned}
&\binom{x_1 + \cdots + x_k + r}{x_1, \ldots, x_k, r} \\
&= \frac{(x_1 + \cdots + x_k)!}{x_1! \ldots x_k!} \frac{(x_1 + \cdots + x_k + r)!}{(x_1 + \ldots + x_k)! r!} \\
&= \frac{(x_1 + \cdots + x_k)!}{x_1! \ldots x_k!} \left\{ \frac{(x_1 + \cdots + x_k + r + 1)!}{(x_1 + \cdots + x_k + 1)! r!} - \frac{(x_1 + \cdots + x_k + r)!}{(x_1 + \cdots + x_k + 1)!(r-1)!} \right\} \\
&= \frac{r+1}{x_1 + \cdots + x_k + 1} \binom{x_1 + \cdots + x_k + r + 1}{x_1, \ldots, x_k, r+1} \\
&\qquad - \frac{r}{x_1 + \cdots + x_k + 1} \binom{x_1 + \cdots + x_k + r}{x_1, \ldots, x_k, r}.
\end{aligned}$$

Therefore, for $x = 0, 1, \ldots,$ we get

$$\begin{aligned}
\sum_{r=0}^{x} \binom{x_1 + \cdots + x_k + r}{x_1, \ldots, x_k, r} &= \sum_{r=0}^{x} \frac{r+1}{x_1 + \cdots + x_k + 1} \binom{x_1 + \cdots + x_k + r + 1}{x_1, \ldots, x_k, r+1} \\
&\quad - \sum_{r=0}^{x} \frac{r}{x_1 + \cdots + x_k + 1} \binom{x_1 + \cdots + x_k + r}{x_1, \ldots, x_k, r} \\
&= \frac{x+1}{x_1 + \cdots + x_k + 1} \binom{x_1 + \cdots + x_k + x + 1}{x_1, \ldots, x_k, x+1}.
\end{aligned}$$

The proof of the lemma is completed.

For $k = 2, 3, \ldots,$ Lemma 2.2 generalizes the "vertical" recurrence relation of the binomial coefficients,

$$\sum_{r=0}^{x} \binom{n+r}{n} = \binom{n+x+1}{n+1} \quad (x, n = 0, 1, \ldots),$$

to which it reduces for $x_1 = n$ and $x_2 = \ldots = x_k = 0$.

We are ready now to give the proof of the theorem.

Proof of Theorem 2.1. For $x = 0, 1, \ldots$, Lemma 2.2 gives

$$\zeta(1,x)\sum_{r=0}^{x-1}\binom{x_1+\cdots+x_k+x-r-1}{x_1,\ldots,x_k,x-r-1} = \frac{x}{x_1+\cdots+x_k+1}\binom{x_1+\cdots+x_k+x}{x_1,\ldots,x_k,x}$$

and

$$\begin{aligned}
&\sum_{r=0}^{x}\binom{x_1+\cdots+x_k+x-r}{x_1,\ldots,x_k,x-r}\\
&= \binom{x_1+\cdots+x_k+x}{x_1,\ldots,x_k,x} + \zeta(1,x)\sum_{r=1}^{x}\binom{x_1+\cdots+x_k+x-r}{x_1,\ldots,x_k,x-r}\\
&= \binom{x_1+\cdots+x_k+x}{x_1,\ldots,x_k,x} + \frac{x}{x_1+\cdots+x_k+1}\binom{x_1+\cdots+x_k+x}{x_1,\ldots,x_k,x}.
\end{aligned}$$

Introducing these expressions into the formula of Lemma 2.1, we get, for $x = 0, 1, \ldots, [n/k]$,

$$\begin{aligned}
&P(N_{n,k}^c = x)\\
&= qp^{n-1}\sum_{i=1}^{k} i \sum\nolimits_1 \binom{x_1+\cdots+x_k+x}{x_1,\ldots,x_k,x}\left(\frac{q}{p}\right)^{x_1+\cdots+x_k}\\
&+ qp^{n-1}\sum_{i=1}^{k} i \sum\nolimits_1 \frac{x}{x_1+\cdots+x_k+1}\binom{x_1+\cdots+x_k+x}{x_1,\ldots,x_k,x}\left(\frac{q}{p}\right)^{x_1+\cdots+x_k}\\
&+ qp^{n-1}\sum_{i=1}^{k}(k-i)\sum\nolimits_1 \frac{x}{x_1+\cdots+x_k+1}\binom{x_1+\cdots+x_k+x}{x_1,\ldots,x_k,x}\left(\frac{q}{p}\right)^{x_1+\cdots+x_k}\\
&+ p^n\delta_{x,[n/k]},
\end{aligned}$$

from which the theorem follows.

We call the probability distribution (2.1) circular binomial distribution of order k, type I, with parameter vector (n, p), and we denote it by $B_{k,I}^c(n,p)$. What is the mean, variance and probability generating function of $B_{k,II}^c(n,p)$? We give the mean in the following.

Proposition 2.1. *Let $N^c_{n,k}$ be a rv as in Theorem 2.1. Then,*

$$E(N^c_{n,k}) = \begin{cases} 0, & \text{if } 1 \le n \le k-1 \ (k \ge 2), \\ nqp^k \frac{1-p^{n-s}}{1-p^k} + [n/k]p^n, & \text{if } n = [n/k]k + s,\ 1 \le s \le k-1 \ (k \ge 2), \\ nqp^k \frac{1-p^{n-k}}{1-p^k} + [n/k]p^n, & \text{if } n = [n/k]k \ (k \ge 1). \end{cases}$$

Proof. Obviously, for $1 \le n \le k-1$ $(k \ge 2)$, $E(N^c_{n,k}) = 0$. If $n = [n/k]k + s$, $1 \le s \le k-1$ $(k \ge 2)$, let E_{ij} be the event that the ith trial $(1 \le i \le n)$ results in the jkth success of a success run of length jk $(1 \le j \le [n/k])$ which is preceded by a failure, and let $E^*_{i,[n/k]}$ be the event that the ith trial $(1 \le i \le [n/k])$ results in the $[n/k]k$th success of a success run of length $[n/k]k$ preceded by s successes. If $n = [n/k]k$ $(k \ge 1)$, let E_{ij} $(1 \le i \le n,\ 1 \le j \le [n/k]-1)$ be as above and let $E^{**}_{i,[n/k]}$ be the event that the ith trial $(1 \le i \le [n/k])$ results in the $[n/k]k$th success of a success run of length $[n/k]k$. We define rvs Y_i, $1 \le i \le n$, as follows:

(a_1) For $n = [n/k]k + s$, $1 \le s \le k-1$ $(k \ge 2)$, and $1 \le i \le [n/k]$,

$$Y_i = \begin{cases} 1, & \text{if } \left(\bigcup_{j=1}^{[n/k]} E_{ij} \right) \bigcup E^*_{i,[n/k]} \text{ occurs}, \\ 0, & \text{otherwise}; \end{cases}$$

(a_2) for $n = [n/k]k + s$, $1 \le s \le k-1$ $(k \ge 2)$, and $[n/k]+1 \le i \le n$,

$$Y_i = \begin{cases} 1, & \text{if } \bigcup_{j=1}^{[n/k]} E_{ij} \text{ occurs}, \\ 0, & \text{otherwise}; \end{cases}$$

(b_1) for $n = [n/k]k$ $(k \ge 1)$ and $1 \le i \le [n/k]$,

$$Y_i = \begin{cases} 1, & \text{if } \left(\bigcup_{j=1}^{[n/k]-1} E_{ij} \right) \bigcup E^{**}_{i,[n/k]} \text{ occurs}, \\ 0, & \text{otherwise}; \end{cases}$$

(b_2) for $n = [n/k]k$ $(k \ge 1)$ and $[n/k]+1 \le i \le n$,

$$Y_i = \begin{cases} 1, & \text{if } \bigcup_{j=1}^{[n/k]-1} E_{ij} \text{ occurs}, \\ 0, & \text{otherwise}. \end{cases}$$

Then, in either case, (a) or (b), $N^c_{n,k} = \sum_{i=1}^n Y_i$, so that

$$E(N^c_{n,k}) = \begin{cases} \sum_{i=1}^{n}\sum_{j=1}^{[n/k]} P(E_{ij}) + \sum_{i=1}^{[n/k]} P(E^*_{i,[n/k]}), \\ \qquad\qquad \text{if } n=[n/k]k+s,\ 1\le s\le k-1\ (k\ge 2), \\ \sum_{i=1}^{n}\sum_{j=1}^{[n/k]-1} P(E_{ij}) + \sum_{i=1}^{[n/k]} P(E^{**}_{i,[n/k]}), \quad \text{if } n=[n/k]k\ (k\ge 1), \end{cases}$$

$$= \begin{cases} \sum_{i=1}^{n}\sum_{j=1}^{[n/k]} qp^{jk} + [n/k]p^n, \\ \qquad\qquad \text{if } n=[n/k]k+s,\ 1\le s\le k-1\ (k\ge 2), \\ \sum_{i=1}^{n}\sum_{j=1}^{[n/k]-1} qp^{jk} + [n/k]p^n, \quad \text{if } n=[n/k]k\ (k\ge 1). \end{cases}$$

The proof of the proposition follows.

Theorem 2.2. *Let $M^c_{n,k}$ be a rv denoting the number of (possibly) overlapping success runs of length k (≥ 1) in n (≥ 1) Bernoulli trias ordered circularly. Then, $P(M^c_{n,k} = x) = p^n$ if $x = n$, 0 if $n-k+1 \leq x \leq n-1$, and*

(2.4)
$$P(M^c_{n,k} = x) = qp^{n-1}\sum_{i=1}^{x+k} i \sum\nolimits_2 \binom{x_1+\cdots+x_n}{x_1,\ldots,x_n}\left(\frac{q}{p}\right)^{x_1+\cdots+x_n} \quad (0\leq x\leq n-k)$$

where the summation $\sum_2$ is taken over all non-negative integers $x_1,\ldots,x_n$ satisfying the conditions $\sum_{j=1}^n jx_j = n-i$ $(1\leq i\leq x+k)$ and $max\{0, i-k\} + \sum_{j=k+1}^n (j-k)x_j = x$ $(0\leq x\leq n-k)$.

Proof. Let 1 denote success and 0 denote failure. We first observe that, for $n-k+1 \leq x \leq n-1$, $P(M^c_{n,k} = x) = P(\emptyset) = 0$, and $P(M^c_{n,k} = n) = P(\underbrace{11\ldots1}_{n}) = p^n$.

For $0 \leq x \leq n-k$, a typical element of the event $(M^c_{n,k} = x)$ is a cyclic arrangement

$$\underbrace{11\ldots1}_{\alpha}0a_1a_2\ldots a_{x_1+\cdots+x_n}\underbrace{11\ldots1}_{\beta}$$

such that x_j of the a's are $e_j = \underbrace{11\ldots1}_{j-1}0$ $(1\leq j\leq n)$, and

(2.5) $\quad 0\leq\alpha\leq x+k-1,\ 0\leq\beta\leq x+k-1$ and $\alpha+\beta\leq x+k-1$;

(2.6) $\sum_{j=1}^{n} jx_j = n-1-(\alpha+\beta)$ and $\max\{0, \alpha+\beta-k+1\} + \sum_{j=k+1}^{n} (j-k)x_j = x.$

If x_j $(1 \leq j \leq n)$, α and β are kept fixed, the number of the above arrangements is

$$\binom{x_1 + \cdots + x_n}{x_1, \ldots, x_n}$$

and each one of them has probability $qp^{n-1}(q/p)^{x_1+\cdots+x_n}$. But α and β may vary subject to (2.5), and the non-negative integers may vary subject to (2.6). Therefore, for $0 \leq x \leq n-k$, and denoting by $\sum_2{}'$ summation over all non-negative integers $x_1, \ldots, x_n$ satisfying (2.5) and (2.6), we have

$$P(M_{n,k}^c = x) = qp^{n-1} \sum_{\substack{\alpha=0 \\ \alpha+\beta \leq x+k-1}}^{x+k-1} \sum_{\beta=0}^{x+k-1} \sum_2{}' \binom{x_1 + \cdots + x_n}{x_1, \ldots, x_n} \left(\frac{q}{p}\right)^{x_1+\cdots+x_n}$$

$$= qp^{n-1} \sum_{i=1}^{x+k} i \sum_2 \binom{x_1 + \cdots + x_n}{x_1, \ldots, x_n} \left(\frac{q}{p}\right)^{x_1+\cdots+x_n}$$

which completes the proof of the theorem.

We call the probability distribution established in Theorem 2.2 circular binomial distribution of order k, type II, with parameter vector (n, p), and we denote it by $B_{k,II}^c(n,p)$. What is the mean, variance and probability generating function of $B_{k,II}^c(n,p)$? We give the mean in the following.

Proposition 2.2. *Let $M_{n,k}^c$ be a rv as in Theorem 2.2. Then its mean is given by $E(M_{n,k}^c) = np^k$.*

Proof. Let $X_1, \ldots, X_n$ be independent rvs with probability distribution function

(2.7) $\qquad P(X_j = x) = p^x(1-p)^{1-x} \quad (x = 0, 1; 0 < p < 1; 1 \leq j \leq n).$

Modifying an idea of Ling (1988), we define random variables Y_i $(1 \leq i \leq n)$ as follows:

$$Y_i = \begin{cases} \prod_{j=i}^{i+k-1} X_j & (1 \leq i \leq n-k+1) \\ \prod_{j=i}^{n} X_j \prod_{j=1}^{i-n+k-1} X_j & (n-k+2 \leq i \leq n). \end{cases} \tag{2.8}$$

Then, $M_{n,k}^c = \sum_{i=1}^n Y_i$. It follows that

$$E(M_{n,k}^c) = \sum_{i=1}^{n} E(Y_i) = np^k \quad \text{(by (2.7) and (2.8))}$$

which was to be shown.

To either one of Theorems 2.1 and 2.2, we have the following corollary.

Corollary 2.1. *Assume that the components of a consecutive-k-out-of-n:F system are ordered circularly and function independently with probability p. Denote the reliability of the system by $R_C(p; k, n)$. Then, for $n \geq k$,*

$$R_C(p; k, n) = pq^{n-1} \sum_{i=1}^{k} i \sum\nolimits_3 \binom{x_1 + \cdots + x_k}{x_1, \ldots, x_k} \left(\frac{p}{q}\right)^{x_1+\cdots+x_k}, \tag{2.9}$$

where the summation $\sum_3$ is taken over all non-negative integers $x_1, \ldots, x_k$ such that $\sum_{j=1}^k jx_j = n - i$, and

$$R_C(p; k, n) = -q^n + \sum_{j=0}^{[n/(k+1)]} (-1)^j \frac{n}{n-jk} \binom{n-jk}{j} p^j q^{jk}. \tag{2.10}$$

Proof. Relation (2.9) readily follows from either one of Theorems 2.1 and 2.2 by setting $x = 0$ and interchanging p and q. Relation (2.10) follows from (2.9), by means of Lemma 2.2 of Philippou and Makri (1985) and Theorems 2.3 and 2.4 of Charalambides (1991). Formula (2.10) was also derived by Charalambides (1991) and, essentially, by Lambiris and Papastavridis (1985, Theorem 2).

3. Relation between circular and linear binomial distributions of order k

In the present section we relate the two types of circular binomial distributions of order k introduced in Section 2 to the usual (linear) binomial distributions of the same order, by relating $N^c_{n,k}$ to $N_{n,k}$ and $M^c_{n,k}$ to $M_{n,k}$.

Theorem 3.1. *Let $N^c_{n,k}$ and $N_{n,k}$ be two rvs denoting the number of nonoverlapping success runs of length k ($k \geq 1$) in n ($n \geq 1$) Bernoulli trials ordered on a circle and on a line, respectively. Set $m = min\{k(x+1)-1, n-2\}$. Then, for $x = 0, 1, \ldots, [n/k]$,*

(3.1)

$$P(N^c_{n,k} = x)$$

$$= \sum_{i=0}^{m} (i+1)q^2p^iP(N_{n-2-i,k} = x - [i/k]) + nqp^{n-1}\delta_{x,[(n-1)/k]} + p^n\delta_{x,[n/k]},$$

where δ is the Kronecker delta function.

Proof. Let $x = 0, 1, \ldots, [n/k]$. We define the events A_j ="j 1's precede the first 0", ($j = 0, 1, \ldots, m$) and B_l ="l 1's follow the last 0", ($l = 0, 1, \ldots, m$), $0 \leq j + l \leq m$, so that

$$(3.2) \qquad A_j = (X_1 = X_2 = \ldots = X_j = 1, X_{j+1} = 0), \quad (j = 0, 1, \ldots, m)$$

and

$$(3.3) \quad B_l = (X_{n-l} = 0, X_{n-l+1} = \ldots = X_{n-1} = X_n = 1), \quad (l = 0, 1, \ldots, m),$$

$j + l \leq m$. Furthermore, we set

(3.4)

$$C = \bigcup_{i=1}^{n} \{X_i = 0, X_j = 1 \ (1 \leq j \neq i \leq n)\} \text{ and } D = \{X_1 = X_2 = \ldots = X_n = 1\}$$

We observe now, by (3.4), that $P(C) = nqp^{n-1}$ and $P(D) = p^n$, while $(N^c_{n,k} \neq$

$[(n-1)/k]) \cap C = (N^c_{n,k} \neq [n/k]) \cap D = \emptyset$. Relations (3.2)-(3.4) imply

(3.5)

$$(N^c_{n,k}=x)=\bigcup_{\substack{j=0\\ j+l\le m}}^{m}\bigcup_{l=0}^{m}[(N^c_{n,k}=x)\cap A_j\cap B_l]\cup[(N^c_{n,k}=x)\cap C]\cup[(N^c_{n,k}=x)\cap D]$$

from which we get

$$(3.6)\qquad P(N^c_{n,k}=x)=\sum_{j=0}^{m}\sum_{\substack{l=0\\ j+l\le m}}^{m}P(A_j)P(B_l)P[(N^c_{n,k}=x)|A_j\cap B_l]$$

$$+P(C)P[(N^c_{n,k}=x)|C]+P(D)P[(N^c_{n,k}=x)|D]$$

$$=\sum_{j+l=0}^{m}(j+l+1)p^j qqp^l P(N_{n-2-j-l,k}=x-[(j+l)/k])$$

$$+nqp^{n-1}\delta_{x,[(n-1)/k]}+p^n\delta_{x,[n/k]},$$

and this establishes the theorem.

Theorem 3.2. *Let $M^c_{n,k}$ and $M_{n,k}$ be two rvs denoting the number of (possibly) overlapping success runs of length k (≥ 1) in n (≥ 1) Bernoulli trials ordered on a circle and on a line, respectively, and set $M = max\{0, i-k+1\}$. Then,*

(3.7)

$$P(M^c_{n,k}=x)=\begin{cases}\sum_{i=0}^{x+k-1}(i+1)p^iq^2P(M_{n-2-i,k}=x-M) & (0\le x\le n-k-1)\\ nqp^{n-1} & (x=n-k(>0))\\ 0 & (n-k+1\le x\le n-1)\\ p^n & (x=n).\end{cases}$$

Proof. Let X_i $(1 \leq i \leq n)$ be as in Proposition 2.2. For $0 \leq x \leq n-k-1$, we consider the events A_j $(j = 0, 1, \ldots, x+k-1)$ and B_l $(l = 0, 1, \ldots, x+k-1)$, $j+l \leq x+k-1$, as in Theorem 3.1. Then,

$$(3.8)\qquad (M^c_{n,k}=x)=\bigcup_{\substack{j=0\\ j+l\le x+k-1}}^{x+k-1}\bigcup_{l=0}^{x+k-1}[(M^c_{n,k}=x)\cap A_j\cap B_l]$$

Therefore, for $0 \leq x \leq n-k-1$,

$$P(M_{n,k}^c = x) = \sum_{\substack{j=0 \\ j+l \leq x+k-1}}^{x+k-1} \sum_{l=0}^{x+k-1} P(A_j)P(B_l)P[(M_{n,k}^c = x)|A_j \cap B_l] \tag{3.9}$$

$$= \sum_{j+l=0}^{x+k-1} (j+l+1)p^j qqp^l P(M_{n-2-j-l,k} = x - \max\{0, j+l-k+1\}).$$

We also observe that

$$P(M_{n,k}^c = n-k) = nqp^{n-1} \qquad (n > k), \tag{3.10}$$

since $M_{n,k}^c = n-k(> 0)$ if and only if $X_1 = 0, X_2 = \ldots = X_n = 1$, or $X_2 = 0, X_3 = \ldots = X_n = X_1 = 1, \ldots,$ or $X_n = 0, X_1 = \ldots = X_{n-1} = 1$. We aslo note that

$$P(M_{n,k}^c = x) = P(\emptyset) = 0 \qquad (n-k+1 \leq x \leq n-1) \tag{3.11}$$

and

$$P(M_{n,k}^c = n) = P(X_1 = \ldots = X_n = 1) = p^n. \tag{3.12}$$

Relations (3.9)-(3.12) establish the theorem.

Setting $x = 0$ and interchanging p and q in either one of Theorems 3.1 and 3.2 we readily get the following corollary, which is one of the main theorems of Derman et al. (1982).

Corollary 3.1. *Let $R_C(p; k, n)$ and $R_L(p; k, n)$ denote the reliability of a circular and linear consecutive-k-out-of-n:F system respectively. Then, for $n \geq k+1$,*

$$R_C(p; k, n) = p^2 \sum_{i=0}^{k-1} (i+1)q^i R_L(p; k, n-i-2). \tag{3.13}$$

Remark 3.1. After the paper has been submitted it was brought to our attention that the cyclic case of overlapping success runs has been treated by Chryssaphinou et al. (1993). Their results and methodology, however, do not overlap with ours.

Acknowledgement

The authors wish to thank two referees for their helpful comments which led to improvements of the paper. They are indebted, in particular, to one of them who suggested the present version of Theorem 2.1 and corrected a previous version of Proposition 2.1.

References

Aki, S. and Hirano, K. (1988). Some characteristics of the binomial distribution of order k and related distributions. In *Statistical Theory and Data Analysis* **II**, 211-222, ed. K. Matusita. Elsevier, Amsterdam.

Charalambides, Ch.A. (1991). Lucas numbers and polynomials of order k and the length of the longest circular success run. *Fibonacci Quart.* **29**, 290-297.

Chryssaphinou, O., Papastavridis, S. and Tsapelas, T. (1993). On the number of overlapping success runs in a sequence of independent Bernoulli trials. In *Applications of Fibonacci Numbers*, vol. **5**, ed. G.E. Bergum et al., Kluwer Academic Publishers, Dordrecht (to appear).

Derman, C. Lieberman, G.J. and Ross, S.M. (1982). On the consecutive-k-out-of-n:F system. *IEEE Trans. Reliability* **31**, 57-63.

Feller, W. (1968). *An Introduction to Probability Theory and its Applications*, Vol. 1, 3rd edn., Wiley, New York.

Godbole, A.P. (1990a). Specific formulae for some success run distributions. *Statist. Prob. Lett.* **10**, 119-124.

Godbole, A.P. (1990b). Degenerate and Poisson convergence criteria for success runs. *Statist. Prob. Lett.* **10**, 247-255.

Hirano, K. (1986). Some properties of the distributions of order k. In *Fibonacci Numbers and Their Applications*, 43-53, ed. A.N. Philippou et al., Reidel, Dordrecht.

Lambiris, M. and Papastavridis, S. (1985). Exact reliability formulas for linear and circular consecutive-k-out-of-n:F systems. *IEEE Trans. Reliability* **R-34**, 124-126.

Ling, K.D. (1988). On binomial distributions of order k. *Statist. Prob. Lett.* **6**, 247-250.

Papastavridis, S.G. (1990). m-consecutive-k-out-of-n:F systems. *IEEE Trans. Reliability* **R-39**, 386-388.

Philippou, A.N. (1986). Distributions and Fibonacci polynomials of order k, longest runs, and reliability of consecutive-k-out-of-n:F systems. In *Fibonacci Numbers and Their Applications*, 203-227, ed. A.N. Philippou et al., Reidel, Dordrecht.

Philippou, A.N. and Makri, F.S. (1985). Longest success runs and Fibonacci-type polynomials. *Fibonacci Quart.* **23**, 338-346.

Philippou, A.N. and Makri, F.S. (1986). Successes, runs and longest runs. *Statist. Prob. Lett.* **4**, 211-215.

Philippou, A.N. and Muwafi, A.A. (1982). Waiting for the kth consecutive success and the Fibonacci sequence of order k. *Fibonacci Quart.* **20**, 28-32.

LENGTH OF THE LONGEST NON-DECREASING SUBSEQUENCE ON TWO SYMBOLS

Arthur O. Pittenger*

University of Maryland Baltimore County

Let L_n denote the length of the longest, non-decreasing subsequence of an n-long sequence of zeroes and ones. Assuming independent, identically distributed observations and also assuming $p_0 \neq p_1$, it is shown that

$$P[L_n = k] = \frac{p_0^{k+1}p_1^{n-k} - p_1^{k+1}p_0^{n-k}}{p_0 - p_1}\left[\binom{n}{k} - \binom{n}{k+1}\right]$$

for $n/2 \leq k \leq n$. The corresponding expression for $p_0 = p_1 = 1/2$ can be obtained as a limit.

The approach leading to the expression above applies when more than two symbols are involved, but does not appear to lead to an easily computable formula.

COMBINATORIAL PROBABILITY, NONDECREASING SUBSEQUENCES.

1. Introduction and Summary of Results.

Consider a sequence of n independent, identically distributed random variables with values in $\{0, 1, 2, \cdots, d\}$, where $d \geq 1$, and suppose one is interested in the distribution of the maximal length of non-decreasing subsequences. If the random variables are uniformly distributed, a combinatorial approach is feasible and leads to the expression in the referenced paper by Schensted.

In this note we use a probabilistic approach to give an exact expression for $P[L_n = k]$ when $d = 1$ and the basic probabilities p_0 and $p_1 = 1 - p_0$ are not equal. Specifically, when $n/2 \leq k \leq n$,

$$P[L_n = k] = \frac{{p_0}^{k+1}{p_1}^{n-k} - {p_1}^{k+1}{p_0}^{n-k}}{p_0 - p_1}\left[\binom{n}{k} - \binom{n}{k+1}\right] \tag{1.1}$$

* Research supported in part by NSA Grant MDA 904-88-H-2019

A. P. Godbole and S. G. Papastavridis (eds.), Runs and Patterns in Probability, 83–89.

using $\binom{n}{n+1} = 0$. Obviously, $P[L_n = k]$ equals zero otherwise. Schensted's result for $d = 1$ can be obtained as a limiting case of (1.1):

$$P[L_n = k] = \frac{2k+1-n}{2^n}\left[\binom{n}{k} - \binom{n}{k+1}\right] \tag{1.2}$$

when $p_0 = p_1 = 1/2$ and $n/2 \le k \le n$.

Equations (1.1) and (1.2) are the only computationally useful results obtained from the approach developed in Section 2, although the ideas used there are valid for $d \ge 1$. The main result is Lemma 2.2 which has as a corollary the precursor to (1.1). Sections 3 and 4 give some of the technical arguments needed to justify (2.3) and to derive (1.1).

2. Notation and Strategy

In keeping with the probabilistic approach used here, let Ω_n denote n-long sequences of elements selected from the set $\{0, 1, \cdots, d\}$ and let $L_n(\omega)$ denote the length of the longest, non-decreasing subsequence of an ω in Ω_n. (When n is obvious, we may omit it from the expressions, and, as usual, we may also suppress the argument ω itself.) Thus, for example, $L_4((0212)) = 3$ where d necessarily is at least 2. The probability used throughout will be that based on the assumption of independent, identically distributed trials with "basic" probabilities $\{p_0, p_1, \cdots, p_d\}$.

Now suppose we look at $(n+1)$-long sequences. Let L'_n denote the length of the longest, non-decreasing subsequence within the first n components of ω in Ω_{n+1}. Let L''_n denote the analogous value using the last n components of ω. Then, after a little reflection, it is intuitively reasonable to expect the following to be true.

Lemma 2.1.

$$P_{n+1}[L'_n < L_{n+1} = k+1] = P_{n+1}[L''_n < L_{n+1} = k+1] \tag{2.1}$$

Unfortunately a short intuitive proof seems elusive, and a recursive argument is presented in section 4.

Assuming the validity of Lemma 2.1, we can decompose each side to obtain a useful representation of the desired probability. First, set:

$L_n(i_1, \cdots, i_r)$ = length of longest non-decreasing subsequence of ω in Ω_{n+1}, restricting attention to entries with values $i_1, i_2, \cdots, i_r$. Thus, if d=1 and $\omega = (0011)$, then $L_4(0)$ is 2 while L_4 itself is 4.

Since the last component of ω has to be included in the longest, non-decreasing subsequence for those ω in $\{L'_n < L_{n+1}\}$, it follows that

$$P_{n+1}[L'_n < L_{n+1} = k+1] = \sum_{r=0}^{d} P_{n+1}[L'_n = L'_n(0, \cdots, r) = k, \omega_{n+1} = r]$$

By the assumption of independence, we can conclude

$$P_{n+1}[L'_n < L_{n+1} = k+1] = \sum_{r=0}^{d} P_n[L_n = L_n(0, \cdots, r) = k]p_r$$

and analogous reasoning using values of ω_1 produces

$$P_{n+1}[L''_n < L_{n+1} = k+1] = \sum_{r=0}^{d} p_r P_n[L_n = L_n(r, \cdots, d) = k].$$

Thus, assuming Lemma 2.1, we have

$$\begin{aligned}(p_0 - p_d)P_n[L_n = k] = &\sum_{r=0}^{d-1} P_n[L_n = L_n(0, \cdots, r) = k]p_r \\ &- \sum_{r=1}^{d} p_r P_n[L_n = L_n(r, \cdots, d) = k].\end{aligned} \tag{2.2}$$

Next define a 1-1 mapping T, $T : \Omega_n \to \Omega_n$, by time reversal followed by spatial complementation. That is, $T\omega$ is obtained by reversing the order of the entries of ω and then complementing each entry with respect to d. Hence, if d=3,

$$T((21001)) = (23321).$$

Note that $L_n(\omega) = L_n(T\omega)$ and that if $\tilde{P}$ is constructed using $\tilde{p}_k = p_{d-k}$ then

$$P(\{\omega\}) = \tilde{P}(\{T\omega\})$$

Further, since $L_n(d-r,\cdots,d)(\omega) = L_n(0,\cdots,r)(T\omega)$, we have

$$P[L_n = L_n(d-r,\cdots,d) = k] = P[L_n(T\omega) = L_n(0,\cdots,r)(T\omega) = k]$$
$$= \tilde{P}[L_n = L_n(0,\cdots,r) = k].$$

Putting these observations together with (2.2) gives a lemma equivalent to (2.1).

Lemma 2.2.

$$(p_0 - p_d)P[L_n = k] = \sum_{r=0}^{d-1}\{p_r P[L_n = L_n(0,\cdots,r) = k] - \tilde{p}_r\tilde{P}[L_n = L_n(0,\cdots,r) = k]\} \tag{2.3}$$

The point of all these manipulations is that we have decomposed the calculation of $P[L_n = k]$ into calculations of analogous quantities which depend on proper subsets of $\{0,\cdots,d\}$. Further, the introduction of $\tilde{P}$ shows that if we can compute the positive terms, then the negative terms can be obtained by using the same functional form. In particular, when d=1, we have the following result.

Corollary 2.3.

$$(p_0 - p_1)P[L_n = k] = p_0 P[L_n = L_n(0) = k] - p_1\tilde{P}[L_n = L_n(0) = k]. \tag{2.4}$$

Thus, in this case the calculation of $P[L_n = k]$ reduces to the calculation of the cases when there is a maximal non-decreasing subsequence using only one symbol.

We will show how to compute that special case in the next section, and it would be nice to compute other cases in closed form. Unfortunately, matters get involved rather quickly when $d > 1$, and it is not readily apparent how to sort things out efficiently using this approach.

3. Calculation of $P[L_n = L_n(0) = k]$ when d=1.

Observe first that if $\lceil n/2 \rceil$ denotes the smallest integer bigger than or equal to $n/2$,

$$P[L_n = k] = 0, \qquad k < \lceil n/2 \rceil. \tag{3.1}$$

If $n = 2r$ and $k = r$, then

$$P[L_{2r} = L_{2r}(0) = r] = p_1 P[L_{2r-1} = L_{2r-1}(0) = r], \tag{3.2}$$

since if ω in Ω_{2r} begins with a zero and has r ones among its components, then $L_{2r}(\omega) \geq r + 1$. Hence, each ω in $\{L_{2r} = L_{2r}(0) = r\}$ begins with a one, and (3.2) is a consequence.

In all other cases with $n/2 < k \leq n$, we assert that

$$\begin{aligned} P[L_n = L_n(0) = k] =& p_0 P[L_{n-1} = L_{n-1}(0) = k - 1] \\ &+ p_1 P[L_{n-1} = L_{n-1}(0) = k]. \end{aligned} \tag{3.3}$$

Here's the reasoning. If $L_n = L_n(0) = k$ and ω begins with a zero, every maximal length, non-decreasing subsequence includes that first zero, and that observation leads to the first term on right of (3.3).

Now suppose the first element is a one. Since $k > n/2$ and there are k zeroes in ω, there have to be strictly fewer than k ones in ω and exactly k zeroes in components 2 through n. Hence both L_n and $L_n(0)$ can be computed using only entries two through n. Conversely, appending a lead one to such an $(n-1)$-long sequence gives an ω with a lead one and with $L_n = L_n(0) = k$, explaining the second term on the right and completing the argument.

This same approach works for $k > n/2$ for $d > 1$, but leads to more complicated expressions when $k < n/2$. It is for this reason that the approach developed in Section 2 does not lead to a simple analysis when $d > 1$.

Equation (3.3) is reminiscent of a relationship among binomial coefficients, and all three equations taken together give a "recursion with boundary conditions" when one includes

$$P[L_2 = L_2(0) = 1] = p_1 p_0 \tag{3.4}$$

and

$$P[L_n = L_n(0) = n] = p_0^n. \tag{3.5}$$

This suggests looking at expressions of the form

$$P[L_n = L_n(0) = k] = p_0^k p_1^{n-k} d(n,k)$$

where $d(n,k)$ involves binomial coefficients. Subsequent analysis then produces the following result.

Lemma 3.1. *When $n/2 \le k \le n$,*

$$P[L_n = L_n(0) = k] = p_0^k p_1^{n-k}\left[\binom{n}{k} - \binom{n}{k+1}\right] \tag{3.6}$$

using the convention $\binom{n}{j} = 0$ when $j > n$.

The proof is a matter of checking that this expression satisfies conditions (3.1) through (3.5), and we omit the details.

4. Proof of Lemma (2.1)

Observe first that if L_n denotes the restriction to the first n components of an $(n+1)$-long ω,

$$\begin{aligned} P[L_{n+1} = k+1, \omega_{n+1} = r] =& P[L_n = L_n(0,\cdots,r) = k, \omega_{n+1} = r] \\ &+ P[L_n = k+1, \omega_{n+1} = r] \\ &- P[L_n = L_n(0,\cdots,r) = k+1, \omega_{n+1} = r]. \end{aligned}$$

Hence,

$$\begin{aligned} P[L_{n+1} = k+1] =& \sum_{r=0}^{d} P[L_n = L_n(0,\cdots,r) = k]p_r + P[L_n = k+1] \\ &- \sum_{r=0}^{d} P[L_n = L_n(0,\cdots,r) = k+1]p_r \end{aligned} \tag{4.1}$$

Analogous reasoning using the first entry in ω produces:

$$\begin{aligned} P[L_{n+1} = k+1] =& \sum_{r=0}^{d} p_r P[L_n = L_n(r,\cdots,d) = k] + P[L_n = k+1] \\ &- \sum_{r=0}^{d} p_r P[L_n = L_n(r,\cdots,d) = k+1]. \end{aligned} \tag{4.2}$$

Subtracting (4.2) from (4.1) and rearranging terms give the following result.

$$\sum_{r=0}^{d}\{P[L_n = L_n(0,\cdots,r) = k]p_r - p_r P[L_n = L_n(r,\cdots,d) = k]\}$$
$$= \sum_{r=0}^{d}\{P[L_n = L_n(0,\cdots,r) = k+1]p_r - p_r P[L_n = L_n(r,\cdots,d) = k+1]\},$$

or recalling the discussion in Section 2,

$$\begin{aligned} &P[L'_n < L_{n+1} = k] - P[L''_n < L_{n+1} = k] \\ &= P[L'_n < L_{n+1} = k+1] - P[L''_n < L_{n+1} = k+1]. \end{aligned} \tag{4.3}$$

It follows by recursion that the proof of (2.1) reduces to verifying that

$$P[L'_n < L_{n+1} = n+1] = P[L''_n < L_{n+1} = n+1] \tag{4.4}$$

But since the components of ω in $\{L_{n+1} = n+1\}$ form an $(n+1)$-long, non-decreasing sequence, $L'_n = L''_n = n$. Hence (4.4) is immediate, and the proof is complete.

Reference

C. Schensted, "Longest increasing and decreasing subsequences," *Canadian Journal of Mathematics* **13** (1961), 179-191.

EXCHANGEABILITY AND RECURSION IN THE CONDITIONAL DISTRIBUTION THEORY OF NUMBER AND LENGTH OF RUNS

Eugene F. Schuster[1]

The University of Texas at El Paso

Let $\vec{s}$ be an ordered sequence of length $N = m + n$ from a two letter alphabet consisting of m letters of type F and n letters of type S where m and n are given integers. We are interested in the conditional distributions of the number and length of runs of selected or combined type under the null hypothesis H_0 of random ordering in the sequence $\vec{s}$ when m and n are known. We represent these statistics as functions of exchangeable random variables in a form which facilitates the study of their distribution theory. We give recursive formulas for the distributions of the number of runs as well as the longest run statistics. The computational utility of these recursive algorithms was tested via the software system *Mathematica*[2]. We correct three errors in the most comprehensive table available on the probability density of the longest run M of either type, Bateman's 1948 table, and give a table of the cdf of M for all pairs n, m, $n \leq m$, for $n + m = N = 2(1)25$ and $N = 30(5)50$.

1 Introduction and Summary

Let $\vec{s}$ be an ordered sequence of N objects of two types, say n S's and m F's where n and m are given integers. We are interested in the distributions of the number and length of runs of selected or combined type under the null hypothesis H_0 of random ordering in the sequence $\vec{s}$ for various applications including: testing the hypothesis of random ordering, median type testing for a random sample, and two sample DNA type sequence comparisons.

Among the best known and easiest to apply of the tests for random versus non-

[1] Partially supported by NIH Grant RCMI 1 G12 RR08124-01

[2] *Mathematica* is a registered trademark of Wolfram Research, Inc.

A. P. Godbole and S. G. Papastavridis (eds.), Runs and Patterns in Probability, 91–118.

random ordering in $\vec{s}$ are the classical runs tests based on the number or lengths of runs. A *run* in $\vec{s}$ is defined as a string of one or more identical symbols which are followed by and preceded by a different symbol or no symbol at all. The *length* of a run is the number of identical symbols comprising the run. Both the total number of runs R and the length of the longest run may give clues to lack of randomness. In a test based on R, both too few and too many runs indicate a nonrandom ordering. In a manufacturing process, long sequences of success runs would be desirable, but long failure runs might suggest that the process is out of control. In DNA type sequence comparisons, one might be interested in the distribution of $M = M(n, m)$, the longest run of matches or mismatches.

A historical summary with related references on the distribution theory of runs follows in Section 2. Parts of Sections 3 and 4, follow Schuster (1991) in representing the number and length of runs of selected or combined type under the null hypothesis H_0 of random ordering as functions of exchangeable random variables. In Section 3, we give recursion formulas for the pdf of the number of runs of individual or combined type. In Section 4, we represent the length of the longest run of one type as the maximum of exchangeable random variables and discuss algorithms, including recursion, for computing the corresponding distribution functions or p-values. Finally, in Section 5, we also give new recursive algorithms for evaluating the distribution function of the longest run of either type. The computational utility of these recursive algorithms is demonstrated via the software system *Mathematica*. We correct three errors in Bateman's (1948) table on the probability density of the longest run of either type. In Table 1, we give rounded, four decimal digit approximations to a where $a = P(M \leq x) = F_M(x; n, m)$ for every x with $0.00005 \leq a < 0.99995$ for all pairs n, m, $n \leq m$ with $N = n + m = 2(1)25$ and $N = 30(5)50$.

2 Historical Background

The classical paper by Mood (1940) reviews the "stormy" history of the distribution theory of runs from its apparent origin in the late nineteenth century until 1940. In this paper, Mood derives moments as well as the finite and asymptotic distributions of runs of given length in the case of random arrangements of fixed numbers of symbols of two or more types (referred to as the conditional problem) and in the case of a random sample of fixed length of symbols obtained by sampling a binomial or multinomial population (the unconditional problem). Mood shows that the asymptotic distribution of suitable normalized versions of these statistics are all normal. In this paper, we treat the conditional problems and distributions.

Wald and Wolfowitz (1940) use the conditional runs test in the context of testing for a common continuous distribution in the two sample problem. Tables for the distribution function of the conditional distribution of the number of runs R in the case of two symbols with $n \leq m \leq 20$ are given in Swed and Eisenhart (1943) and reprinted in part in Koskoska and Nevison (1989). Guenther (1978) expresses the distribution of R in terms of mixtures of hypergeometric distributions.

Tables of the probability mass function of the length of the longest run of either type in the conditional problem are given in Bateman (1948) for select n, m, $n+m \leq 20$. At the end of Section 5, we correct 3 errors in Bateman's Table. Mosteller (1941) gives tables when $n = m$ for $2n = 10(10)50$ in the quality control context of testing randomness in a sample by analyzing the number and lengths of runs above and below the median in a sample. Power functions against simple Markov alternatives are discussed in Bateman (1948) and David (1947). Additional references citing related work in the conditional distribution theory of the longest run are given at the beginning of Section 4. Schuster (1991) first uses the theory of exchangeability for finite sequences of random variables in studying the distribution theory of runs. The nonparametrics book by Gibbon's (1985) contains an introductory chapter on "tests based on the theory of runs". Kingman (1978) gives an introduction to the theory of exchangeability with a long list of references.

3 Exchangeable Representation of R

Our first result is an exchangeable representation of the total number of runs R. Perhaps the simplest textbook representation of R is (see Gibbons (1985), page 56):

$$R = 1 + \sum_{k=2}^{N} J_k \quad \text{where} \quad J_k = \begin{cases} 1 & \text{if the } k^{th} \text{ element in } \vec{s} \neq k\text{-}1^{\ th} \\ 0 & \text{otherwise.} \end{cases}$$

Then under the null hypothesis H_0 of random ordering in $\vec{s}$, J_2, ..., J_N are identically distributed Bernoulli random variables with parameter $p = nm/\binom{N}{2}$. From this representation the null distribution computation $E(R) = 1 + (N-1)nm/\binom{N}{2} = 1 + 2nm/N$ is easy. However, since $J_2, \ldots, J_n$ are not exchangeable, i.e., the covariance of J_i and J_j, $i \neq j$, depends on i and j, the computation of Var(R) is tedious. In the following, we represent $R-1$ as a linear combination of Bernoulli random variables which are exchangeable under H_0.

In this direction, consider a random experiment ξ which produces the randomly ordered sequence (or permutation) $\vec{s}$ of N objects of two types, say n S's (successes) and m F's (failures) where n and m are given integers. Let R, R_F, and R_S be the number of runs, success runs, and failure runs, respectively and let M, M_F, and M_S be the length of the longest run of either type, the longest failure run, and the longest success run, respectively. Here and in the remainder of the paper, we assume the null hypothesis distribution of random ordering, i.e., we continue sampling without replacement until we exhaust the population of N. As a sample space Ω for the random experiment ξ, we take all $(N+1)$ - tuples $\vec{\omega} = (s_1, \ldots, s_N, F)$ where $\vec{s} = (s_1, \ldots, s_N)$ is an ordered arrangement of n S's and m F's. For reasons (of symmetry) to become clear in the following, we add the final F to the ordered outcome $\vec{s}$.

Let us then define a random vector $\vec{L} = (L_1, \ldots, L_{m+1})$ over Ω as follows: L_1 records the number of S's before the first F in $\vec{\omega}$, and for $i = 2, \ldots, m+1$, L_i records the number of S's between the $(i-1)^{th}$ F and the ith F in $\vec{\omega}$. If consecutive F's is interpreted as an S run of 0 length between the F's, then L_i

can be interpreted as the length of the ith S run. For each $\vec{\omega}$ in Ω, let us write $\vec{\omega}=(\omega_1,\ldots,\omega_{m+1})$ where each ω_i is a sequence of zero or more S's ending with an F. It then follows that every permutation $\vec{\omega}^* = (\omega_{i_1},\ldots,\omega_{i_{m+1}})$ of $\vec{\omega}$ is also in Ω. Thus one can easily argue that $L_1,\ldots,L_{m+1}$ are identically distributed and the joint distribution of $L_{i_1},\ldots,L_{i_k}, 1 \leq k \leq m+1$, does not depend on the permutation $i_1,\ldots, i_k$ from 1, 2,...,m+1, i.e., $L_1,\ldots,L_{m+1}$ are exchangeable random variables which are sufficient for the data. We will freely use this in the following. In fact, $P(L_1 = l_1,\cdots,L_{m+1} = l_{m+1}) = \binom{N}{n}^{-1}$ for nonnegative integers l_i, $i = 1,\cdots,m+1$ which sum to n. Define $Y_{(0)} = 0$, $Y_{(m+1)} = N+1$, and $Y_{(i)} = i + \sum_{k=1}^{i} L_i$, $i = 1,\cdots,m$. Then $Y_{(1)},\cdots,Y_{(m)}$ have the same joint distribution as the order statistics of a simple random sample of size m without replacement from the integers $1,2,\cdots,N$. In this case, $S_{(i)} = Y_{(i)} - Y_{(i-1)} = L_i + 1$ can be interpreted as the ith spacing, $i = 1,\cdots,m+1$. The marginals distributions of $\vec{L}$ can be studied or simulated by using the

Beta-Binomial representation of the sampling distribution of $\vec{L}$: *First sample $\Theta \sim Beta(1,m)$, observing θ_1 and then obtain the observed value l_1 of L_1 by sampling $(L_1|\Theta = \theta_1) \sim Binomial(n,\theta_1)$. For fixed integer k, $1 < k \leq m$, suppose $l_1,\cdots,l_{k-1}$ are the observed values of $L_1,\cdots,L_{k-1}$. Then obtain the observed value l_k of L_k, given $l_1,\cdots,l_{k-1}$, by first sampling $\Theta \sim Beta(1,m-k+1)$, observing, say θ_k and then sampling $(L_k|L_1 = l_1,\cdots,L_{k-1} = l_{k-1},\Theta = \theta_k) \sim Binomial(n-\sum_{i=1}^{k-1} l_i,\theta_k)$ (Note that for any positive integer I, $1-U^{1/I} \sim Beta(1,I)$ when U is uniform over (0,1)).*

The pdf of L_i is easily seen to be $p(k;n,m) = \binom{N-k-1}{m-1}/\binom{N}{m}$ for $k = 0,1,\cdots,n$. Since the L_i's are exchangeable and $\sum_{i=1}^{m+1} L_i = n$, it follows that $E(L_i) = n/(m+1)$ and for $i \neq j$ that $Cov(L_i,L_j) = -Var(L_1)/m$, i.e. the correlation coefficient $\rho(L_i,L_j) = -1/m$. This is the smallest correlation that a finite set of $m+1$ exchangeable random variables can have (Kingman (1978)). The Beta-Binomial representation indicates that $Var(L_i) = E(Var(L_i|\Theta)) + Var(E(L_i|\Theta)) = nE\Theta(1-\Theta) + n^2 Var\Theta = \frac{nm(N+1)}{(m+1)^2(m+2)}$. Thus $E(Y_{(r)}) = r + rn/(m+1)$ and $Var(Y_{(r)}) = rVar(L_1)(1-\frac{r-1}{m})$. These and other properties of the random variables $L_1, \ldots, L_{m+1}$ and the negative hypergeometric random variables $Y_{(r)}$ are discussed in Schuster and Sype (1987). In the following we express the number of runs as a function of $\vec{L}$.

We begin with an elementary (urn model) property of exchangeable random variables with values in {0,1} which will be used in the following.

Lemma 3.1 *Let $I_1, \ldots, I_W$ be a finite sequence of exchangeable Bernoulli random variables. Let j, k, r, s be integers with $0 \leq r \leq s$, $0 \leq r \leq j \leq k \leq W$, and $P(\sum_{i=1}^{k} I_i = s) > 0$. Then $P(\sum_{i=1}^{j} I_i = r | \sum_{i=1}^{k} I_i = s) = \binom{j}{r}\binom{k-j}{s-r}/\binom{k}{s}$.*

Proof. Let $p_{s,k} = P(I_1 = 1, \ldots, I_s = 1, I_{s+1} = 0, \ldots, I_k = 0)$. Using the exchangeability of the I_i's, we see that $P(\sum_{i=1}^{k} I_i = s) = \binom{k}{s} p_{s,k}$ and $P(\sum_{i=1}^{j} I_i = r, \sum_{i=1}^{k} I_i = s) = P(\sum_{i=1}^{j} I_i = r, \sum_{i=j+1}^{k} I_i = s-r) = \binom{j}{r}\binom{k-j}{s-r} p_{s,k}$. The conclusion in the lemma now follows from the definition of conditional probability.

Let $I(x)$ be the indicator function of the set $(0, \infty)$, i.e., $I(x) = 1$ if $0 < x < \infty$ and $I(x) = 0$, otherwise. Define exchangeable Bernoulli random variables $I_1, \ldots, I_{m+1}$, by $I_i = I(L_i)$, i=1,...,m+1. Let $p = n/N$ and $q = 1 - p = m/N$ (recall $N = n + m$). Then:

Lemma 3.2 *$E(I_i) = p$, $Var(I_i) = pq$, $i = 1, \cdots, m+1$, $Cov(I_i, I_j) = -pq/(N-1)$ for $i \neq j$ and $Var(\sum_{i=1}^{m+1} \alpha_i I_i) = pq(\sum_{i=1}^{m+1} \alpha_i^2 - \sum_{i \neq j} \alpha_i \alpha_j /$
$(N-1))$ for any constants $\alpha_1, \ldots, \alpha_{m+1}$.*

Proof. $I_1 = 1$ indicates that the random sequence $\vec{s}$ begins with an S. Thus $E(I_1) = P(I_1 = 1) = p$ and so $Var(I_1) = pq$. Now $I_1 = 1$ and $I_{m+1} = 1$ only if the random sequence $\vec{s}$ begins and ends with an S. Thus $E(I_1 I_{m+1}) = n(n-1)/N(N-1)$. Take $i \neq j$. Then since the I_i's are exchangeable, $Cov(I_i, I_j) = Cov(I_1, I_{m+1}) =$

$E(I_1 I_{m+1}) - E(I_1)E(I_{m+1}) = -pq/(N-1)$ and the proof of the lemma is easily completed.

Let R_S be the number of S runs and R be the total number of runs in the ordered sequence $\vec{s}$. Then

Theorem 3.3 $R_S = \sum_{i=1}^{m+1} I_i$, $R_F = R_S + 1 - I_1 - I_{m+1}$ *and* $R = 1 + \sum_{i=1}^{m+1} \alpha_i I_i$ *where* $\alpha_1 = \alpha_{m+1} = 1$ *and* $\alpha_i = 2$ *for* $2 \le i \le m$.

Proof. Let $\vec{s}$ be the ordered sequence of n S's and m F's. Again we add an F at the end of $\vec{s}$ to obtain $\vec{\omega} = (\omega_1, \cdots, \omega_{m+1})$ where each ω_i is zero or more S's ending with an F. Now $L_i > 0$ indicates that there is an S run of length L_i between the $(i-1)^{th}$ and the ith F and so it follows that $R_S = \sum_{i=1}^{m+1} I_i$ is the number of S runs.

The total number of runs R in $\vec{s}$ equals the sum of the number of F runs R_F and the number of S runs R_S. It is easy to see that R_F will differ from R_S by at most 1. Now $R_S = R_F + 1$ indicates that there is one more S run than there are F runs. This can happen only if $\vec{s}$ begins and ends with an S. But then $I_1 = 1$ and $I_{m+1} = 1$ and so $R_F = R_S - 1 = R_S + 1 - I_1 - I_{m+1}$ and hence $R = 2R_S + 1 - I_1 - I_{m+1}$. The cases $R_S = R_F - 1$ and $R_S = R_F$ follow in similar fashion and the validity of the theorem follows.

If $\alpha_1 = \alpha_{m+1} = 1$ and $\alpha_i = 2$, for $i = 2, \cdots, m$, then

(i) $\sum_{i=1}^{m+1} \alpha_i = 2m$,

(ii) $\sum_{i=i}^{m+1} {\alpha_i}^2 = 4m - 2$, and

(iii) $\sum_{i \neq j}^{m+1} \alpha_i \alpha_j = (\sum_{i=1}^{m+1} \alpha_i)^2 - \sum_{i=1}^{m+1} \alpha_i^2 = 4m^2 - 4m + 2$.

Using (i) - (iii), Lemma 3.2 and Theorem 3.3, we easily obtain:

Corollary 3.4 $E(R_S) = \mu_S = (m+1)p$, $E(R) = \mu_R = 1 + 2mp$, $Var(R_S) = (m+1)pq(n-1)/(N-1) = (\mu_S)(\mu_S - 1)/(N-1)$, *and* $Var(R) = 2pq(2nm - N)/(N-1) = (\mu_R - 1)(\mu_R - 2)/(N-1)$ $(p = n/N$ *and* $q = 1 - p = m/N)$.

Remark 1. If n and m tend to infinity in such a way that the proportion of successes S in the sequence, $p = p_n = n/N$, tends to p_0, then $Var[R_S/(m+1)]$ tends to zero and one can use standard arguments to conclude that $R_S/(m+1)$ converges in probability to p_0.

Remark 2. Theorem 3.3 suggests that much of the difficulty in studying the theoretical properties of R is due to the fact that R is not symmetrically defined in terms of the exchangeable variables $I_1, \ldots, I_{m+1}$. In fact for $i \neq j$, $R_{i,j} = 2R_S + 1 - I_i - I_j$ has the same distribution as $R = R_{1,m+1}$. If we replace R by the average $\bar{R} = \sum_{i \neq j}^{m+1} R_{ij}/\{m(m+1)\}$, then $\bar{R} = 1 + cR_S$ where $c = 2m/(m+1)$. Thus, it would seem that one should simply use the number of success runs R_S (or failure runs) as the test statistic in testing random ordering. Since R_F will differ from R_S by at most 1, it is easy to see that the asymptotic distribution of normalized R would be the same as that of normalized R_S. The asymptotic distribution of normalized $R_S = (R_S - E(R_S))/\sigma(R_S)$ can be established using Stirling's approximation to factorials in Corollary 3.5. In addition, $\boldsymbol{R_S}$ **has a hypergeometric distribution:**

Corollary 3.5 *R_S has the hypergeometric distribution of a random variable which records the number of failures in a random sample of n without replacement from a population of $m+1$ failures and $n-1$ successes, i.e. R_S has density $f(r) = \binom{m+1}{r}\binom{n-1}{n-r}/\binom{N}{n}$ for*
$r = 1, \ldots, min\{m+1, n\}$.

Proof. $P(R_S = r) = P(\sum_{i=1}^{m+1} I_i = r) = \binom{m+1}{r}P(A)$ where $A = \{I_1 = 1, \ldots, I_r = 1, I_{r+1} = 0, \ldots, I_{m+1} = 0\}$. Again we take $\vec{\omega} = (\omega_1, \ldots, \omega_{m+1})$ where each ω_i is the sequence of zero or more S's preceding the ith F. Now $\vec{\omega} \epsilon A$, implies $\omega_i = F$, $i = r+1, \cdots, m+1$. The n S's in $\vec{\omega}$ must be placed in (the beginnings of) the $\omega_1, \ldots, \omega_r$ strings with at least one S in each string. But n S's (balls) can be placed in the r strings (cells) in $\binom{n-1}{r-1}$ ways if every string (cell) receives at least one S (ball). Thus $P(A) = \binom{n-1}{r-1}/\binom{N}{n} = \binom{n-1}{n-r}/\binom{N}{n}$ and the proof is complete.

Remark 3. Note that the pdf $f(r)$ of R_S can be recursively generated:

$$f(r+1) = (n-r)(m+1-r)f(r)/\{r(r+1)\}.$$

Of course, R_F is also hypergeometric and so R is the sum of two highly correlated hypergeometrics (see Corollary 3.6). Let $r = (n+1)(m+2)/(N+1)$. Since R_S is hypergeometric, it follows from Corollary 3.5 (or from the recursion formula) that the most probable value of R_S occurs at both $r-1$ and r if r is a positive integer and at $[r]$ (the greatest integer in r) otherwise. In either case, the most probable value of R_S is close to the mean $E(R_S) = n(m+1)/N$.

Using the representation $R_F = R_S + 1 - I_1 - I_{m+1}$ of Theorem 3.3 and the exchangeability of the I_i's, one can compute the correlation coefficient $\rho(R_S, R_F)$ without using the joint density (or marginals) of (R_S, R_F).

Corollary 3.6 $\rho = \rho(R_S, R_F) = \sqrt{\{1-2/(m+1)\}\{1-2/(n+1)\}}$.

Proof. $Cov(R_S, R_F) = Cov(R_S, R_S + 1 - I_1 - I_{m+1})$

$$\begin{aligned}
&= Cov(R_S, R_S) - 2Cov(R_S, I_1) \\
&= Var(R_S) - 2(m+1)Cov(R_S, I_1)/(m+1) \\
&= Var(R_S) - 2Cov(R_S, \sum_{i=1}^{m+1} I_i)/(m+1) \\
&= Var(R_S)\{1-2/(m+1)\} \\
&= Var(R_F)\{1-2/(n+1)\} \text{ (by symmetry.)}
\end{aligned}$$

Thus $\rho^2 = \{1-2/(m+1)\}\{1-2/(n+1)\}$. Since $Cov(R_S, R_F) \geq 0$, the corollary follows.

For completeness, we sketch our proof of:

Corollary 3.7 *The probability density function of R is*

$$f(2r) = 2\frac{\binom{m-1}{r-1}\binom{n-1}{r-1}}{\binom{N}{n}} \text{ and } f(2r+1) = \frac{\binom{m-1}{r}\binom{n-1}{r-1} + \binom{m-1}{r-1}\binom{n-1}{r}}{\binom{N}{n}}$$

for $r = 1, \ldots, min\{n, m\}$ *(here* $f(2r+1) = 0$ *if* $n = m = r$*).*

Proof. Let $f(\cdot)$ be the pdf of R_S as given in Corollary 3.5 and let $u = min\{m+1, n\}$. Then we can use the represention $R = 2R_S + 1 - I_1 - I_{m+1}$ of Theorem 3.3, condition on R_S, and use the exchangeability of the I_i's to arrive

$$\begin{aligned} P(R=k) &= \sum_{j=1}^{u} P(R=k, R_S=j)) \\ &= \sum_{j=1}^{u} P(2R_S + 1 - I_1 - I_{m+1} = k, R_S = j) \\ &= \sum_{j=1}^{u} P(I_1 + I_2 = 2j + 1 - k | \sum_{i=1}^{m+1} I_i = j) f(j). \end{aligned}$$

Observing that $I_1 + I_2 \in \{0, 1, 2\}$, the Corollary follows from Lemma 3.1 by considering the cases, k odd and k even.

Remark 4. We note here that the pdf $f(r) = f(r; n, m)$ of R can be recursively generated:
$f(2) = 2/\binom{n+m}{n}$, $f(2r+1) = (n+m-2r)f(2r)/(2r)$, and
$f(2r+2) = (n-r)(m-r)f(2r)/r^2$.

As an alternative to R_S or R, one could test using $R_M = max\{R_S, R_F\}$. In this case:

Corollary 3.8 *The probability density function of* R_M *is*

$$f(r) = \left\{\binom{m}{r-1}\binom{n-1}{r-1} + \binom{n}{r-1}\binom{m-1}{r-1}\right\} / \binom{N}{n} \text{ for } r = 1, 2, \ldots, min\{n, m\}.$$

Proof. Again we take $\vec{s}$ to be our ordered arrangement. Now
$P(R_M = r) = P(R_M = r, s_1 = S) + P(R_M = r, s_1 = F)$. Using the exchangeability of the I_i's and Lemma 3.1, we see that

$$P(R_M = r, s_1 = S) = P(R_S = r, s_1 = S)$$

$$\begin{aligned}
&= P(R_S = r, I_1 = 1) \\
&= P(R_S = r)P(I_1 = 1|R_S = r) \\
&= P(R_S = r)\frac{\binom{m}{r-1}}{\binom{m+1}{r}} \\
&= \frac{\binom{m}{r-1}\binom{n-1}{r-1}}{\binom{N}{n}}.
\end{aligned}$$

By a symmetric argument (one can interchange the names of S and F), one can conclude that $P(R_M = r, s_1 = F) = \binom{n}{r-1}\binom{m-1}{r-1}/\binom{N}{n}$ and the corollary follows.

4 Distribution of the Longest S Run

We consider the distribution of the longest run of one type, say $M_S = M_S(n,m)$, the longest S run in a random ordering of n S's and m F's. Recall that as the sample space Ω for the random experiment of randomly ordering the n S's and m F's, we took all $(N+1)$ - tuples $\vec{\omega} = (s_1, \ldots, s_N, F)$ where $\vec{s} = (s_1, \ldots, s_N)$ is an ordered arrangement of n S's and m F's. As in Section 3, L_i is the number of S's immediately preceding the ith F in the random ordering and $M_S = M_S(n,m) = max\{L_1, \ldots, L_{m+1}\}$.

Exact formulas for the distribution of M_S can be found in Burr and Cane (1961), Gibbons (1985), and Philippou and Makri (1986). However, we have found some of these results cumbersome to use in computations (see, for example, computations in the case $l = 2, n = 5, m = 6$, considered on page 61 of Gibbons(1985)), and present alternate approaches in the following.

As indicated in Burr and Cane, one can present a combinatorial argument as in Freund and Pozner (1956) to arrive at our first representation of the cdf of M_S in Theorem 4.1. We sketch our proof using exchangeability and a generating function technique. The basic representation used is: the distribution of a random sample of size $m+1$ from a negative binomial with parameters 1 and θ, conditioned on the sample summing to n has the same distribution as $\vec{L}$.

Theorem 4.1 *For positive integer l, $1 \leq l \leq n$,*

$$F_S(l,n,m) = Prob(M_S \leq l) = a_{m,n}(l)/\binom{n+m}{n}$$

where $a_{m,n}(l)$ is the coefficient of θ^n in the expansion of $\{(1-\theta^{l+1})/(1-\theta)\}^{m+1}$.

Proof. Let l be a nonnegative integer, $\{X_i\}$ be a sequence of iid Bernoulli random variables with parameter $P(X_i = 1) = \theta$, and for i=1,2,$\cdots$, let L_i^* be the number of ones immediately preceding the ith zero in $(X_1, X_2, \cdots)$. Then L_i^* has cdf $F(l) = (1-\theta^{l+1})$ and $\sum_{i=1}^{m+1} L_i^*$ has the negative binomial distribution. Next, observe that the conditional distribution of $(L_1^*, \cdots, L_{m+1}^*) | \sum_{i=1}^{m+1} L_i^* = n$ is the same as that of $(L_1, \cdots, L_{m+1})$. It then follows that

$$\begin{aligned}
(1-\theta^{l+1})^{m+1} &= P(max\{L_1^*, \cdots, L_{m+1}^*\} \leq l) \\
&= \sum_{k=0}^{\infty} P(max\{L_1^*, \cdots, L_{m+1}^*\} \leq l | \sum_{i=1}^{m+1} L_i^* = k) \cdot \\
&\qquad\qquad P(\sum_{i=1}^{m+1} L_i^* = k) \\
&= \sum_{k=0}^{\infty} F_S(l,k,m)\binom{k+m}{k}\theta^k(1-\theta)^{m+1}.
\end{aligned}$$

Dividing the equation by $(1-\theta)^{m+1}$ and equating like coefficients, the theorem follows.

Corollary 4.2 *For positive integers n, m, l, with $1 \leq l \leq n$, take $u = min\{m+1, [n/(l+1)]\}$, then*
$F_S(l,n,m) = Prob(M_S \leq l) = \sum_{i=0}^{u}(-1)^i\binom{m+1}{i}\binom{n+m-i(l+1)}{m}/\binom{n+m}{n}$.

Proof. As indicated in Theorem 4.1, for positive integer l, $F_S(l,n,m) = Prob(M_S \leq l) = a_{m,n}(l)/\binom{n+m}{n}$ where $a_{m,n}(l)$ is the coefficient of θ^n in the expansion of $\{(1-$

$\theta^{l+1})/(1-\theta)\}^{m+1}$. Following Burr and Cane (1961) we multiply the binomial expansion of $(1-\theta^{l+1})^{m+1}$ and the negative binomial expansion of $(1-\theta)^{-(m+1)}$, to see that $a_{m,n}(l)$ is the coefficient of θ^n in $\{\sum_{i=0}^{m+1}(-1)^i\binom{m+1}{i}\theta^{i(l+1)}\}\{\sum_{j=0}^{\infty}\binom{m+j}{m}\theta^j\} = \sum_{i=0}^{m+1}(-1)^i\binom{m+1}{i}\{\sum_{j=0}^{\infty}\binom{m+j}{m}\theta^{i(l+1)+j}\}$. The theorem then follows by noting that for each fixed i, $i(l+1)+j=n$, implies that $j=n-i(l+1)$. Approaching the cdf of M_S through recursion we obtain:

Corollary 4.3 *For positive integers* n, m, l, $1 \le l \le n$,

$$F_S(l,n,m) = Prob(M_S \le l) = \sum_{i=n-l}^{n} F_S(l,i,m-1)\binom{i+m-1}{m-1}\Big/\binom{n+m}{n}.$$

Proof. Now $\omega = (\vec{s}, F) \in A_i = \{L_{m+1} = i\}$ implies that ω begins with a string of $n-i$ S's mixed with $m-1$ F's and ends with a string initiated by an F, followed by i S's and terminating with an F. Using the marginal distributions and exchangeability of the L_i's it then follows that

$$\begin{aligned} F_S(l,n,m) &= \sum_{i=0}^{l} Prob(M_S \le l, L_{m+1} = i) \\ &= \sum_{i=0}^{l} Prob(M_S|A_i)Prob(A_i) \\ &= \sum_{i=0}^{l} F_S(l,n-i,m-1)\binom{n-i+m-1}{m-1}\Big/\binom{n+m}{m} \end{aligned}$$

and the corollary easily follows.

For moderate n and m (and small l), one can easily program the recursion (the boundary conditions when at least one of n or m are zero are easy to compute) of Corollary 4.3 in BASIC, FORTRAN, or hand compute to obtain $F_S(l,n,m)$. In particular, we note that the recursion formula indicates that $a_{m,n}(l) = F_S(l,n,m)\binom{n+m}{n}$ $= A_{m+1,n+1}(l+1)$ where $A_{i,j}(k)$ is the entry in row i and column j of **Pascal's triangle of order k** (see Philippou and Georghiou (1986)). Pascal's triangle of order k is generated by starting with an infinite row beginning with k ones and zeros thereafter. The entries in row i (≥ 2) are recursively generated with the entry in column j obtained by adding the k entries $A_{i-1,j}(k), \cdots, A_{i-1,j-k+1}(k)$ (consider $A_{i-1,x}(k) = 0$ if $x \leq 0$) above it in row $i-1$. The case $k = 2$ corresponds to the usual Pascal's triangle. In the following, we illustrate the computation of $F_S(l, n, m)$ using Pascal's triangle of order 3 in Gibbon's example when $n = 5$, $m = 6$, and $l = 2$.

Example 4.1 *Find the p-value corresponding to an observation of a longest S run of $M_S(5,6) = 3$ under the hypothesis of random ordering of the 6 F's and 5 S's, i.e., find $P(M_S(5,6) \geq 3) = 1 - F_S(2,5,6)$.*

Solution. Pascal's triangle of order $l + 1 = 3$ begins:

$$\left(\begin{array}{cccccc|ccccccccc}
1 & 1 & 1 & 0 & 0 & 0 & 0 & 0 & 0 & 0 & 0 & 0 & 0 & \cdots \\
1 & 2 & 3 & 2 & 1 & 0 & 0 & 0 & 0 & 0 & 0 & 0 & 0 & \cdots \\
1 & 3 & 6 & 7 & 6 & 3 & 1 & 0 & 0 & 0 & 0 & 0 & 0 & \cdots \\
1 & 4 & 10 & 16 & 19 & 16 & 10 & 4 & 1 & 0 & 0 & 0 & 0 & \cdots \\
1 & 5 & 15 & 30 & 45 & 51 & 45 & 30 & 15 & 5 & 1 & 0 & 0 & \cdots \\
1 & 6 & 21 & 50 & 90 & 126 & 141 & 126 & 90 & 50 & 21 & 6 & 1 & \cdots \\
1 & 7 & 28 & 77 & 161 & \mathbf{266} & 357 & 393 & 357 & 266 & 161 & 77 & 28 & \cdots
\end{array}\right)$$

From the table, we see that $a_{6,5}(2) = A_{7,6}(3) = 266$. Since $\binom{11}{5} = 462$, $F_S(2,5,6) = 266/462 = 0.576$ and the p-value is $196/462 = 0.424$.

Remark 5. Note that in computing $a_{m,n}(l) = A_{m+1,n+1}(l+1)$, one could save storage locations by iterating $m+1$ times on a one dimensional array of dimension $n+1$. The successive iterations would produce the rows in the first $n+1$ columns of Pascal's triangle of order $l+1$. In Example 4.1, these iterations are the rows to the left of the partition line in the solution matrix.

Remark 6. Although the expansion in Corollary 4.2 has been around for quite some time, it has been mainly of mathematical interest and useful only for small values of n and m. Terms in the finite sum alternate in sign and have magnitudes which require some ingenuity to approximate accurately. For example, the binomial coefficient $\binom{1000}{500}$ is $O(300)$. However, with the advent of software systems with exact arithmetic capabilities like *Mathematica* (see Wolfram (1991)), formulas like these are computationally feasible for reasonably large N. For example, *Mathematica*, Version 2.0, running on a SUN Sparc 1^+ has a remarkably fast algorithm for computing binomial coefficients (exactly) and can use Corollary 4.2 to compute $F_S(15, 500, 500) = 0.993250$ (or to any desired accuracy) in less than 7 seconds with a one line function statement. Additional examples and discussion of computation of moments of M_S can be found in Schuster (1991).

5 Recursion for the Longest Run

Let us now turn our attention to algorithms for computing the cdf for the longest run M of either type in a random ordering of n S's and m F's. In this case we are not aware of any recursive algorithms or closed form type formulas for the pdf or cdf of M such as that given in Corollaries 4.2 and 4.3 for the case of the longest run of one type. In the following we develop recursion formulas for computing the cdf of M.

For fixed integer l, we begin by recursively defining a function $F_C(l, n, m)$ and then express the cdf of M in terms of this function:

$$F_C(l,n,m) \equiv \sum_{i=max\{0,n-l\}}^{n-1} \sum_{j=max\{0,m-l\}}^{m-1} F_C(l,i,j)\binom{i+j}{j} / \binom{n+m}{n},$$

$1 \leq l < max\{n, m+1\}$, with *boundary conditions*:

$$\begin{aligned}
F_C(l,n,m) &= 0 \text{ if } l < 1 \\
F_C(l,0,m) &= \begin{cases} 1 & \text{if } l \geq m+1 \\ 0 & \text{otherwise} \end{cases} \\
F_C(l,n,0) &= \begin{cases} 1 & \text{if } l \geq n \\ 0 & \text{otherwise} \end{cases} \\
F_C(l,n,m) &= 1 \text{ if } l \geq max\{n, m+1\}.
\end{aligned}$$

As the sample space for the random ordering of the n S's and m F's, we took all

$(N+1)$ - tuples $\vec{\omega} = (s_1, \ldots, s_N, F)$ where $\vec{s} = (s_1, \ldots, s_N)$ is an ordered arrangement of n S's and m F's. Let $M_C = M_C(n, m)$ be the random variable which records the length of the longest run in the random ordering in the vector $(\vec{s}, F)$ where the final F placed at the end in the sample space counts in analyzing the length of the longest F run. Note that the r.v. M does not count the final F. The r.v. M_C may be of independent interest in that it has the same (conditional) distribution as the length of the longest run of successes or failures in a random sample of size $m+1$ from a geometric distribution which records the number of trials until (including the trial of) the first F in repeated, independent, Bernoulli (S-F) trials with $Prob(S) = \theta$, $0 < \theta < 1$, conditioned on the sample summing to $N+1$.

Theorem 5.1 *M_C has cdf $F_C(\cdot\,, n, m)$.*

Proof. Recall that L_i is the number of S's immediately before the ith F, $i = 1, 2, \cdots, m+1$. For positive integer l and nonnegative integer k, take

$$A_{(i,k)} = \{L_{m+1-0} = \cdots = L_{m+1-(i-2)} = 0, L_{m+1-(i-1)} = k\}.$$

Then since the last F counts, $\omega = (\vec{s}, F) \in A_{(i,0)}$ implies ω ends with $i+1$ F's. Thus

$$Prob(M_C \leq l, L_{m+1} = \cdots = L_{m+1-(i-1)} = 0) = Prob(M_C \leq l, A_{(i,0)}) = 0$$

for $i > min\{l, m\}$. Now $Prob(M_C \leq l) = Prob(M_C \leq l, L_{m+1} = 0) + \sum_{k=1}^{n} Prob$ $(M_C \leq l, L_{m+1} = k)$ and $Prob(M_C \leq l, L_{m+1} = 0) = Prob(M_C \leq l, L_{m+1} = 0, L_m = 0) + \sum_{k=1}^{min\{l,n\}} Prob(M_C \leq l, A_{(2,k)})$. Continuing in this fashion, and using $Prob(M_C \leq l, A_{(i,0)}) = 0$ for $i > min\{l, m\}$, we see that

$$\begin{aligned} Prob(M_C \leq l) &= \sum_{i=1}^{min\{l,m\}} \sum_{k=1}^{min\{l,n\}} Prob(M_C \leq l, A_{(i,k)}) \\ &= \sum_{i=1}^{min\{l,m\}} \sum_{k=1}^{min\{l,n\}} Prob(M_C \leq l | A_{(i,k)}) Prob(A_{(i,k)}). \end{aligned}$$

Now $\omega = (\vec{s}, F) \in A_{(i,k)}$ implies that ω ends with a string initiated by an F, followed by k S's, and terminating in i F's. Thus whenever $A_{(i,k)}$ occurs, there are $n-k$

free S's and $m-i$ free F's. Let $F(l,i,j) = Prob(M_C(i,j) \leq l)$ for $0 \leq i \leq n$ and $0 \leq j \leq m$. Using the exchangeability of the L_i's, it then follows that

$$\begin{aligned} F(l,n,m) &= \sum_{i=1}^{min\{l,m\}} \sum_{k=1}^{min\{l,n\}} F(l,n-k,m-i)\frac{\binom{n-k+m-i}{n-k}}{\binom{n+m}{n}} \\ &= \sum_{i=max\{0,n-l\}}^{n-1} \sum_{j=max\{0,m-l\}}^{m-1} F(l,i,j)\frac{\binom{i+j}{j}}{\binom{n+m}{n}}, \end{aligned}$$

and $F = F_C$ follows by checking boundary conditions.

Using the interpretation of F_C given in Theorem 5.1 we can write a single sum recursion formula for F_C:

Corollary 5.2 *For positive integers n, m, l, $l \leq max\{n, m+1\}$,*

$$F_C(l,n,m) = Prob(M_C \leq l) = \sum_{i=max\{0,m+1-l\}}^{m} F_C(l,i,n-1)\frac{\binom{n+i-1}{i}}{\binom{n+m}{n}}.$$

Proof. Take

$$A_i = \{L_{m+1-0} = \cdots = L_{m+1-(i-2)} = 0, L_{m+1-(i-1)} > 0\}.$$

Then $\omega = (\vec{s}, F) \in A_i$ implies that ω ends with a string initiated by an S and terminating in i F's. Thus whenever A_i occurs, there are $n-1$ free S's and $m-i+1$ free F's which end with an S. Interchanging the roles of the F's and S's, noting the ending S now counts, and using the exchangeability of the L_i's, we can proceed as in the proof of Theorem 5.1 to see that

$$\begin{aligned} F_C(l,n,m) &= \sum_{i=1}^{min\{l,m+1\}} Prob(M_C \leq l, A_i) \\ &= \sum_{i=1}^{min\{l,m+1\}} Prob(M_C \leq l | A_i))Prob(A_i) \\ &= \sum_{i=1}^{min\{l,m+1\}} F_C(l,m-i+1,n-1)\frac{\binom{n-1+m+1-i}{m+1-i}}{\binom{n+m}{n}} \\ &= \sum_{i=max\{0,m+1-l\}}^{m} F_C(l,i,n-1)\frac{\binom{n+i-1}{i}}{\binom{n+m}{n}}. \end{aligned}$$

Remark 7. Note that $F_C(l,n,m)$ is not symmetrically defined in terms of n and m since the boundary conditions depend on the order of n and m. One could also take

$$D_C(l,n,m) = F_C(l,n,m)\binom{n+m}{n}$$

and restate the recursion formulas in Theorem 5.1 and Corollary 5.2 in the simplified form

$$D_C(l,n,m) = \sum_{i=max\{0,n-l\}}^{n-1} \sum_{j=max\{0,m-l\}}^{m-1} D_C(l,i,j)$$

and

$$D_C(l,n,m) = \sum_{i=max\{0,m+1-l\}}^{m} D_C(l,i,n-1)$$

with corresponding boundary conditions.

We can now express the cdf of $M = M(n,m)$ in terms of the cdf $F_C(\cdot,n,m)$ of M_C.

Theorem 5.3 *(CDF of the Longest Run of Either Type)For positive integers n,m and l, $l \leq max\{n,m\}$,*

$$F_M(l,n,m) = Prob(M \leq l) = \sum_{i=max\{0,n-l\}}^{n-1} \sum_{j=max\{0,m-1-l\}}^{m-1} F_C(l,i,j)\frac{\binom{i+j}{j}}{\binom{n+m}{n}}.$$

Proof. We use the same notation as in the proof of Theorem 5.1. However, in the present case, the final F in $\omega = (\vec{s}, F)$ does not count in determining the longest run in ω. Recall that for positive integer i, $\omega = (\vec{s}, F) \in A_{(i,0)}$ implies ω ends with $i+1$ F's. But then

$$Prob(M \leq l, L_{m+1} = \cdots = L_{m+1-(i-1)} = 0) = Prob(M \leq l, A_{(i,0)}) = 0$$

for all $i \geq l+1$. Proceeding as in the proof of Theorem 5.1, we observe that

$$F_M(l,n,m) \quad = \quad \sum_{i=1}^{min\{m,l+1\}} \sum_{k=1}^{min\{l,n\}} Prob(M \leq l, A_{(i,k)})$$

$$\begin{aligned}
&= \sum_{i=1}^{min\{m,l+1\}} \sum_{k=1}^{min\{l,n\}} Prob(M \le l | A_{(i,k)}) Prob(A_{(i,k)}) \\
&= \sum_{i=1}^{min\{m,l+1\}} \sum_{k=1}^{min\{l,n\}} F_C(l, n-k, m-i) \frac{\binom{n-k+m-i}{n-k}}{\binom{n+m}{n}} \\
&= \sum_{i=max\{0,n-l\}}^{n-1} \sum_{j=max\{0,m-1-l\}}^{m-1} F_C(l,i,j) \frac{\binom{i+j}{j}}{\binom{n+m}{n}}.
\end{aligned}$$

In the following we give an alternate single sum expression for the cdf of M in terms of the cdf F_C.

Corollary 5.4 *For positive integers n, m, l, with $l \le max\{n, m\}$,*

$$\begin{aligned}
F_M(l,n,m) &= \sum_{i=max\{0,m-l\}}^{m} F_C(l,i,n-1)\binom{n+i-1}{i}\Big/\binom{n+m}{n} \\
&= \sum_{i=max\{0,n-l\}}^{n} F_C(l,i,m-1)\binom{m+i-1}{i}\Big/\binom{n+m}{n}.
\end{aligned}$$

Proof. Again take

$$A_i = \{L_{m+1-0} = \cdots = L_{m+1-(i-2)} = 0, L_{m+1-(i-1)} > 0\}.$$

Then $F_M(l,n,m) = \sum_{i=1}^{min\{l+1,m+1\}} Prob(M \le l, A_i)$. Condition on A_i, interchange the roles of F and S and proceed as in the proof of Corollary 5.2.

Remark 8. One can use the formula of Corollary 5.2 to iterate in the single sum formulas to arrive at the double sum formulas. In particular, one can use this approach in the two representations of $F_M(l,n,m)$ in Corollary 5.4 to prove that the double sum formula of Theorem 5.3 is symmetric in n and m. The index i can take the value $m(n)$ in the single sum formula of Corollary 5.4 (Corollary 5.2) but not in the double sum formula of Theorem 5.3 (Theorem 5.1). In addition, the double sum recursion appears to be better "organized" than the single sum recursion. Hence, the single sum formulas may require more work in some cases than the double sum algorithms. This has been our experience for small to moderate sized n and m. For large n and m, the single sum algorithm seems to be faster.

Errors in Bateman's table. We programmed the double sum recursion formula for the cdf of F_C in *Mathematica* running on a SUN Sparc 1^+ and computed (in less than 60 seconds) all values of the pdf of M in the most comprehensive table available, namely Bateman's (1948) table. Three discrepancies were found: For $n = 7$, $m = 4$, Bateman gives $P(M(7,4) = 3) = 107/330 = 0.324242$, and $P(M(7,4) = 2) = 48/330 = 0.145455$. For $n = m = 10$, he gives $P(M(10,10) = 6) = 13180/184756 = 0.0713373$. Our algorithms give $P(M(7,4) = 3) = 125/330 = 0.378788$, $P(M(7,4) = 2) = 30/330 = 0.0909091$, and $P(M(10,10) = 6) = 13810/184756 = 0.0747472$. We verified our answers when $n = 7$ and $m = 4$ by listing the 330 sample points in this case. In the case of $n = m = 10$, our computations agree with Bateman except for the count of 13180 when $l = 6$. On can easily observe that Bateman has at least one error in his table when $n = m = 10$, since the table entries do not sum to $\binom{20}{10} = 184756$, i.e. the probabilities do not sum to one. Hence, it appears that Bateman's count of 13180 in the case of $n = m = 10$, instead of 13810, is a simple typing error of transposition of the two digits, 8 and 1.

Table 1. CDF of Longest Run M. In Table 1, we give rounded, four decimal digit approximations to a where $a = F_M(z, n, m)$ for every z with $0.00005 \leq a < 0.99995$ for all pairs n, m with $N = n + m = 2(1)25$ and $N = 30(5)50$. To find a in the table, first locate the appropriate $n + m$ in column 1. Then move to the row(s) corresponding to the value of n. Located to the right of n is the value of x. We then find $a = F_M(z, n, m)$ in column $+i$ where $z = x + i$. For example, when $n + m = 50$, $n = 20$, we find that $x = 2$, i.e. the cdf $F_M(l, n, m)$ rounds to 0.0000 for $l \leq x = 2$. Adding $+i = +5$ to x, the table gives $F_M(x+5, 20, 30) = F_M(7, 20, 30) = F_M(7, 30, 20) = 0.7736$. Note that some values of N and n require a second line in the table. In these cases, we leave the N and n entries in the second blank. The second line x value is 18 plus the first line x value. For example, when $n + m = 50$ and $n = 12$, we find two lines in the table. The second line x value is 22. Since $x + 2 = 24$, $F_M(24, 12, 48)$ rounds to 0.9994.

Table 1. CDF of Longest Run M of Either Type.
Table entries give $a = F_M(x+i, n, m)$ for $x+i$ with $0.00005 \leq a < 0.99995$.

N	n	x	+1	+2	+3	+4	+5	+6	+7	+8	+9	+10	+11	+12	+13	+14	+15	+16	+17	+18
2	1	0																		
3	1	0	.3333																	
4	1	1	.5000																	
4	2	0	.3333																	
5	1	1	.2000	.6000																
5	2	0	.1000	.7000																
6	1	2	.3333	.6667																
6	2	1	.4000	.8000																
6	3	0	.1000	.7000																
7	1	2	.1429	.4286	.7143															
7	2	1	.1429	.5714	.8571															
7	3	0	.0286	.5143	.8857															
8	1	3	.2500	.5000	.7500															
8	2	1	.0357	.3571	.6786	.8929														
8	3	1	.2857	.7143	.9286															
8	4	0	.0286	.4857	.8857															
9	1	3	.1111	.3333	.5556	.7778														
9	2	2	.1667	.5000	.7500	.9167														
9	3	1	.1190	.5238	.8095	.9524														
9	4	0	.0079	.3571	.7857	.9603														
10	1	4	.2000	.4000	.6000	.8000														
10	2	2	.0667	.3333	.6000	.8000	.9333													
10	3	1	.0333	.3333	.6667	.8667	.9667													
10	4	1	.2048	.6381	.8810	.9762														
10	5	0	.0079	.3333	.7698	.9603														
11	1	4	.0909	.2727	.4545	.6364	.8182													
11	2	2	.0182	.1818	.4545	.6727	.8364	.9455												
11	3	1	.0061	.1879	.5152	.7576	.9030	.9758												
11	4	1	.0909	.4697	.7727	.9242	.9848													
11	5	0	.0022	.2446	.6861	.9156	.9870													
12	1	5	.1667	.3333	.5000	.6667	.8333													
12	2	3	.0909	.3182	.5455	.7273	.8636	.9545												
12	3	2	.0909	.3636	.6364	.8182	.9273	.9818												
12	4	1	.0303	.3131	.6465	.8485	.9495	.9899												
12	5	1	.1439	.5606	.8384	.9545	.9924													
12	6	0	.0022	.2251	.6667	.9091	.9870													
13	1	5	.0769	.2308	.3846	.5385	.6923	.8462												
13	2	3	.0385	.1923	.4231	.6154	.7692	.8846	.9615											
13	3	2	.0350	.2378	.5105	.7203	.8601	.9441	.9860											
13	4	1	.0070	.1888	.5105	.7552	.8951	.9650	.9930											
13	5	1	.0676	.4196	.7382	.9021	.9720	.9953												
13	6	0	.0006	.1661	.5944	.8660	.9691	.9959												
14	1	6	.1429	.2857	.4286	.5714	.7143	.8571												
14	2	3	.0110	.1099	.3077	.5055	.6703	.8022	.9011	.9670										
14	3	2	.0110	.1429	.3846	.6154	.7802	.8901	.9560	.9890										
14	4	1	.0010	.1009	.3806	.6503	.8252	.9251	.9750	.9950										
14	5	1	.0250	.2887	.6224	.8322	.9371	.9820	.9970											
14	6	1	.0999	.4885	.7945	.9337	.9837	.9977												
14	7	0	.0006	.1509	.5746	.8566	.9668	.9959												
15	1	6	.0667	.2000	.3333	.4667	.6000	.7333	.8667											
15	2	4	.0571	.2000	.4000	.5714	.7143	.8286	.9143	.9714										
15	3	2	.0022	.0769	.2747	.5077	.6923	.8242	.9121	.9648	.9912									
15	4	2	.0476	.2674	.5385	.7436	.8718	.9451	.9817	.9963										
15	5	1	.0070	.1818	.5015	.7483	.8881	.9580	.9880	.9980										
15	6	1	.0492	.3716	.7023	.8821	.9608	.9902	.9986											
15	7	0	.0002	.1119	.5122	.8159	.9473	.9893	.9988											
16	1	7	.1250	.2500	.3750	.5000	.6250	.7500	.8750											
16	2	4	.0250	.1250	.3000	.4750	.6250	.7500	.8500	.9250	.9750									
16	3	3	.0357	.1857	.4000	.6000	.7500	.8571	.9286	.9714	.9929									
16	4	2	.0192	.1758	.4286	.6538	.8077	.9038	.9588	.9863	.9973									
16	5	1	.0014	.1044	.3860	.6538	.8269	.9231	.9712	.9918	.9986									

Table entries give $a = F_M(x+i, n, m)$ for $x+i$ with $0.00005 \le a < 0.99995$.

N	n	x	+1	+2	+3	+4	+5	+6	+7	+8	+9	+10	+11	+12	+13	+14	+15	+16	+17	+18
16	6	1	.0196	.2620	.5973	.8163	.9266	.9755	.9939	.9991										
16	7	1	.0687	.4238	.7500	.9119	.9745	.9944	.9993											
16	8	0	.0002	.1007	.4929	.8050	.9436	.9885	.9988											
17	1	7	.0588	.1765	.2941	.4118	.5294	.6471	.7647	.8824										
17	2	4	.0074	.0735	.2059	.3824	.5368	.6691	.7794	.8676	.9338	.9779								
17	3	3	.0147	.1176	.3029	.5059	.6706	.7941	.8824	.9412	.9765	.9941								
17	4	2	.0063	.1071	.3277	.5588	.7353	.8529	.9265	.9685	.9895	.9979								
17	5	1	.0002	.0543	.2830	.5545	.7557	.8778	.9457	.9796	.9942	.9990								
17	6	1	.0062	.1710	.4888	.7387	.8812	.9525	.9842	.9960	.9994									
17	7	1	.0351	.3267	.6658	.8621	.9505	.9852	.9967	.9996										
17	8	1	.0750	.4398	.7666	.9237	.9805	.9964	.9996											
18	1	8	.1111	.2222	.3333	.4444	.5556	.6667	.7778	.8889										
18	2	5	.0392	.1373	.2941	.4510	.5882	.7059	.8039	.8824	.9412	.9804								
18	3	3	.0049	.0686	.2206	.4118	.5882	.7255	.8284	.9020	.9510	.9804	.9951							
18	4	2	.0016	.0605	.2402	.4641	.6569	.7941	.8856	.9428	.9755	.9918	.9984							
18	5	2	.0252	.1968	.4559	.6765	.8235	.9118	.9608	.9853	.9958	.9993								
18	6	1	.0015	.1032	.3845	.6527	.8258	.9208	.9683	.9894	.9974	.9996								
18	7	1	.0149	.2350	.5710	.8000	.9170	.9698	.9910	.9980	.9997									
18	8	1	.0469	.3662	.7060	.8891	.9645	.9906	.9981	.9998										
18	9	1	.0669	.4216	.7548	.9190	.9791	.9962	.9996											
19	1	8	.0526	.1579	.2632	.3684	.4737	.5789	.6842	.7895	.8947									
19	2	5	.0175	.0877	.2105	.3684	.5088	.6316	.7368	.8246	.8947	.9474	.9825							
19	3	3	.0010	.0361	.1538	.3251	.5046	.6533	.7688	.8555	.9174	.9587	.9835	.9959						
19	4	2	.0003	.0312	.1680	.3744	.5743	.7291	.8375	.9097	.9549	.9807	.9936	.9987						
19	5	2	.0103	.1295	.3630	.5926	.7616	.8700	.9350	.9711	.9892	.9969	.9995							
19	6	1	.0003	.0573	.2902	.5627	.7616	.8808	.9458	.9783	.9928	.9982	.9997							
19	7	1	.0052	.1579	.4731	.7277	.8742	.9476	.9809	.9943	.9987	.9998								
19	8	1	.0248	.2856	.6294	.8416	.9402	.9803	.9946	.9989	.9999									
19	9	1	.0501	.3767	.7188	.8992	.9703	.9931	.9988	.9999										
20	1	9	.1000	.2000	.3000	.4000	.5000	.6000	.7000	.8000	.9000									
20	2	5	.0053	.0526	.1474	.2895	.4316	.5579	.6684	.7632	.8421	.9053	.9526	.9842						
20	3	4	.0175	.1018	.2491	.4211	.5789	.7053	.8035	.8772	.9298	.9649	.9860	.9965						
20	4	3	.0144	.1115	.2931	.4912	.6594	.7833	.8700	.9278	.9639	.9845	.9948	.9990						
20	5	2	.0036	.0804	.2794	.5077	.6935	.8212	.9025	.9512	.9783	.9919	.9977	.9996						
20	6	2	.0291	.2099	.4728	.6906	.8331	.9166	.9621	.9848	.9949	.9987	.9998							
20	7	1	.0014	.0989	.3785	.6486	.8229	.9183	.9659	.9876	.9963	.9992	.9999							
20	8	1	.0110	.2091	.5438	.7829	.9076	.9646	.9882	.9968	.9994	.9999								
20	9	1	.0318	.3155	.6632	.8658	.9538	.9864	.9967	.9994	.9999									
20	10	1	.0444	.3599	.7066	.8937	.9684	.9926	.9988	.9999										
21	1	9	.0476	.1429	.2381	.3333	.4286	.5238	.6190	.7143	.8095	.9048								
21	2	6	.0286	.1000	.2143	.3571	.4857	.6000	.7000	.7857	.8571	.9143	.9571	.9857						
21	3	4	.0075	.0632	.1850	.3429	.5038	.6391	.7474	.8316	.8947	.9398	.9699	.9880	.9970					
21	4	3	.0058	.0702	.2222	.4110	.5865	.7243	.8246	.8947	.9415	.9708	.9875	.9958	.9992					
21	5	2	.0010	.0467	.2074	.4252	.6213	.7665	.8638	.9257	.9628	.9835	.9938	.9982	.9997					
21	6	2	.0134	.1451	.3869	.6153	.7786	.8808	.9404	.9729	.9892	.9964	.9991	.9999						
21	7	1	.0003	.0577	.2921	.5659	.7641	.8819	.9455	.9773	.9917	.9975	.9994	.9999						
21	8	1	.0041	.1438	.4553	.7155	.8670	.9431	.9781	.9927	.9980	.9996								
21	9	1	.0173	.2486	.5938	.8205	.9297	.9754	.9925	.9981	.9997									
21	10	1	.0333	.3219	.6729	.8742	.9592	.9889	.9976	.9996										
22	1	10	.0909	.1818	.2727	.3636	.4545	.5455	.6364	.7273	.8182	.9091								
22	2	6	.0130	.0649	.1558	.2857	.4156	.5325	.6364	.7273	.8052	.8701	.9221	.9610	.9870					
22	3	4	.0026	.0364	.1325	.2727	.4286	.5714	.6883	.7818	.8545	.9091	.9481	.9740	.9896	.9974				
22	4	3	.0021	.0417	.1627	.3363	.5126	.6617	.7744	.8565	.9139	.9522	.9761	.9897	.9966	.9993				
22	5	2	.0002	.0253	.1483	.3477	.5473	.7068	.8195	.8947	.9426	.9713	.9872	.9952	.9986	.9998				
22	6	2	.0055	.0957	.3078	.5383	.7185	.8390	.9133	.9567	.9803	.9921	.9974	.9993	.9999					
22	7	2	.0312	.2174	.4831	.6995	.8390	.9195	.9628	.9845	.9944	.9983	.9996							
22	8	1	.0013	.0929	.3693	.6421	.8189	.9155	.9638	.9861	.9954	.9987	.9997							
22	9	1	.0080	.1847	.5163	.7653	.8981	.9596	.9856	.9956	.9989	.9998								
22	10	1	.0215	.2710	.6221	.8422	.9426	.9817	.9950	.9989	.9998									
22	11	1	.0293	.3067	.6605	.8682	.9569	.9883	.9975	.9996										

Table entries give $a = F_M(x+i, n, m)$ for $x+i$ with $0.00005 \le a < 0.99995$.

N	n	x	+1	+2	+3	+4	+5	+6	+7	+8	+9	+10	+11	+12	+13	+14	+15	+16	+17	+18
23	1	10	.0435	.1304	.2174	.3043	.3913	.4783	.5652	.6522	.7391	.8261	.9130							
23	2	6	.0040	.0395	.1107	.2174	.3478	.4664	.5731	.6680	.7510	.8221	.8814	.9289	.9644	.9881				
23	3	4	.0006	.0198	.0909	.2117	.3574	.5031	.6273	.7290	.8103	.8735	.9209	.9548	.9774	.9910	.9977			
23	4	3	.0006	.0232	.1146	.2688	.4404	.5963	.7205	.8137	.8814	.9289	.9605	.9802	.9915	.9972	.9994			
23	5	3	.0127	.1020	.2773	.4739	.6435	.7705	.8588	.9176	.9551	.9775	.9900	.9963	.9989	.9998				
23	6	2	.0020	.0600	.2379	.4622	.6544	.7918	.8810	.9359	.9680	.9854	.9942	.9981	.9995	.9999				
23	7	2	.0156	.1557	.4031	.6308	.7901	.8880	.9440	.9742	.9892	.9961	.9988	.9997						
23	8	1	.0003	.0563	.2902	.5656	.7644	.8819	.9449	.9764	.9909	.9970	.9992	.9998						
23	9	1	.0032	.1296	.4361	.7022	.8593	.9387	.9755	.9913	.9973	.9993	.9999							
23	10	1	.0120	.2155	.5591	.7992	.9189	.9703	.9903	.9972	.9994	.9999								
23	11	1	.0221	.2747	.6292	.8491	.9475	.9842	.9960	.9992	.9999									
24	1	11	.0833	.1667	.2500	.3333	.4167	.5000	.5833	.6667	.7500	.8333	.9167							
24	2	7	.0217	.0761	.1630	.2826	.4022	.5109	.6087	.6957	.7717	.8370	.8913	.9348	.9674	.9891				
24	3	5	.0099	.0593	.1601	.2925	.4348	.5652	.6739	.7628	.8340	.8893	.9308	.9605	.9802	.9921	.9980			
24	4	3	.0001	.0119	.0777	.2095	.3718	.5299	.6636	.7671	.8447	.9012	.9407	.9671	.9835	.9929	.9976	.9995		
24	5	3	.0058	.0672	.2154	.4032	.5782	.7174	.8183	.8882	.9348	.9644	.9822	.9921	.9970	.9992	.9999			
24	6	2	.0006	.0356	.1784	.3891	.5879	.7399	.8438	.9108	.9519	.9760	.9891	.9956	.9985	.9996	.9999			
24	7	2	.0071	.1071	.3285	.5602	.7362	.8513	.9207	.9603	.9817	.9924	.9972	.9992	.9998					
24	8	1	.0001	.0320	.2207	.4886	.7047	.8426	.9213	.9633	.9843	.9939	.9980	.9994	.9999					
24	9	1	.0011	.0858	.3579	.6339	.8140	.9125	.9617	.9847	.9945	.9983	.9996	.9999						
24	10	1	.0058	.1622	.4889	.7472	.8884	.9546	.9831	.9944	.9984	.9996	.9999							
24	11	1	.0145	.2324	.5828	.8185	.9310	.9766	.9930	.9982	.9996	.9999								
24	12	1	.0194	.2611	.6168	.8427	.9449	.9834	.9958	.9992	.9999									
25	1	11	.0400	.1200	.2000	.2800	.3600	.4400	.5200	.6000	.6800	.7600	.8400	.9200						
25	2	7	.0100	.0500	.1200	.2200	.3400	.4500	.5500	.6400	.7200	.7900	.8500	.9000	.9400	.9700	.9900			
25	3	5	.0043	.0365	.1174	.2348	.3696	.5026	.6174	.7130	.7913	.8539	.9026	.9391	.9652	.9826	.9930	.9983		
25	4	4	.0055	.0506	.1589	.3083	.4644	.6043	.7174	.8043	.8696	.9170	.9502	.9723	.9862	.9941	.9980	.9996		
25	5	3	.0024	.0423	.1628	.3368	.5126	.6612	.7739	.8547	.9106	.9478	.9715	.9858	.9937	.9976	.9993	.9999		
25	6	2	.0002	.0200	.1297	.3210	.5208	.6843	.8022	.8813	.9322	.9635	.9817	.9917	.9967	.9989	.9997			
25	7	2	.0030	.0707	.2613	.4896	.6784	.8097	.8929	.9429	.9714	.9868	.9945	.9980	.9994	.9999				
25	8	2	.0169	.1623	.4138	.6414	.7980	.8929	.9465	.9750	.9893	.9959	.9986	.9996	.9999					
25	9	1	.0003	.0537	.2853	.5626	.7630	.8810	.9440	.9755	.9902	.9965	.9989	.9997						
25	10	1	.0024	.1158	.4162	.6881	.8513	.9343	.9730	.9899	.9966	.9990	.9998							
25	11	1	.0082	.1863	.5256	.7777	.9079	.9651	.9880	.9963	.9990	.9998								
25	12	1	.0147	.2341	.5878	.8242	.9353	.9791	.9941	.9986	.9997									
30	1	14	.0667	.1333	.2000	.2667	.3333	.4000	.4667	.5333	.6000	.6667	.7333	.8000	.8667	.9333				
30	2	9	.0138	.0483	.1034	.1793	.2759	.3724	.4621	.5448	.6207	.6897	.7517	.8069	.8552	.8966	.9310	.9586	.9793	.9931
30	3	6	.0010	.0138	.0542	.1281	.2256	.3360	.4483	.5517	.6414	.7182	.7833	.8374	.8818	.9172	.9448	.9655	.9803	.9901
		24	.9961	.9990																
30	4	5	.0026	.0252	.0887	.1927	.3184	.4472	.5661	.6679	.7510	.8174	.8695	.9097	.9398	.9617	.9770	.9872	.9936	.9973
		23	.9991	.9998																
30	5	4	.0018	.0253	.1011	.2266	.3738	.5163	.6399	.7395	.8161	.8736	.9157	.9458	.9667	.9805	.9894	.9947	.9976	.9991
		22	.9997																	
30	6	3	.0004	.0154	.0882	.2265	.3930	.5505	.6811	.7812	.8541	.9056	.9410	.9646	.9798	.9891	.9946	.9975	.9990	.9997
		21	.9999																	
30	7	3	.0043	.0558	.1930	.3770	.5540	.6970	.8021	.8749	.9236	.9550	.9747	.9865	.9933	.9969	.9987	.9995	.9999	
30	8	2	.0002	.0203	.1311	.3251	.5268	.6901	.8066	.8838	.9327	.9626	.9802	.9901	.9954	.9980	.9992	.9997	.9999	
30	9	2	.0020	.0585	.2372	.4648	.6586	.7952	.8826	.9354	.9660	.9830	.9920	.9965	.9986	.9995	.9998			
30	10	2	.0096	.1245	.3613	.5957	.7651	.8710	.9324	.9662	.9840	.9929	.9971	.9989	.9996	.9999				
30	11	1	.0001	.0291	.2149	.4878	.7075	.8453	.9225	.9631	.9834	.9930	.9973	.9990	.9997	.9999				
30	12	1	.0005	.0636	.3166	.6018	.7953	.9021	.9557	.9810	.9924	.9972	.9991	.9997	.9999					
30	13	1	.0020	.1073	.4103	.6915	.8577	.9392	.9756	.9908	.9968	.9990	.9997	.9999						
30	14	1	.0043	.1452	.4765	.7484	.8947	.9599	.9860	.9955	.9987	.9997	.9999							
30	15	1	.0055	.1603	.5003	.7680	.9070	.9666	.9892	.9969	.9992	.9998								
35	1	16	.0286	.0857	.1429	.2000	.2571	.3143	.3714	.4286	.4857	.5429	.6000	.6571	.7143	.7714	.8286	.8857	.9429	
35	2	10	.0017	.0168	.0471	.0924	.1529	.2286	.3143	.3950	.4706	.5412	.6067	.6672	.7227	.7731	.8185	.8588	.8941	.9244
		28	.9496	.9697	.9849	.9950														
35	3	7	.0002	.0053	.0252	.0689	.1358	.2193	.3125	.4087	.5013	.5844	.6578	.7219	.7775	.8252	.8655	.8992	.9267	.9487
		25	.9658	.9786	.9878	.9939	.9976	.9994												
35	4	6	.0013	.0136	.0518	.1218	.2174	.3255	.4352	.5383	.6299	.7078	.7727	.8262	.8697	.9044	.9317	.9527	.9685	.9799
		24	.9880	.9933	.9967	.9986	.9995	.9999												

Table entries give $a = F_M(x+i, n, m)$ for $x+i$ with $0.00005 \le a < 0.99995$.

N	n	x	+1	+2	+3	+4	+5	+6	+7	+8	+9	+10	+11	+12	+13	+14	+15	+16	+17	+18
35	5	5	.0014	.0167	.0667	.1567	.2739	.3994	.5191	.6249	.7135	.7851	.8416	.8856	.9193	.9445	.9630	.9762	.9854	.9915
		23	.9953	.9977	.9990	.9996	.9999													
35	6	4	.0006	.0127	.0645	.1666	.3010	.4417	.5706	.6793	.7661	.8328	.8830	.9199	.9466	.9655	.9784	.9870	.9926	.9960
		22	.9980	.9991	.9996	.9999														
35	7	4	.0055	.0470	.1506	.2983	.4549	.5954	.7097	.7973	.8617	.9078	.9401	.9621	.9769	.9864	.9923	.9959	.9980	.9991
		22	.9996	.9999																
35	8	3	.0008	.0228	.1120	.2661	.4398	.5963	.7207	.8128	.8777	.9222	.9518	.9711	.9833	.9907	.9951	.9975	.9989	.9995
		21	.9998	.9999																
35	9	3	.0051	.0613	.2057	.3950	.5730	.7138	.8153	.8843	.9296	.9584	.9762	.9869	.9931	.9966	.9984	.9993	.9997	.9999
35	10	2	.0002	.0184	.1251	.3184	.5222	.6875	.8050	.8826	.9315	.9613	.9789	.9889	.9945	.9974	.9988	.9995	.9998	.9999
35	11	2	.0013	.0469	.2122	.4379	.6370	.7797	.8720	.9282	.9611	.9797	.9899	.9952	.9978	.9991	.9996	.9999		
35	12	2	.0054	.0950	.3149	.5531	.7338	.8501	.9190	.9579	.9789	.9899	.9954	.9980	.9992	.9997	.9999			
35	13	2	.0157	.1617	.4228	.6559	.8110	.9014	.9507	.9763	.9892	.9953	.9981	.9993	.9997	.9999				
35	14	1	.0001	.0347	.2399	.5248	.7413	.8694	.9372	.9711	.9873	.9947	.9979	.9992	.9997	.9999				
35	15	1	.0005	.0608	.3173	.6107	.8064	.9107	.9608	.9836	.9935	.9976	.9992	.9997	.9999					
35	16	1	.0012	.0872	.3796	.6725	.8501	.9369	.9750	.9907	.9967	.9989	.9997	.9999						
35	17	1	.0018	.1040	.4144	.7049	.8720	.9496	.9816	.9938	.9981	.9995	.9999							
40	1	19	.0500	.1000	.1500	.2000	.2500	.3000	.3500	.4000	.4500	.5000	.5500	.6000	.6500	.7000	.7500	.8000	.8500	.9000
		37	.9500																	
40	2	12	.0038	.0192	.0462	.0846	.1346	.1962	.2692	.3423	.4115	.4769	.5385	.5962	.6500	.7000	.7462	.7885	.8269	.8615
		30	.8923	.9192	.9423	.9615	.9769	.9885	.9962											
40	3	9	.0020	.0121	.0368	.0810	.1417	.2146	.2951	.3789	.4615	.5385	.6077	.6696	.7247	.7733	.8158	.8526	.8842	.9109
		27	.9332	.9514	.9660	.9773	.9858	.9919	.9960	.9984	.9996									
40	4	7	.0008	.0078	.0316	.0788	.1494	.2362	.3309	.4263	.5172	.5999	.6726	.7349	.7879	.8326	.8698	.9004	.9253	.9452
		25	.9609	.9729	.9819	.9885	.9931	.9962	.9981	.9992	.9997	.9999								
40	5	6	.0012	.0119	.0463	.1118	.2034	.3095	.4182	.5213	.6137	.6933	.7599	.8144	.8586	.8940	.9219	.9436	.9602	.9726
		24	.9817	.9883	.9928	.9958	.9977	.9989	.9995	.9998	.9999									
40	6	5	.0008	.0110	.0498	.1273	.2352	.3568	.4766	.5852	.6782	.7547	.8159	.8639	.9010	.9293	.9505	.9661	.9774	.9854
		23	.9909	.9945	.9969	.9983	.9992	.9996	.9998	.9999										
40	7	4	.0002	.0064	.0412	.1226	.2425	.3781	.5091	.6241	.7189	.7939	.8515	.8948	.9268	.9501	.9667	.9784	.9863	.9917
		22	.9951	.9972	.9985	.9993	.9997	.9999	.9999											
40	8	4	.0019	.0247	.0993	.2251	.3740	.5182	.6423	.7416	.8174	.8734	.9139	.9426	.9626	.9762	.9853	.9912	.9949	.9972
		22	.9985	.9992	.9996	.9998	.9999													
40	9	3	.0002	.0089	.0634	.1844	.3443	.5044	.6417	.7493	.8289	.8858	.9253	.9522	.9701	.9818	.9893	.9939	.9966	.9982
		21	.9991	.9996	.9998	.9999														
40	10	3	.0012	.0271	.1255	.2885	.4663	.6221	.7429	.8302	.8906	.9311	.9576	.9745	.9852	.9916	.9954	.9976	.9988	.9994
		21	.9997	.9999																
40	11	3	.0052	.0618	.2088	.4011	.5806	.7212	.8213	.8887	.9323	.9599	.9769	.9870	.9930	.9963	.9982	.9991	.9996	.9998
		21	.9999																	
40	12	2	.0001	.0157	.1159	.3066	.5123	.6807	.8005	.8795	.9292	.9596	.9775	.9879	.9937	.9969	.9985	.9993	.9997	.9999
40	13	2	.0008	.0370	.1882	.4108	.6147	.7638	.8613	.9211	.9564	.9767	.9879	.9939	.9971	.9987	.9994	.9998	.9999	
40	14	2	.0030	.0722	.2741	.5134	.7039	.8300	.9062	.9499	.9741	.9870	.9938	.9971	.9987	.9995	.9998	.9999		
40	15	2	.0085	.1216	.3664	.6080	.7780	.8807	.9383	.9691	.9851	.9931	.9969	.9987	.9995	.9998	.9999			
40	16	2	.0189	.1814	.4567	.6898	.8368	.9182	.9605	.9816	.9918	.9965	.9986	.9994	.9998	.9999				
40	17	1	.0001	.0341	.2438	.5370	.7557	.8809	.9447	.9753	.9894	.9956	.9983	.9994	.9998	.9999				
40	18	1	.0003	.0512	.2988	.5998	.8037	.9114	.9621	.9845	.9939	.9977	.9992	.9997	.9999					
40	19	1	.0006	.0651	.3367	.6398	.8328	.9292	.9719	.9895	.9963	.9988	.9996	.9999						
40	20	1	.0007	.0704	.3502	.6536	.8425	.9350	.9751	.9910	.9970	.9991	.9997	.9999						
45	1	21	.0222	.0667	.1111	.1556	.2000	.2444	.2889	.3333	.3778	.4222	.4667	.5111	.5556	.6000	.6444	.6889	.7333	.7778
		39	.8222	.8667	.9111	.9556														
45	2	14	.0061	.0212	.0455	.0788	.1212	.1727	.2333	.3000	.3636	.4242	.4818	.5364	.5879	.6364	.6818	.7242	.7636	.8000
		32	.8333	.8636	.8909	.9152	.9364	.9545	.9697	.9818	.9909	.9970								
45	3	10	.0007	.0059	.0202	.0479	.0909	.1463	.2110	.2819	.3559	.4299	.5008	.5659	.6251	.6786	.7268	.7700	.8083	.8421
		28	.8717	.8974	.9194	.9380	.9535	.9662	.9763	.9842	.9901	.9944	.9972	.9989	.9997					
45	4	8	.0005	.0048	.0200	.0524	.1038	.1720	.2509	.3350	.4195	.5007	.5758	.6434	.7028	.7545	.7992	.8374	.8699	.8973
		26	.9201	.9389	.9542	.9664	.9760	.9834	.9889	.9930	.9958	.9977	.9988	.9995	.9998					
45	5	7	.0011	.0089	.0337	.0822	.1537	.2415	.3370	.4326	.5230	.6051	.6772	.7391	.7913	.8348	.8707	.9001	.9239	.9429
		25	.9579	.9696	.9785	.9853	.9902	.9937	.9961	.9977	.9988	.9994	.9997	.9999						
45	6	6	.0010	.0098	.0401	.1005	.1877	.2912	.3991	.5026	.5962	.6774	.7458	.8022	.8478	.8843	.9132	.9359	.9534	.9667
		24	.9767	.9840	.9894	.9931	.9957	.9974	.9985	.9992	.9996	.9998	.9999							
45	7	5	.0004	.0072	.0371	.1032	.2019	.3186	.4377	.5484	.6451	.7259	.7915	.8435	.8840	.9153	.9390	.9568	.9699	.9795
		23	.9863	.9911	.9944	.9966	.9980	.9989	.9994	.9997	.9999	.9999								

Table entries give $a = F_M(x+i, n, m)$ for $x+i$ with $0.00005 \leq a < 0.99995$.

N	n	x	+1	+2	+3	+4	+5	+6	+7	+8	+9	+10	+11	+12	+13	+14	+15	+16	+17	+18
45	8	4	.0001	.0032	.0263	.0903	.1954	.3236	.4543	.5735	.6747	.7567	.8210	.8703	.9073	.9348	.9548	.9693	.9795	.9866
		22	.9915	.9947	.9968	.9982	.9990	.9995	.9997	.9999	.9999									
45	9	4	.0007	.0129	.0651	.1690	.3063	.4496	.5797	.6882	.7737	.8388	.8870	.9221	.9471	.9647	.9769	.9852	.9908	.9944
		22	.9967	.9981	.9990	.9995	.9997	.9999	.9999											
45	10	4	.0034	.0352	.1260	.2667	.4232	.5676	.6869	.7791	.8474	.8965	.9309	.9548	.9709	.9817	.9887	.9932	.9961	.9978
		22	.9988	.9994	.9997	.9998	.9999													
45	11	3	.0002	.0115	.0745	.2065	.3738	.5357	.6706	.7735	.8480	.9000	.9354	.9591	.9746	.9846	.9909	.9947	.9970	.9984
		21	.9992	.9996	.9998	.9999														
45	12	3	.0013	.0290	.1322	.3005	.4812	.6369	.7558	.8404	.8981	.9362	.9609	.9765	.9862	.9921	.9956	.9976	.9988	.9994
		21	.9997	.9999	.9999															
45	13	3	.0047	.0596	.2063	.4004	.5818	.7234	.8235	.8903	.9334	.9605	.9770	.9870	.9928	.9962	.9980	.9990	.9995	.9998
		21	.9999																	
45	14	2	.0001	.0129	.1054	.2922	.4995	.6714	.7943	.8754	.9265	.9576	.9762	.9869	.9930	.9964	.9982	.9991	.9996	.9998
		20	.9999																	
45	15	2	.0005	.0288	.1660	.3843	.5923	.7477	.8504	.9141	.9519	.9738	.9861	.9928	.9964	.9983	.9992	.9996	.9998	.9999
45	16	2	.0016	.0548	.2383	.4765	.6752	.8105	.8937	.9421	.9693	.9842	.9921	.9962	.9982	.9992	.9997	.9999	.9999	
45	17	2	.0046	.0914	.3174	.5635	.7463	.8605	.9260	.9619	.9810	.9908	.9957	.9980	.9992	.9996	.9999	.9999		
45	18	2	.0103	.1369	.3970	.6413	.8048	.8990	.9495	.9755	.9885	.9948	.9977	.9990	.9996	.9999	.9999			
45	19	2	.0190	.1864	.4706	.7065	.8507	.9276	.9661	.9846	.9932	.9971	.9988	.9996	.9998	.9999				
45	20	1	.0001	.0296	.2332	.5320	.7570	.8843	.9477	.9772	.9904	.9961	.9985	.9994	.9998	.9999				
45	21	1	.0002	.0396	.2698	.5760	.7914	.9064	.9603	.9840	.9938	.9977	.9992	.9997	.9999					
45	22	1	.0002	.0457	.2899	.5990	.8087	.9172	.9665	.9871	.9953	.9984	.9995	.9998						
50	1	24	.0400	.0800	.1200	.1600	.2000	.2400	.2800	.3200	.3600	.4000	.4400	.4800	.5200	.5600	.6000	.6400	.6800	.7200
		42	.7600	.8000	.8400	.8800	.9200	.9600												
50	2	15	.0008	.0082	.0229	.0449	.0743	.1110	.1551	.2065	.2653	.3241	.3804	.4343	.4857	.5347	.5812	.6253	.6669	.7061
		33	.7429	.7771	.8090	.8384	.8653	.8898	.9118	.9314	.9486	.9633	.9755	.9853	.9927	.9976				
50	3	11	.0002	.0029	.0112	.0286	.0580	.0992	.1500	.2082	.2714	.3376	.4043	.4694	.5306	.5869	.6386	.6857	.7286	.7673
		29	.8022	.8335	.8612	.8857	.9071	.9257	.9416	.9551	.9663	.9755	.9829	.9886	.9929	.9959	.9980	.9992	.9998	
50	4	9	.0003	.0031	.0132	.0357	.0734	.1260	.1906	.2627	.3384	.4141	.4874	.5561	.6190	.6754	.7254	.7693	.8078	.8412
		27	.8701	.8948	.9158	.9336	.9483	.9605	.9704	.9783	.9845	.9893	.9928	.9954	.9973	.9985	.9992	.9997	.9999	
50	5	8	.0009	.0070	.0254	.0622	.1183	.1904	.2727	.3588	.4439	.5244	.5982	.6641	.7217	.7714	.8137	.8495	.8796	.9047
		26	.9254	.9424	.9561	.9671	.9757	.9825	.9876	.9915	.9943	.9964	.9978	.9987	.9993	.9996	.9998	.9999		
50	6	7	.0011	.0089	.0334	.0815	.1528	.2405	.3362	.4321	.5227	.6048	.6769	.7387	.7908	.8340	.8696	.8986	.9220	.9407
		25	.9555	.9671	.9761	.9829	.9880	.9918	.9945	.9965	.9978	.9987	.9992	.9996	.9998	.9999				
50	7	6	.0007	.0079	.0341	.0891	.1717	.2722	.3792	.4831	.5781	.6610	.7314	.7896	.8370	.8750	.9052	.9289	.9473	.9615
		24	.9723	.9804	.9863	.9907	.9938	.9960	.9975	.9984	.9991	.9995	.9997	.9999	.9999					
50	8	5	.0002	.0047	.0276	.0836	.1733	.2845	.4021	.5142	.6139	.6986	.7681	.8239	.8678	.9019	.9280	.9479	.9628	.9738
		23	.9819	.9877	.9918	.9946	.9966	.9979	.9987	.9993	.9996	.9998	.9999	.9999						
50	9	5	.0016	.0168	.0665	.1575	.2771	.4054	.5269	.6332	.7212	.7915	.8463	.8881	.9195	.9429	.9600	.9724	.9813	.9875
		23	.9918	.9948	.9967	.9980	.9988	.9993	.9996	.9998	.9999									
50	10	4	.0002	.0067	.0425	.1265	.2502	.3892	.5222	.6375	.7313	.8044	.8598	.9009	.9309	.9525	.9678	.9786	.9859	.9910
		22	.9943	.9965	.9979	.9988	.9993	.9996	.9998	.9999										
50	11	4	.0013	.0193	.0853	.2050	.3529	.4994	.6271	.7297	.8082	.8663	.9081	.9378	.9586	.9728	.9825	.9889	.9931	.9958
		22	.9975	.9986	.9992	.9996	.9998	.9999	.9999											
50	12	4	.0047	.0432	.1455	.2961	.4570	.6007	.7163	.8034	.8664	.9107	.9413	.9620	.9758	.9849	.9907	.9944	.9967	.9981
		22	.9990	.9994	.9997	.9999	.9999													
50	13	3	.0003	.0130	.0813	.2206	.3928	.5559	.6892	.7890	.8601	.9090	.9418	.9634	.9774	.9863	.9919	.9953	.9973	.9985
		21	.9992	.9996	.9998	.9999														
50	14	3	.0013	.0292	.1343	.3057	.4887	.6451	.7632	.8464	.9025	.9393	.9629	.9777	.9869	.9925	.9958	.9977	.9988	.9994
		21	.9997	.9998	.9999															
50	15	3	.0041	.0559	.2005	.3956	.5791	.7223	.8233	.8904	.9335	.9604	.9769	.9868	.9926	.9960	.9979	.9989	.9994	.9997
		21	.9999	.9999																
50	16	2	.0001	.0104	.0948	.2767	.4851	.6606	.7870	.8706	.9234	.9556	.9748	.9860	.9924	.9960	.9979	.9990	.9995	.9998
		20	.9999																	
50	17	2	.0003	.0223	.1458	.3587	.5702	.7316	.8396	.9071	.9474	.9709	.9843	.9917	.9957	.9979	.9990	.9995	.9998	.9999
50	18	2	.0009	.0414	.2070	.4420	.6478	.7915	.8814	.9345	.9647	.9814	.9904	.9952	.9977	.9989	.9995	.9998	.9999	
50	19	2	.0025	.0687	.2749	.5223	.7159	.8406	.9138	.9547	.9767	.9884	.9943	.9973	.9988	.9995	.9998	.9999		
50	20	2	.0056	.1032	.3448	.5957	.7736	.8798	.9383	.9692	.9850	.9929	.9967	.9986	.9994	.9997	.9999			
50	21	2	.0105	.1420	.4116	.6594	.8204	.9100	.9563	.9794	.9905	.9958	.9982	.9992	.9997	.9999				
50	22	2	.0169	.1807	.4699	.7111	.8564	.9322	.9691	.9863	.9941	.9975	.9990	.9996	.9999	.9999				
50	23	2	.0236	.2138	.5153	.7490	.8819	.9474	.9775	.9907	.9963	.9986	.9995	.9998	.9999					
50	24	1	.0001	.0288	.2361	.5441	.7722	.8970	.9562	.9823	.9931	.9974	.9991	.9997	.9999					
50	25	1	.0001	.0307	.2440	.5539	.7800	.9020	.9591	.9838	.9939	.9978	.9992	.9998	.9999					

References

Bateman, G. (1948) On the power function of the longest run as a test for randomness in a sequence of alternatives. *Biometrika* **35**, 97-112.

Burr, E. J. and Cane, G. (1961) Longest run of consecutive observations having a specified attribute. *Biometrika* **48**, 461-465.

David, F.N. (1947) A power function for tests for randomness in a sequence of alternatives. *Biometrika* **34**, 335-339.

Freund, John E. and Pozner, Arthur N. (1956) Some results on restricted occupancy theory. *Ann. Math. Statistics* **27**, 537-540.

Gibbons, J. D. (1985) *Nonparametric Statistics.* Marcel Decker, New York.

Guenther, William C. (1978) Some remarks on the runs test and the use of the hypergeometric distribution. *The American Statistician* **32**, 71-73.

Kingman, J. F. C. (1978) Uses of exchangeability. *Ann. Probab.* **6**, 183-197.

Kokoska, S. and Nevison, C (1989) *Statistical Tables and Formulae.* Springer-Verlag, New York.

Mood, A. M. (1940) The distribution theory of runs. *Ann. Math. Statistics* **11**, 367-392.

Mosteller, F. (1941) Note on an Application of runs to Quality Control Charts. *Ann. Math. Statistics* **12**, 228-232.

Philippou, Andreas N. and Makri, Frosso S. (1986) Successes, runs and longest runs. *Statistics and Probability Letters* **4**, 101-105.

Philippou, G. N. and Georghiou, C. (1986) Fibonacci-type polynomials and Pascal triangles of order k., in *Fibonacci Numbers and Their Applications*, edited by A. N. Philippou et al., D. Riedel Publishing Company, 229-233.

Schuster, E. F. and Sype, W. R. (1987) On the negative hypergeometric distribution. *Int. J. Math. Educ. Sci. Technol.* **18**, 453-459.

Schuster, E. F. (1991) Distribution theory of runs via exchangeable random variables. *Statistics and Probability Letters* **11**, 379-386.

Swed, F. S. and Eisenhart C. (1943) Tables for testing randomness of grouping in a sequence of alternatives. *Ann. Math. Statistics* **14**, 66-87.

Wald, A., and Wolfowitz J. (1940) On a test whether two samples are from the same population. *Ann. Math. Statistics* **11**, 147-162.

Wolfram, S. (1991) *Mathematica.* Second edition, Addison-Wesley, New York.

THE LONGEST RUN OF ANY LETTER IN A RANDOMLY GENERATED WORD

Kenneth A. Suman

Winona State University

A word is formed by choosing n letters at random and with replacement from an alphabet of k letters. The probability distribution of the longest run of the same letter is derived.

LONGEST RUN, NEAREST INTEGER FUNCTION, PARTIAL FRACTION APPROXIMATION, LAGRANGE INVERSION THEOREM, GENERATING FUNCTION.

1. Introduction

Let $\mathcal{W}_n$ be the set of all n letter words that can be formed from a k letter alphabet $\mathcal{A}_k$ (allowing repeated use of any letter). Define a *run* of the letter j in the word ω to be a string of consecutive letter j's and let ρ_j to be the length of the longest run of the letter j in the word ω. Then let $\tau = \max(\rho_1, \rho_2, \ldots, \rho_k)$. For example, in the word *abbaaa* from the alphabet $\mathcal{A}_3 = \{a, b, c\}$, we have $\rho_a = 3$, $\rho_b = 2$, $\rho_c = 0$, and $\tau = 3$.

Suppose an n letter word $\omega \in \mathcal{W}_n$ is generated randomly, that is each of the n letters of ω is selected at random and with replacement from $\mathcal{A}_k$. In this paper we will derive the exact distribution of τ for such randomly generated words.

The study of longest runs has a long history. Dunkel (1925) examines various questions about the longest success run in n Bernoulli trials. More recently, Philippou and Makri (1986) and Godbole (1990) have derived closed form solutions for various problems involving success runs. Both Mood (1940) and David and Barton (1962) give nice treatments of the case when the number of each type of letter is fixed. Recent work has focused more on asymptotic results and approximations. In

A. P. Godbole and S. G. Papastavridis (eds.), Runs and Patterns in Probability, 119–130.

this connection see Csörgo (1979), Schilling (1990), and Godbole (1991).

In Theorems 1, 2, and 3 we derive successively simpler solutions to the probability distribution of τ. In Theorem 1 we derive a recursion formula for $N(\mathcal{C}_{r+1,n})$, where $\mathcal{C}_{r+1,n} = \{\omega \in \mathcal{W}_n | \tau < r+1\}$ and then derive a double sum solution of this recurrence relationship. In Theorem 2 we use combinatorial identities to simplify the solution given in Theorem 1 into a single sum solution. The proof of Theorem 3 involves the use of a partial fraction approximation. Theorem 3 gives us the computational tool necessary for calculating the probability distribution of τ for larger values of n.

2. Statement of results

Theorem 1.

$$P(\tau < r+1) = \frac{1}{k^n} \cdot \sum_{j=0}^{n} \sum_{u=0}^{j} k(k-1)^{j-1} \binom{j}{u} (-1)^u \binom{n-u\cdot r-1}{n-u\cdot r-j}.$$

Theorem 2.

$$\begin{aligned} P(\tau < r+1) =& \frac{1}{k^n} \cdot \sum_{u=0}^{n} (-1)^u \cdot \left(\frac{k-1}{k^{r+1}}\right)^u \cdot \binom{n-u\cdot r-1}{n-u\cdot r-1-u} \\ &+ \frac{1}{k^n} \cdot \left(\frac{k}{k-1}\right) \cdot \sum_{u=0}^{n} (-1)^u \cdot \left(\frac{k-1}{k^{r+1}}\right)^u \cdot \binom{n-u\cdot r-1}{n-u\cdot r-u}. \end{aligned}$$

Theorem 3. *There exists an $N < \infty$ such that for all $n > N$,*

$$P(\tau < r+1) = \frac{1}{k^n} \cdot \left\langle \frac{k}{(k-1)^2} \cdot \left(\frac{(\lambda_b)^{r+2} - 2(\lambda_b)^{r+1} + (\lambda_b)^r}{(\lambda_b)^{r+1} - (r+1)(\lambda_b) + r} \right) \cdot (\lambda_b)^n \right\rangle$$

where $\langle x \rangle$ represents the nearest integer to x and where

$$\frac{1}{\lambda_b} = \sum_{j=0}^{\lfloor \frac{w-1}{r+1} \rfloor} \left(\frac{1}{k}\right)^{jr+j+1} \cdot (k-1)^j \cdot \frac{1}{(jr+j+1)} \cdot \binom{jr+j+1}{j} + o(k^{-w}).$$

3. Derivation of results

Proof of Theorem 1.

The set of words $\mathcal{C}_{r+1,n}$ can be partitioned according to the length of the final run of each word. For example, the length of the final run in the word *abbaaa* would be 3. Let $\mathcal{C}_{r+1,n}(v)$ be that subset of $\mathcal{C}_{r+1,n}$ for which the final run has length $v, 1 \le v \le r$. Then we have for $n \ge r+1$,

$$N(\mathcal{C}_{r+1,n}) = N\left(\mathcal{C}_{r+1,n}(1)\right) + N\left(\mathcal{C}_{r+1,n}(2)\right) + \cdots + N\left(\mathcal{C}_{r+1,n}(r)\right).$$

We can construct each word in the set $\mathcal{C}_{r+1,n}(v)$ through the following steps.

Step 1. Select a word from the set $\mathcal{C}_{r+1,n-v}$.

Step 2. Append to this word v copies of the letter y, where y is any letter in the alphabet $\mathcal{A}_k$ *except* the last letter of the word selected in Step 1.

Keeping the first subscript fixed at $r+1$, let $c_j = N(\mathcal{C}_{r+1,j})$. Then there are c_{n-v} ways to carry out Step 1 and there are $(k-1)$ ways to carry out Step 2. Hence, for $n \ge r+1$, we have the recurrence relation,

$$c_n = (k-1)c_{n-1} + (k-1)c_{n-2} + \cdots + (k-1)c_{n-r}.$$

For $n \le r$, $\tau < r+1$ and thus $\mathcal{C}_{r+1,n} \equiv \mathcal{W}_n$ for $n \le r$. Therefor, $c_n = k^n$ for all $n \le r$. Now define the generating function $\mathbf{A}(x) = \sum_{n=0}^{\infty} c_n x^n$. Then

$$\begin{aligned}\mathbf{A}(x) &= \sum_{n=0}^{r} k^n x^n + \sum_{n=r+1}^{\infty} \left((k-1)c_{n-1} + (k-1)c_{n-2} + \cdots + (k-1)c_{n-r}\right) x^n \\ &= \sum_{n=0}^{r} k^n x^n + \left((k-1)x^1 \left(\mathbf{A}(x) - \sum_{n=0}^{r-1} c_n x^n\right)\right) \\ &\qquad + \cdots + \left((k-1)x^r \left(\mathbf{A}(x) - \sum_{n=0}^{0} c_n x^n\right)\right).\end{aligned}$$

Solving for $\mathbf{A}(x)$, we have

$$\mathbf{A}(x) = -\frac{1}{k-1} + \frac{k}{(k-1)} \left(\frac{1}{1-(k-1)(x+x^2+\cdots+x^r)}\right).$$

Upon simplifying we find that

$$c_n = \sum_{j=0}^{n}\sum_{u=0}^{\lambda} k(k-1)^{j-1}\binom{j}{u}(-1)^u\binom{n-u\cdot r-1}{n-u\cdot r-j}$$

where

$$\lambda(n,j,r) = \min\left(\left\lfloor\frac{n-j}{r}\right\rfloor, j\right).$$

We note that if we replace λ with any larger number it will not change the sum because for $u > \lambda$, either

$$\binom{j}{u} = 0 \qquad \text{or} \qquad \binom{n-u\cdot r-1}{n-u\cdot r-j} = 0.$$

In particular, we can replace λ with j. □

Proof of Theorem 2.

It is possible to manipulate the double sum solution of Theorem 1 into just a single sum. In particular,

$$c_n = \sum_{j=0}^{n} \sum_{u=0}^{j} k(k-1)^{j-1} \binom{j}{u} (-1)^u \binom{n-u\cdot r-1}{n-u\cdot r-j}$$

$$= \sum_{u=0}^{n} \frac{k}{(k-1)} (-1)^u \left(\sum_{j=u}^{n} (k-1)^j \binom{j}{u} \binom{n-u\cdot r-1}{n-u\cdot r-j} \right)$$

$$= \sum_{u=0}^{n} k(k-1)^{u-1}(-1)^u \cdot \frac{1}{(n-u\cdot r)} \cdot \binom{n-u\cdot r}{n-u\cdot r-u}$$

$$\cdot \left(\sum_{j=u}^{n-ur-u} (k-1)^j \binom{n-u\cdot r-u}{n-u\cdot r-u-j} (j+u) \right)$$

$$= \sum_{u=0}^{n} k(k-1)^{u-1}(-1)^u \cdot \frac{1}{(n-u\cdot r)} \cdot \binom{n-u\cdot r}{n-u\cdot r-u}$$

$$\cdot \left((k-1)\cdot(n-u\cdot r-u)\cdot k^{(n-ur-u-1)} + u\cdot k^{(n-ur-u)} \right)$$

$$= \sum_{u=0}^{n} (-1)^u \cdot (k-1)^u \cdot k^{(n-ur-u)} \cdot \binom{n-u\cdot r-1}{n-u\cdot r-1-u}$$

$$+ \sum_{u=0}^{n} (-1)^u \cdot (k-1)^{u-1} \cdot k^{(n-ur-u+1)} \cdot \binom{n-u\cdot r-1}{n-u\cdot r-u}.$$

This completes the proof of Theorem 2. □

Proof of Theorem 3.

Let $\lambda_1, \lambda_2, \ldots, \lambda_d$ be the distinct roots $(d \leq r)$ of

$$C(x) = x^r - (k-1)x^{r-1} - (k-1)x^{r-2} - \cdots - (k-1)x^1 - (k-1)x^0 = 0$$

and let m_j be the multiplicity of the root λ_j. Define $h(x) = (x-1)\cdot C(x)$. We note that all the roots of $h(x) = 0$ [and hence of $C(x) = 0$] are simple. Thus, we can write

$$C(x) = (x-\lambda_1)(x-\lambda_2)\cdots(x-\lambda_r).$$

Let $Q(x) = 1-(k-1)(x+x^2+\cdots+x^r)$. We note that

$$Q(x) = x^r C\left(\frac{1}{x}\right) \qquad \text{and} \qquad C(x) = x^r Q\left(\frac{1}{x}\right).$$

Therefore,

$$\begin{aligned}
\mathbf{A}(x) &= -\frac{1}{k-1} + \frac{k}{(k-1)}\left(\frac{1}{x^r C\left(\frac{1}{x}\right)}\right) \\
&= -\frac{1}{k-1} + \frac{k}{(k-1)}\left(\frac{1}{(1-\lambda_1 x)(1-\lambda_2 x)\cdots(1-\lambda_r x)}\right) \\
&= -\frac{1}{k-1} + \frac{k}{(k-1)}\left[\sum_{n=0}^{\infty}\left(\sum_{j=1}^{r}\delta_j(\lambda_j)^n\right)x^n\right]
\end{aligned}$$

where the δ_j are the coefficients of our partial fraction expansion. It follows that for $n \geq 1$,

$$c_n = \frac{k}{(k-1)}\sum_{j=1}^{r}\delta_j(\lambda_j)^n$$

and from the theory of partial fractions we have that

$$\delta_j = \frac{-\lambda_j}{Q'\left(\frac{1}{\lambda_j}\right)}.$$

It is easily seen that

$$Q'(x) = -(k-1) \cdot \frac{\left(rx^{r+1} - (r+1)x^r + 1\right)}{(1-x)^2}.$$

Thus,

$$\delta_j = \frac{1}{(k-1)} \cdot \left(\frac{(\lambda_j)^{r+2} - 2(\lambda_j)^{r+1} + (\lambda_j)^r}{(\lambda_j)^{r+1} - (r+1)(\lambda_j) + r}\right)$$

and

$$c_n = \frac{k}{(k-1)^2} \sum_{j=1}^{r} \left(\frac{(\lambda_j - 1)^2}{(\lambda_j)^{r+1} - (r+1)(\lambda_j) + r}\right) \cdot (\lambda_j)^{n+r}.$$

The key to all further simplifications is the Lagrange Inversion Theorem. The particular form used here is from Whittaker and Watson (1958) p. 133. With some simplification the equation $h(\lambda) = 0$ can be written in the form

$$\psi = \frac{1}{k}\left(1 + (k-1)\psi^{r+1}\right)$$

with $\psi = 1/\lambda$. After verifying that the sufficient conditions of the Lagrange Inversion Theorem are met we can conclude that the equation

$$\psi = \frac{1}{k}\left(1 + (k-1)\psi^{r+1}\right) = \frac{1}{k}\phi(\psi)$$

has exactly one root ($\psi = \psi_0$) in the interior of this circle **C**. This in turn implies that $C(x) = 0$ has exactly one root ($1/\psi_0$) *outside* the unit circle on the complex plane.

Let $\lambda_b = 1/\psi_0$ be the one root of $C(x)$ such that $|\lambda_b| > 1$. Then,

$$\lim_{n\to\infty} \left[\frac{k}{(k-1)^2} \left(\frac{(\lambda_j - 1)^2}{(\lambda_j)^{r+1} - (r+1)(\lambda_j) + r}\right) \cdot (\lambda_j)^{n+r}\right] = 0 \qquad j \neq b$$

and it follows that $\exists N < \infty$ such that for all $n > N$,

$$\left| \frac{k}{(k-1)^2} \sum_{\substack{j=1 \\ j\neq b}}^{r} \left(\frac{(\lambda_j - 1)^2}{(\lambda_j)^{r+1} - (r+1)(\lambda_j) + r}\right) \cdot (\lambda_j)^{n+r} \right| < \frac{1}{2}.$$

But c_n is an integer for all n and thus for $n > N$

$$\begin{aligned} c_n =& \frac{k}{(k-1)^2}\left(\frac{(\lambda_b-1)^2}{(\lambda_b)^{r+1}-(r+1)(\lambda_b)+r}\right)\cdot(\lambda_b)^{n+r} \\ &+\frac{k}{(k-1)^2}\sum_{\substack{j=1\\ j\neq b}}^{r}\left(\frac{(\lambda_j-1)^2}{(\lambda_j)^{r+1}-(r+1)(\lambda_j)+r}\right)\cdot(\lambda_j)^{n+r} \\ =&\left\langle\frac{k}{(k-1)^2}\cdot\left(\frac{(\lambda_b)^{r+2}-2(\lambda_b)^{r+1}+(\lambda_b)^r}{(\lambda_b)^{r+1}-(r+1)(\lambda_b)+r}\right)\cdot(\lambda_b)^n\right\rangle \end{aligned}$$

where $\langle x\rangle$ represents the nearest integer to x. From Bürmann's Theorem, [see Takács (1967) p. 209], we have that

$$\psi_0=\sum_{n=1}^{w}\left(\frac{1}{k}\right)^n\cdot\frac{1}{n!}\cdot\left(\frac{d^{n-1}}{d\psi^{n-1}}(\phi(\psi)^n)\Big|_{\psi=0}\right)+o(k^{-w}).$$

Simplifying, we find

$$\begin{aligned} \psi_0 =& \sum_{n=1}^{w}\left(\frac{1}{k}\right)^n\cdot\frac{1}{n!}\cdot\left(\sum_{j=0}^{n}\binom{n}{j}(k-1)^j(n-1)!\,I_{\{(r+1)\cdot j\}}(n-1)\right)+o(k^{-w}) \\ =&\sum_{j=0}^{\lfloor\frac{w-1}{r+1}\rfloor}\left(\frac{1}{k}\right)^{jr+j+1}\cdot(k-1)^j\cdot\frac{1}{(jr+j+1)}\cdot\binom{jr+j+1}{j}+o(k^{-w}). \end{aligned}$$

Hence, for $n > N$, we have

$$c_n=\left\langle\frac{k}{(k-1)^2}\cdot\left(\frac{(\lambda_b)^{r+2}-2(\lambda_b)^{r+1}+(\lambda_b)^r}{(\lambda_b)^{r+1}-(r+1)(\lambda_b)+r}\right)\cdot(\lambda_b)^n\right\rangle$$

where

$$\frac{1}{\lambda_b}=\sum_{j=0}^{\lfloor\frac{w-1}{r+1}\rfloor}\left(\frac{1}{k}\right)^{jr+j+1}\cdot(k-1)^j\cdot\frac{1}{(jr+j+1)}\cdot\binom{jr+j+1}{j}+o(k^{-w}),$$

proving Theorem 3. □

4. Notes

(i) The condition $r \geq 2$ assumed in Theorem 3 was necessary in order to satisfy the sufficient conditions of the Lagrange Inversion Theorem. However, the conclusions drawn from the Lagrange Inversion Theorem do remain valid for the case $r = 1$. In this case, $\lambda_b = k - 1$ and

$$c_n = \left\langle \frac{k}{(k-1)^2} \cdot \left(\frac{(k-1) \cdot k^2}{k^2} \right) \cdot (k-1)^n \right\rangle = k \cdot (k-1)^{n-1}.$$

We recognize that this is the correct solution for the case $r = 1$. [There are k ways to fill the first slot and $(k-1)$ ways to fill each of the remaining $n-1$ slots.]

(ii) In computations it will be necessary to approximate λ_b. The direct approach is to take the reciprocal of

$$\sum_{j=0}^{t} \left(\frac{1}{k}\right)^{jr+j+1} \cdot (k-1)^j \cdot \frac{1}{(jr+j+1)} \cdot \binom{jr+j+1}{j}$$

with some "large" value of t. An alternative approach is based on realizing that

$$\lim_{n \to \infty} \left(\frac{c_{n+1}}{c_n} \right) = \lambda_b.$$

This follows immediately from the fact that

$$c_n = \langle \text{constant} \cdot (\lambda_b)^n \rangle \qquad \text{for sufficiently large } n.$$

We can use the recurrence relation

$$c_n = (k-1)c_{n-1} + (k-1)c_{n-2} + \cdots + (k-1)c_{n-r} \qquad n \geq r+1$$

with initial conditions $c_n = k^n$ for all $n \leq r$ to find c_t and c_{t+1} for "large" t and then take

$$\lambda_b \simeq \frac{c_{t+1}}{c_t}.$$

With this approach for estimating λ_b the value of t must be "large" relative to r in order to get an accurate approximation. However it is straightforward to compute an accurate value of c_t even for large t.

(iii) Numerical computations comparing the asymptotic result of Theorem 3 with the exact solution derived in Theorem 2 suggest that the phrase "for all $n > N$" can be replaced with the phrase "for all $n \geq 1$".

The author has examined the cases for $n \leq 50$, $r \leq n-1$, and $k \leq n$. In each of these cases the error in the approximation, namely

$$\left| \frac{k}{(k-1)^2} \sum_{\substack{j=1 \\ j \neq b}}^{r} \left(\frac{(\lambda_j - 1)^2}{(\lambda_j)^{r+1} - (r+1)(\lambda_j) + r} \right) \cdot (\lambda_j)^{n+r} \right|$$

is very much less than 1/2 and hence the nearest integer approximation is in fact exact. For another example where the use of a similar approximation turns out to give an exact result see Wilf (1987). Wilf is able to establish analytically that the error is less than 1/2 for all $n \geq 1$ for the problem considered in his paper. However this author has not been able to apply Wilf's arguments with any success.

(iv) Suppose ω is an n letter word generated randomly from a k letter alphabet $\mathcal{A}_k$. Let τ be the length of the longest run of any of the letters in $\mathcal{A}_k$ in the word ω.

Now, suppose δ is an $n-1$ letter word generated from the 2 letter alphabet $\{S, F\}$ in a manner such that each of the $n-1$ letters are chosen independently and at each of the $n-1$ stages the letter S is chosen with probability $1/k$. Let ρ_S be the length of the longest run of the letter S in the word δ. Then,

$$P(\tau = r) = P(\rho_S = r - 1).$$

This is easily seen to follow from an example. Suppose the letters in our alphabet $\mathcal{A}_k$ were $\{a, b, c, d\}$ and suppose that we randomly generated the 10 letter word *abcbaabbdc*. Under each pair of adjacent letters write an S if the letters are the same and write an F if the letters are different.

$$\begin{array}{ccccccccccccccccccc} a & & b & & c & & b & & a & & a & & b & & b & & d & & c \\ & F & & F & & F & & F & & S & & F & & S & & F & & F & \end{array}$$

This generates an $n-1$ letter word from the two letter alphabet $\{S, F\}$ with the properties that each of the $n-1$ letters are chosen independently and at each of the $n-1$ stages the letter S is chosen with probability $1/k$.

Furthermore, it is clear that $\tau = r \Leftrightarrow \rho_S = r-1$. Hence, $P(\tau = r) = P(\rho_S = r-1)$.

There has been a considerable amount of work concerning the limiting behavior (as $n \to \infty$) of ρ_S. From the relation above we see how this work is immediately applicable to the limiting behavior of τ. Two important results from this work are that

$$E(\tau) \approx \log_k \left(\frac{(n-1)(k-1)}{k} \right) + \frac{\gamma}{\ln(k)} + \frac{1}{2}$$

and

$$\mathrm{Var}(\tau) \approx \frac{\pi^2}{2\,(\ln(k))^2} + \frac{1}{12}$$

where $\gamma = 0.577\cdots$ is Euler's constant. For a survey and a bibliography of these and other related asymptotic results see Schilling (1990).

5. Application

As an illustrative example we consider finding the probability distribution of τ, the length of the longest run of any letter, for the case for $n = 100$ and $k = 10$. We have that

$P(\tau = 1) = 0.00002951266$	$P(\tau = 8) = 0.00000755097$
$P(\tau = 2) = 0.40519676821$	$P(\tau = 9) = 0.00000074699$
$P(\tau = 3) = 0.51083523051$	$P(\tau = 10) = 0.00000007389$
$P(\tau = 4) = 0.07532239692$	$P(\tau = 11) = 0.00000000730$
$P(\tau = 5) = 0.00776041680$	$P(\tau = 12) = 0.00000000072$
$P(\tau = 6) = 0.00077097797$	$P(\tau = 13) = 0.00000000007$
$P(\tau = 7) = 0.00007631692$	$P(\tau = 14) = 0.00000000000$

The identical probabilities (to the listed precision) were determined using the exact results of Theorem 1 and the approximate results of Theorem 3. For each value of r we used

$$\lambda_b(r) \simeq \frac{c_{100}}{c_{99}}.$$

References

Csörgo, S. (1979). Erdös-Rényi laws, *Annals of Statistics* **7**, 772-787.

David, F. N. and Barton, D. E. (1962). *Combinatorial Chance.* Griffin, London.

Dunkel, O. (1925). Solutions of a probability difference equation. *American Mathematical Monthly*, **32**, 354-370.

Godbole, A. P. (1990). Specific formulae for some success run distributions. *Statistics & Probability Letters* **10**, 119-124.

Godbole, A. P. (1991). Poisson approximations for runs and patterns of rare events. *Advances in Applied Probability* **23**, 851-865.

Mood, A. (1940). The distribution theory of runs. *Annals of Mathematical Statistics* **11**, 367-392.

Philippou, A. N. and Makri, F. (1986). Successes, runs and longest runs. *Statistics and Probability Letters* **4**, 211-215.

Schilling, M. F. (1990). The longest run of heads. *The College Mathematics Journal* **3**, 196-207.

Takács, L. (1967). *Combinatorial Methods in the Theory of Stochastic Processes*, Wiley.

Whittaker, E. T. and Watson, G. N. (1958). *Modern Analysis*, Cambridge.

Wilf, H. S. (1987). The Editor's Corner: Strings, substrings, and the 'Nearest Integer' Function. *American Mathematical Monthly*, 855-860.

APPLICATIONS OF CONSECUTIVE SYSTEM RELIABILITY IN SELECTING ACCEPTANCE SAMPLING STRATEGIES

Hae-Il Ahn
Iowa State University

Way Kuo
Texas A&M University

Analyses of lot-by-lot acceptance sampling by attributes call for much computational and analytic endeavor. In light of recent advancements in the area of reliability evaluation, however, the sampling procedure would better be regarded as being implemented much like a consecutive structure of reliability systems. A customized design of lot-by-lot acceptance sampling strategy for stipulated Type I and Type II risks is discussed with a view to an application of analytic results available in the literature. In order to enhance the effectiveness of the design, diversification of switching rules are taken into consideration.

1. Introduction

Standard sampling procedures for inspection by attributes were developed more than half a century ago. The methodology for monitoring a process and estimating the process parameters has been applied to statistical process control (SPC) and acceptance sampling systems. A conspicuous application of SPC would be the control chart. The most widely used acceptance sampling system for attributes in the world today would be MIL-STD-105 and its civilian versions such as ANSI/ASQCZ1.4 and ISO 2859.

The background of MIL-STD-105D and its development out of the 105 series are given in the papers by Pabst [15] and Keefe [13]. Both of them explain some of the intricacies of the system and its development. The story behind the structure of the MIL-STD-105D tables is well presented in a paper by Hill [10]. An exten-

A. P. Godbole and S. G. Papastavridis (eds.), Runs and Patterns in Probability, 131–162.

sive and informative investigation of the properties of MIL-STD-105D schemes is presented in the papers by Stephens and Larson [24] and Hahn and Schilling [7]. Scheme properties are also investigated by Schilling and Sheesley [20,21] and measures of performance tabulated. A set of plans indexed by limiting quality (LQ) and compatible with MIL-STD-105D in the same lot-size classes and sample sizes has been proposed by Duncan, Mundel, Godfrey and Patridge[3]. Their tables, which simplify the selection of a limiting quality plan, can be used independently or in conjunction with MIL-STD-105D and other related standards.

Fundamentally, there are several modes of acceptance sampling plans. Probably, the most commonly used designations for acceptance sampling attribute plans are AQL (Acceptable Quality Level), LTPD (Lot Tolerance Percent Defective), and AOQL (Average Outgoing Quality Limit). According to Dodge [2], the choice among them depends primarily on what requirements one wishes to emphasize with respect to the quality of product under consideration. It makes little difference which of the three types of plan one uses, of whether one designs his own plan, so long as the rules are followed closely, the sampling operation is fair and both producer and consumer understand the principles involved.

As a matter of fact, MIL-STD-105 is indexed by AQL and lot size. An LQL (Limiting Quality Level) indexed scheme is also available in Duncan *et. al.* [3]. Since the scheme is prepared in table format, there are some rather significant gaps between sample sizes and AQL's. Besides, the control over the discriminatory power of the sampling scheme is only through the choice of inspection level. For this reason, the tables of MIL-STD-105 fail to provide a flexible sampling scheme more often than not, whereas the determination of acceptance sampling plans that provide desired levels of protection for both producer and consumer is frequently faced by the quality control personnel and it provides fair ground for compromise. A customization of switching system is worthy of study.

Analyses of switching systems require much of analytic and computational power. Since the standards were established, signaling or switching rules have been studied by many authors in the area of statistics and reliability. Moreover, if we take advantage of recent development in computation, the problem gets tractable. As a matter of fact, modern quality control system is being harnessed by electronic devices which provides immense computing power. This paper is concerned with customized design of a sampling scheme for lot-by-lot acceptance sampling by attributes on the basis of stipulated consumer's and producer's risks.

1.1. Notation and Nomenclature

AQL $\equiv p_1$	Acceptable Quality Level
AOQ	Average Outgoing Quality
RQL $\equiv p_2$	Rejectable Quality Level (or LTPD, LQL, LQ)
s, N	sample size, lot size
s_N, s_T	sample size of normal, tightened inspection respectively
P	probability of concern ($Q = 1 - P$)
P	rejection probability on Normal Inspection
P	acceptance probability on Tightened Inspection
P_N	acceptance probability on normal inspection
$P_{N_1} = P_N(p_1)$	acceptance probability on normal inspection at AQL
P_T	acceptance probability on tightened inspection
$P_{T_2} = P_T(p_2)$	acceptance probability on tightened inspection at RQL
π_N, π_T, π_R	probability the system is run on each inspection
$f(i; s, p)$	probability function
$B(c, s, p)$	cumulative distribution function of $f(i; s, p)$
$\bar{N} \equiv E(X)$	mean recurrence time on normal inspection
$\bar{T} \equiv E(Y)$	mean recurrence time on tightened inspection
$\mathbf{r}_j$	a vector of r_j
$\mathbf{f}_j$	a vector of f_j

$E(\mathbf{X})$	a vector of $E(X)$
$F(z)$	probability generating function of f_j
$R(z)$	probability generating function of r_j

1.2. Lot Sentencing Process

The use of sampling plans for inspecting a continuing stream of lots can be regarded as a Bernoulli trial. For each sampling inspection of a lot, the probability of obtaining x defectives in a random sample of s items drawn without replacement from a lot of size N containing X defectives is

$$f(x;N;X,s) = \frac{\binom{X}{x}\binom{N-X}{s-x}}{\binom{N}{s}} \quad N = 0,1,\cdots \quad 0 \leq x \leq N \quad 0 \leq s \leq N$$

and

$$\max[s - N + x, 0] \leq x \leq \min[X, s].$$

If N is large, $f(x;N,X,s)$ reduces to $b(x,s,p) = \binom{s}{x}p^x(1-p)^{s-x}, x = 0,1,\cdots s$. For large s and small p, $b(x,s,p)$ approximates to $d(s,p)$, where d is a Poisson distribution.

In a lot-by-lot acceptance sampling attribute plan, the probability of acceptance of a lot is calculated as a cumulative probability based on assumed probability functions such as hypergeometric, binomial, Poisson and others.

The operating characteristic (OC) curve of a sampling plan is the acceptance probability at various percent defective. If we regard the probability of acceptance (rejection) as a success (failure) probability, the sampling process follows the two-state recurrent Bernoulli trial. The rejection of a lot is analogous to a failure of a component and thus the acceptance is equivalent to success of a component in reliability systems.

Define

E_j	event of k successes out of m consecutive trials occur for the first time on trial $j, j = 1,\cdots,\infty$

f_j $\quad$ $Prob\{E_j \text{ occurs }\}$

$r_j = r(k, m, j)$ $\quad$ $Prob\{\text{ the event } E_j \text{ occurs after } j\text{-th trial}\}$

Let X and Y be discrete random variables that follow the above probability distribution. The variables X and Y are the number of trials on which k-out-of-m consecutive trials yielding the outcome concerned occur. In view of the reliability system, f_j represents the failure of system at j-th component and r_j is analogous to survival function (or system reliability). It is interesting to note that

$$f_j = Prob(E_j) = r(k, m, j-1) - r(k, m, j) = r_{j-1} - r_j \tag{1}$$

$$E(X) = \sum_{j=0}^{\infty} j f_j = \sum_{j=1}^{\infty} j(r_{j-1} - r_j) = \sum_{j=1}^{\infty} r_j \quad \text{where } r_0 = 1 \tag{2}$$

The $E(X)$ compares to the mean-time-to-failure (MTTF) of reliability systems. Accordingly, if r_j is known, the identification of the other measures are readily available. The r_j can be obtained in several ways such as Markov chain, probability generating function technique and combinatorial approach. Even though the expressions vary depending on the methods used, their results are supposed to be exactly the same, which will be discussed in later sections.

1.3. Monitoring of a Process

In view of the control chart and acceptance sampling system, we recall that most of signaling and switching rules in classical statistical quality control (SQC) procedures take the form of "signal or switch as soon as k out of m consecutive trials result in the occurrence of an event of interest". This procedure is appealing and hence widely accepted because the criterion depends solely on the probability with which certain statistics fall on the designated warning zone or an alarming event occurs, without specifying peculiar configuration for a run. The sampling process in a signaling or switching system can be regarded as two-state recurrent Bernoulli trial (or binary process) in which k outcomes within m consecutive trials are of interest. By varying the values of k and m, it is possible to regulate the "false alarming" of a control chart and to design the "scheme performance" of an acceptance sampling system.

1.4. Signaling in a Control Chart

In a control chart, a warning zone is designated for monitoring the production process. An assignable cause is sought whenever k out of m consecutive points fall on the warning zone. Let the warning zone be one-sided above the mean plus predetermined standard deviations. The probability of the sample mean falling on this warning zone given no change in the process average, say α, is Type I error. There exist two kinds of Type I and Type II errors. The first kind is associated with the sampling plan for each lot. The Type I and Type II error of the second kind are associated with "false alarming" of the process.

If the average run length ($E(X)$) of samples between false alarms is given, then the Type I error can be expressed as $1/E(X)$. Calculation of the Type II error can be done analogously, as a function of the magnitude of the shift in the mean.

Monitoring of a production process requires the knowledge not only of the first kind, but also of the second kind of errors. In order to have control over Type I and Type II errors of a process, it is essential to know the average run length (or the mean recurrence time) of the event concerned.

1.5. Switching in a Sampling System

The most famous acceptance sampling system that employs the switching rule would be the MIL-STD-105 series and its emulating versions such as ANSI/ASQ-CZ1.4, ABC-STD-105 and ISO 2859. These inspection systems are based on the AQL concept and use basically three sampling plans, called Normal, Tightened and Reduced. Switching between tightened and normal plan is made mandatory by the standards, while use of reduced plans is optional. Tightened Inspection protects the consumer, in the case that submitted quality gets significantly worse than the AQL.

The invocation of Tightened Inspection causes greater rejections of lots, and forces the producer to improve his quality. The switching from one to the other provides a composite system OC curve more close to the ideal OC curve than the individual OC curve is.

1.5.1. Switching Rules

The convention is of shifting from Normal to Tightened Inspection when two out of any five consecutive lots or batches are rejected on Normal Inspection. Normal Inspection is resumed when five consecutive lots are accepted, while Tightened Inspection is under way. See Figure 1. The rule is originally suggested by Dodge [2] and later revised by Paul Martel.

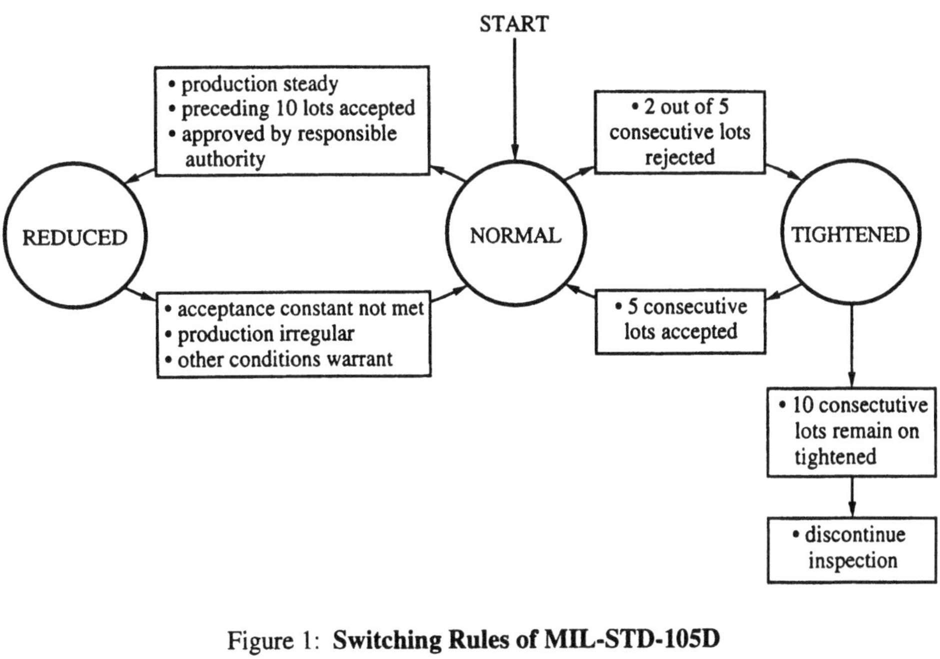

Figure 1: **Switching Rules of MIL-STD-105D**

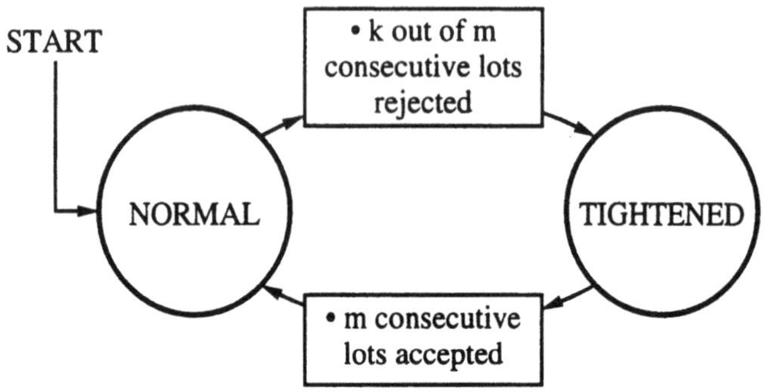

Figure 2: **Switching Rules for NTN Scheme**

1.5.2. Composite System OC Curve

The composite measure P_C represents the OC function of the set of sampling plans used as a "system" in accordance with the rules outlined above. The composite acceptance probability P_C is determined as follows:

$$P_C = \pi_N \cdot P_N + \pi_T \cdot P_T (+\pi_R \cdot P_R) \tag{3}$$

where π_N,π_T,π_R are the probabilities the system is run on normal, tightened, and reduced inspection, respectively. Since the reduced inspection is optional, the major part of the system is the switching between Normal and Tightened Inspection. The scheme that only Normal and Tightened Inspection is involved is what we call "NTN scheme" as in Figure 2 and Figure 3.

The acceptance probability of a lot on each inspection takes a different form depending on the assumed probability distributions. For any type of distribution,

$$\begin{aligned} P_C &= \pi_N \cdot B(c_N, s_N, p) + \pi_T \cdot B(c_T, s_N, p) \\ &= \pi_N \cdot \sum_{i=0}^{c_N} f(i, s_N, p) + \pi_T \cdot \sum_{i=0}^{c_T} f(i, s_N, p) \end{aligned} \tag{4}$$

According to Dodge [2] and Stephens and Larson [24], the calculation of π_N and π_T can be evaluated via Markov chain approach. In case of large k and m, however, it gets tremendously complex.

A simpler way of doing this is the use of average run length on each inspection. Since the transition probabilities are reciprocals of average run lengths on each inspection, the following equations hold when the system reaches the steady state.

$$\pi_N \cdot \frac{1}{E(X)} = \pi_T \cdot \frac{1}{E(Y)} \tag{5}$$

$$\pi_N + \pi_T = 1 \tag{6}$$

By solving these equations with respect to π_N and π_T, we get

$$\pi_N = \frac{E(X)}{E(X)+E(Y)} \quad \text{and} \quad \pi_T = \frac{E(Y)}{E(X)+E(Y)} \tag{7}$$

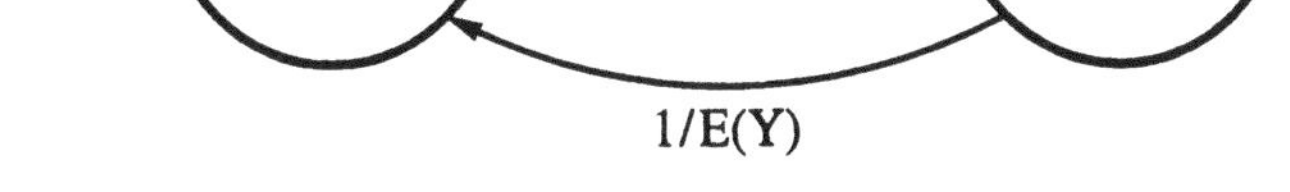

Figure 3: **Transition Diagram for NTN Scheme**

If the average run lengths for both Normal and Tightened Inspection are known, the expected portion of lots inspected on each inspection can be obtained as follows:

$$\pi_N = \frac{\bar{N}}{\bar{N}+\bar{T}} \quad \text{and} \quad \pi_T = \frac{\bar{T}}{\bar{N}+\bar{T}} \tag{8}$$

where π_N and π_T serve as weights assigned to each inspection and $\bar{N}$ and $\bar{T}$ are average run lengths of each inspection. Since they play a key role in calculating composite system OC curve, the study is more concerned with computation of average run length (on mean recurrence time of the event) on each inspection.

2. Formulas for Switching Rules

The early studies of this problem such as Roberts' [16] date as far back as the 1950's. Later, the study has been succeeded by Greenberg [5] and Saperstein [17]. On the other hand, there have been attempts to solve the problem in the context of consecutive k-out-of-m-from-n:F reliability system. The consecutive k-out-of-m-from-n:F system is a system of n components ordered in sequence, and the system fails if and only if out of m consecutive components at least k fail. Usually n is assumed to be quite large. This problem lends itself to a number of interesting applications. An application is in radar detection problem as in the minimal-order model of Nelson [14]. Later, Saperstein [17] dealt with the same problem in connection with quality control and inspection procedures. Recently, Sfakianakis *et.al.*[23] presented a closed form solution for the case of $k = 2$ and very close bounds for the cases of k greater than 2. The solution of those problems is still known to be unwieldy, especially when k and m take on large values.

2.1. Markov Chain

In connection with generalization of consecutive k-out-of-n system, Griffith [6] via a Markov chain approach devised recursive ways to compute the failure probability of consecutive k-out-of-m-from-n systems with i.i.d components. In fact, there could exist better ways of applying Markov chain theory as in Nelson [14]. Among the relevant formulas out of Markov chain theory, the following would be the simplest and it complies with reliability theory.

In a Bernoulli trial, it is possible to write out all possible sequences of success and failure of length $m-1$ with at most $k-1$ successes and to think of the evolution of the sequence of successes and failures as defined by Markov chain with $\sum_{i=0}^{k-1}\binom{m-1}{i}+1$ states indexed by these sequences.

If the last row and column of the one step transition matrix correspond to the absorbing (system failure) state, then partition the transition matrix $\mathbf{T}$ as

$$\mathbf{T}=\begin{bmatrix}\mathbf{A} & \mathbf{q}\\ \mathbf{0} & 1\end{bmatrix}, \text{where} \begin{cases}\mathbf{A}:(l\times l) & \text{transitions among non-absorbing states}\\ \mathbf{q}:(l\times l) & \text{transitions to absorbing state}\\ \mathbf{0}:(l\times l) & \text{transitions from absorbing state}\end{cases}$$

The l is given as $\sum\limits_{i=0}^{k-1}\binom{m-1}{i}$. For example, given that $k=2$ and $m=3$, the non-absorbing states are $\{SS, SF, FS\}$. The transition matrix $\mathbf{T}$ is

$$\mathbf{T}=\begin{matrix} & SS & SF & FS & (*)\\ SS & P & 1-P & 0 & 0\\ SF & 0 & 0 & P & 1-P\\ FS & P & 1-P & 0 & 0\\ (*) & 0 & 0 & 0 & 1\end{matrix} \tag{9}$$

The entries of $\mathbf{A}^j\cdot\mathbf{1}$ gives the probability of survival until j steps, given starts at various states. The entries of $\mathbf{A}^{j-1}(\mathbf{I}-\mathbf{A})\cdot\mathbf{1}$ gives the probabilities of absorption in j steps, given starts at various states. Then the $m\times 1$ vector $(\mathbf{I}-\mathbf{A})^{-1}\cdot\mathbf{1}$ has as its entries the mean times to absorption beginning at the corresponding non-absorbing states. The algebraic relations are shown as follows:

$$\mathbf{r}_j=\mathbf{A}^j\cdot\mathbf{1}\quad\text{for } j=0,1,2,\cdots \tag{10}$$

$$\mathbf{f}_j=\mathbf{r}_{j-1}-\mathbf{r}_j=(\mathbf{A}^{j-1}-\mathbf{A}^j)\cdot\mathbf{1}=\mathbf{A}^{j-1}(\mathbf{I}-\mathbf{A})\cdot\mathbf{1} \tag{11}$$

$$E[\mathbf{X}]=\sum_{j=0}^{\infty}\mathbf{r}_j=(\sum_{j=0}^{\infty}\mathbf{A}_j)\cdot\mathbf{1}=(\mathbf{I}-\mathbf{A})^{-1}\cdot\mathbf{1} \tag{12}$$

The r_j,f_j and $E[X]$ are the first entries of $\mathbf{r}_j$, $\mathbf{f}_j$ and $E[\mathbf{X}]$, respectively. For example, when $k=2$ and $m=3$,

$$E(X)=\frac{1+2P-P^2}{P(1-(1-P)^2)}=\frac{2-Q^2}{P(1-Q^2)} \tag{13}$$

In general, for $k = 2$, it can be shown that

$$E(X) = \frac{2 - Q^{m-1}}{P(1 - Q^{m-1})} \tag{14}$$

As the values of k and m become larger, it gets unwieldy to derive a symbolic equation for mean recurrence time (or average run length). If the problem size is moderate, however, the method is useful for computational purposes.

2.2. Probability Generating Function

A more systematic way of obtaining average run length (or mean recurrence time) is the probability generating function method. The general background of generating function has been studied and developed by many authors such as Roberts [16], Greengberg [5] and Saperstein [17] and Nelson [14]. Their studies were focused on reduction of the complexity of the problem.

For given k and m, define

E	event that at least k out of consecutive m failure occurs
f_j	Prob$\{E$ occurs for the first time on trial $j\}$
r_j	Prob$\{E$ occurs after j-th trial $\}$
$r_{j,i}$	Prob$\{E$ does not occur on first j trials, trial $j-i$ was a failure, last i trials were successes $\}$
P	rejection probability on Normal Inspection
P	acceptance probability on Tightened Inspection

Let $F(z)$ and $R(z)$ be the probability generating function of f_j and r_j, respectively. From the fact that $f_j = r_{j-1} - r_j$, and f_j is the probability that the k-th failure in consecutive m trials occurs on trial j, we know that $1 - F(z) = (1-z)R(z)$, for $|z| < 1$, since

$$F(z) = \sum_{j=0}^{\infty} f_j z^j = \sum_{j=0}^{\infty} r_{j-1} z^j - \sum_{j=0}^{\infty} r_j z^j \tag{15}$$

$$= 1 + zR(z) - R(z), \text{ where } r_j = 1 \text{ if } j \leq 0 \tag{16}$$

$F(z)$ and $R(z)$ are obtainable using the following probabilities.

$$f_0 = f_1 = \cdots = f_{k-1} = 0$$

$$f_j = \binom{j-1}{k-1} Q^k P^{j-k}, \quad \text{for } j = k, k+1, \cdots, m \tag{17}$$

$$r_0 = r_1 = \cdots = r_{k-1} = 1$$

$$r_j = 1 - \sum_{i=k}^{j} \binom{j-1}{k-1} Q^k P^{j-k}, \quad \text{for } j = k, k+1, \cdots, m \tag{18}$$

$$r_{j,0} = P \cdot r_{j-1}$$

$$r_{j,k-1} = Q^{k-1} P^{m-k+1} \cdot r_{j-k}$$

$$r_j = r_{j,0} + r_{j,1} + \cdots + r_{j,k-1}, \quad \text{for } j \geq m \tag{19}$$

Since $R(z) = \sum_{j=0}^{\infty} r_j z^j$, the average run length (or mean recurrence time) can be given as $R(1) = \sum_{j=0}^{\infty} r_j$. For instance, when $m = 5$,

$$R(1) = \begin{cases} \dfrac{1}{1-P}, & k{=}1 \quad (20) \\ \dfrac{2-P^4}{Q(1-P^4)}, & k{=}2 \quad (21) \\ \dfrac{1+PQ(1+QP+QP^2)-Q^3P^2(1+P)^2+Q^3P^6}{Q(1-P^2-2QP^3+Q^2P^5+Q^3P^6)}, & k{=}3 \quad (22) \\ \dfrac{1+Q+2Q^2+2Q^3P-Q^4P-Q^4P^2-Q^6P^2-Q^6P^3}{Q^3(1+P-2P^2+QP^3+Q^3P^4)}, & k{=}4 \quad (23) \\ \dfrac{1-Q^5}{PQ^5}, & k{=}5 \quad (24) \end{cases}$$

When $k = 1$, the mean recurrence time $R(1)$ corresponds to the mean of the geometric distribution. It takes a rather simple form when $k = 1, 2$ and m. For instance, when $k = 2$ and for any m

$$E(X) = \frac{2 - Q^{m-1}}{P(1 - Q^{m-1})} \tag{25}$$

When $k = m$, the system is equivalent to the consecutive-k-out-of-n system with $n \to \infty$ and gives

$$E(X) = \frac{1 - Q^m}{PQ^m} \tag{26}$$

However, it becomes unwieldy, when $3 \leq k \leq m-1$ as shown in eqns (20)-(24). The probability generating function technique is not much different from Markovian approach, except for the fact that the r_j's and f_j's are obtainable via z-transform.

2.3. Combinatorial Approach

In the area of reliability, studies on consecutive systems have been done. The above procedure is equivalent to the k-out-of-m-from-n system. As a special case, if $k = m$, then the system reduces to consecutive k-out-of-n system. According to Kounias and Sfakianakis [26], a closed form solution is obtainable for some cases of k and m. For $k = 2$, notice that

$$r(2, m, n) = \sum_{j=0}^{[\frac{n+m-1}{m}]} \binom{n-(j-1)(m-1)}{j} Q^j P^{n-j} \quad \text{for } n = 1, 2, \cdots \tag{27}$$

$$r(2, m, n) = \begin{cases} P \cdot r(2, m, n-1) + QP^{m-1} \cdot r(2, m, n-m) & \text{for } n \geq 2m \\ P \cdot r(2, m, n-1) + QP^{m-1} \cdot r(2, m, n-m) & \text{for } n \leq 2m \end{cases}$$

For $n = m + \lambda$, $\lambda \leq m$,

$$r(k, m, n) = \sum_{i=1}^{k} r(i, \lambda, 2\lambda) \binom{m-\lambda}{k-i} P^{m-\lambda-k+i} Q^{k-i} \quad \text{where } r(\cdot) \equiv 1, \text{ if } i > \lambda. \tag{28}$$

For some other cases, very close upper and lower bounds are available. The consecutive k-out-of-n:F system has been studied by many authors. A recursive formula for consecutive k-out-of-n system is available in Zuo and Kuo [25]. In case of i.i.d probabilities, a closed form solution is available in Hwang [11]. The mean recurrence time is given as $E(X) = \sum_{x=1}^{\infty} x \cdot r(k, m, x)$.

3. Operating Characteristics

In connection with designing a switching system, three kinds or Type I and Type II errors are involved. The first kind is associated with the sampling plan of each lot. The second kind is characterized by the probabilities of misswitching to and from Tightened Inspection. The third kind is associated with discontinuation probability of inspection procedure. Among those, the errors of second kind are of most interest.

The probabilities that characterize a switching system are evaluated following the next steps. The following example is based on the switching rules when $k = 2$ and $m = 5$, just as is given in MIL-STD-105D.

3.1. Probability of Acceptance

The Type I and Type II errors of first kind can be calculated on the basis of these equations. For example,

(1) The process is at AQL.

$$1 - \text{Type I error} \equiv P_{N_1} \equiv 1 - \alpha_{N_1} = \sum_{i=0}^{c_N} f(i; s_N, p_1) \tag{29}$$

$$P_{T_1} \equiv 1 - \alpha_{T_1} = \sum_{i=0}^{c_T} f(i; s_T, p_1) \tag{30}$$

(2) The process is at RQL.

$$P_{N_2} \equiv \beta_{N_2} = \sum_{i=0}^{c_N} f(i; s_N, p_2) \tag{31}$$

$$\text{Type II error} \equiv P_{T_2} \equiv \beta_{T_2} = \sum_{i=0}^{c_T} f(i; s_T, p_2) \tag{32}$$

3.2. Average Run Length

In case of Normal Inspection, the formula for the average run length is represented using the rejection probability of a lot. Therefore, P is substituted by P_N

and Q by $1 - P_N$. Conversely, the probability concerned with Tightened Inspection is the acceptance probability of a lot. Therefore, P is substituted by $1 - P_N$ and Q by P_N. For example, using eqs (25) and (26),

(1) The process is at AQL.

$$E(X_1) \equiv \bar{N}_1 = \frac{2 - P_{N_1}^4}{(1 - P_{N_1})(1 - P_{N_1}^4)} \tag{33}$$

$$E(Y_1) \equiv \bar{T}_1 = \frac{1 - P_{T_1}^5}{(1 - P_{T_1})P_{T_1}^5} \tag{34}$$

(2) The process is at RQL.

$$E(X_2) \equiv \bar{N}_2 = \frac{2 - P_{N_2}^4}{(1 - P_{N_2})(1 - P_{N_2}^4)} \tag{35}$$

$$E(Y_2) \equiv \bar{T}_2 = \frac{1 - P_{T_2}^5}{(1 - P_{T_2})P_{T_2}^5} \tag{36}$$

3.3 True Signal and False Alarm

The probability of false alarming is the probability that the event concerned occurs while the process is at AQL. The false alarming is the Type I error of second kind. Type II error can be defined analogously. The probability of every alarming or signaling can be expressed as a reciprocal of each average run length.

(1) The process is at AQL.

$$\text{Type I error} \equiv \text{Prob}\{\text{misswitching to Tightened}\} = \frac{1}{\bar{N}_1} \tag{37}$$

$$\text{Prob}\{\text{returning from Tightened to Normal}\} = \frac{1}{\bar{T}_1} \tag{38}$$

(2) The process is at RQL.

$$\text{Type II error} \equiv \text{Prob}\{\text{misswitching to Normal}\} = \frac{1}{\bar{T}_2} \tag{39}$$

$$\text{Prob}\{\text{returning from Normal to Tightened}\} = \frac{1}{\bar{N}_2} \tag{40}$$

3.4. State Probability

The probability that the system is run on each inspection at a given percent defective can be evaluated using the equation (8). For example, from eqs (33)-(36),

(1) The process is at AQL.

$$\pi_{N_1} = \frac{\bar{N}_1}{\bar{N}_1 + \bar{T}_1} = \frac{P_T^5(1-P_T)(2-P_N^4)}{P_T^5(1-P_T)(2-P_N^4) + (1-P_N^4)(1-P_T^5)(1-P_N)} \tag{41}$$

$$\pi_{T_1} = \frac{\bar{T}_1}{\bar{N}_1 + \bar{T}_1} = \frac{(1-P_N^4)(1-P_T^5)(1-P_N)}{P_T^5(1-P_T)(2-P_N^4) + (1-P_N^4)(1-P_T^5)(1-P_N)} \tag{42}$$

(2) The process is at RQL.

$$\pi_{N_2} = \frac{\bar{N}_2}{\bar{N}_2 + \bar{T}_2} = \frac{P_T^5(1-P_T)(2-P_N^4)}{P_T^5(1-P_T)(2-P_N^4) + (1-P_N^4)(1-P_T^5)(1-P_N)} \tag{43}$$

$$\pi_{T_2} = \frac{\bar{T}_2}{\bar{N}_2 + \bar{T}_2} = \frac{(1-P_N^4)(1-P_T^5)(1-P_N)}{P_T^5(1-P_T)(2-P_N^4) + (1-P_N^4)(1-P_T^5)(1-P_N)} \tag{44}$$

This result compares to those in Stephens and Larson [24] and Dodge [2].

3.5. Composite OC Curve

From eq (3), a formula for a composite OC curve can be constructed as a weighted sum of acceptance probabilities on both Normal and Tightened Inspections.

$$P_C = \pi_N \cdot P_N + \pi_T \cdot P_T \tag{45}$$

$$= \frac{P_N P_T^5(1-P_T)(2-P_N^4) + P_T(1-P_N^4)(1-P_T^5)(1-P_N)}{P_T^5(1-P_T)(2-P_N^4) + (1-P_N^4)(1-P_T^5)(1-P_N)} \tag{46}$$

3.6. Average Outgoing Quality

The AOQ can be expressed as a weighted sum of AOQ's on each inspection.

$$AOQ = \pi_N P_N \cdot p \cdot \frac{N - s_N}{N} + \pi_T P_T \cdot p \cdot \frac{N - s_T}{N} \tag{47}$$

$$= (\pi_N \cdot P_N + \pi_T \cdot P_T) \cdot p \cdot \frac{N - s}{N} \quad \text{where } s = s_N = s_T \tag{48}$$

$$= P_C \cdot p \cdot \frac{N - s}{N} \tag{49}$$

3.7. Average Sample Number

The ASN can be expressed as a weighted sum of ASN's on each inspection.

$$ASN = s_N \cdot \pi_N + s_T \cdot \pi_T \quad (50)$$

$$= s \cdot (\pi_N + \pi_T) = s \quad \text{where } s = s_N = s_T \quad (51)$$

3.8. Average Total Inspection

The ATI can be expressed as a weighted sum of ATI's on each inspection.

$$ATI = \pi_N\{s_N + (N - s_N)(1 - P_N)\} + \pi_T\{s_T + (N - s_T)(1 - P_T)\} \quad (52)$$

$$= s + (N - s)[\pi_N(1 - P_N) + \pi_T(1 - P_T)] \quad \text{where } s = s_N = s_T \quad (53)$$

4. Effect of Switching Rules

When the MIL-STD-105 system was designed, it seemed to be that an emphasis was placed on simplicity of the rule. Pabst [15] reported that the original convention was adopted following Dodge [2] and in an effort to simplify switching procedures. The convention of shifting from Normal to Tightened Inspection when 2-out-of-5 consecutive lots or batches are rejected on original inspection was established in the earlier versions of MIL-STD-105. In case of MIL-STD-105C, Normal inspection is resumed when 3-out-of-5 consecutive lots are accepted while under tightened inspection. Later, this switching rule has changed to just "5 consecutive lots" in MIL-STD-105D on the ground of simplicity.

In every standard switching system, the same predetermined switching rules are employed, seemingly for simplicity. Since the switching rules can affect the performance of the sampling system, it would be worth trying to diversify the switching rules and examine the effect of various switching rules.

4.1. Various k Values

In studying sensitivity of switching rules with various k values, it is assumed that (1) the switching rule from normal to tightened is k out of 5 and (2) the switching rule from tightened to normal is 5 out of 5. In Figure 4, AOQ is made 10 times as big as it is for visibility. There is a tendency that the composite OC curve gets steeper as k takes on larger value and AOQL goes upward as in Figure 4. The increase of k seems not to be desirable due to the fact that AOQL worsens.

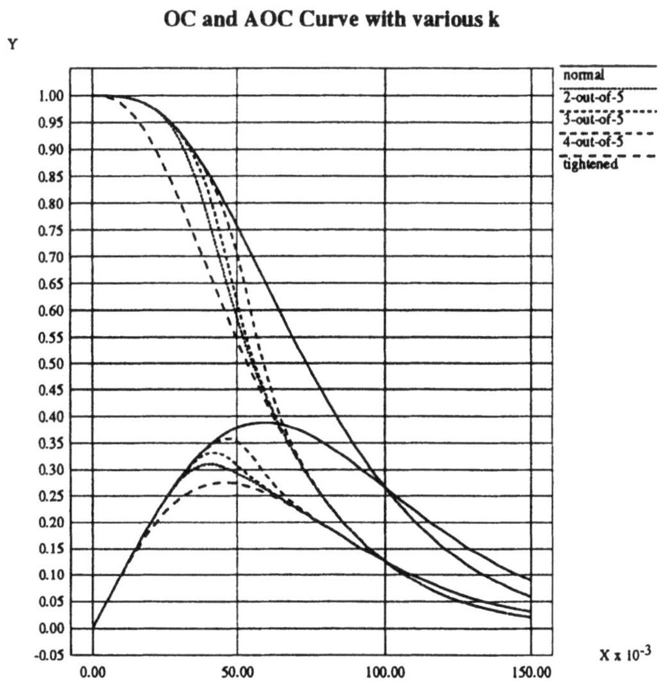

Figure 4: **Shift of Composite OC Curve with different k**

4.2. Various m Values

In this case, it is assumed that (1) the switching rule from normal to tightened is 2-out-of-m and (2) the switching rule from tightened to normal is m-out-of-m. Again, the real values of AOQ is one tenth of the values read from the graph. The composite OC curve shifts right as m gets smaller and hence less protective of consumer against low quality lots as shown in Figure 5.

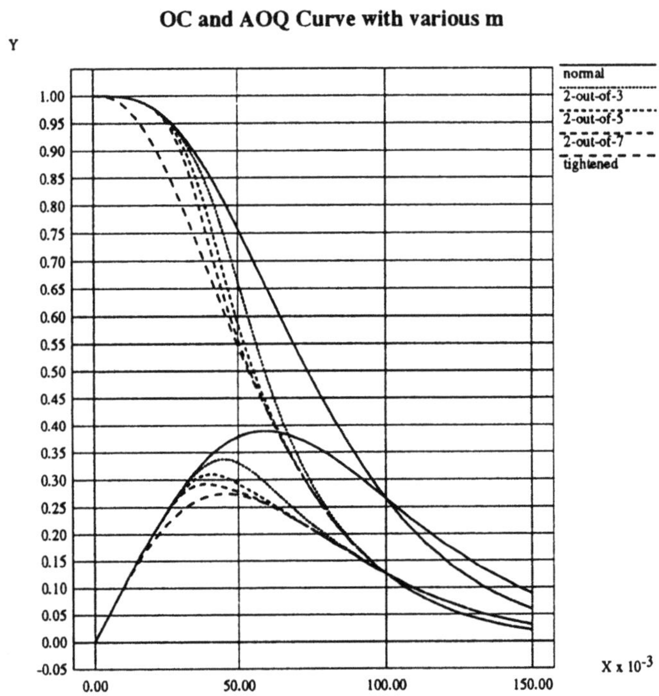

Figure 5: **Shift of Composite OC Curve with different** m

4.3. Various u and v Values

In this case, it is assumed that (1) the switching rule from normal to tightened is 2 out of $v+1$ and (2) the switching rule from tightened to normal is u out of u. If u lots in a row are accepted, switch to normal inspection. Continue at that level until a lot is rejected. If an additional lot is rejected in the next v lots, switch back to tightened inspection; otherwise, continue with normal inspection. From equation (25), (26) and (45), the composite acceptance probability can be given as

$$P_C = \frac{P_N P_T^u (1-P_T)(2-P_N^v) + P_T(1-P_N^v)(1-P_T^u)(1-P_N)}{P_T^u(1-P_T)(2-P_N^v) + (1-P_N^v)(1-P_T^u)(1-P_N)} \tag{54}$$

As seen in Figure 6, there is a tendency that the composite OC curve shifts right and gets steeper as u of switching rule takes on a larger value and the AOQL goes upward.

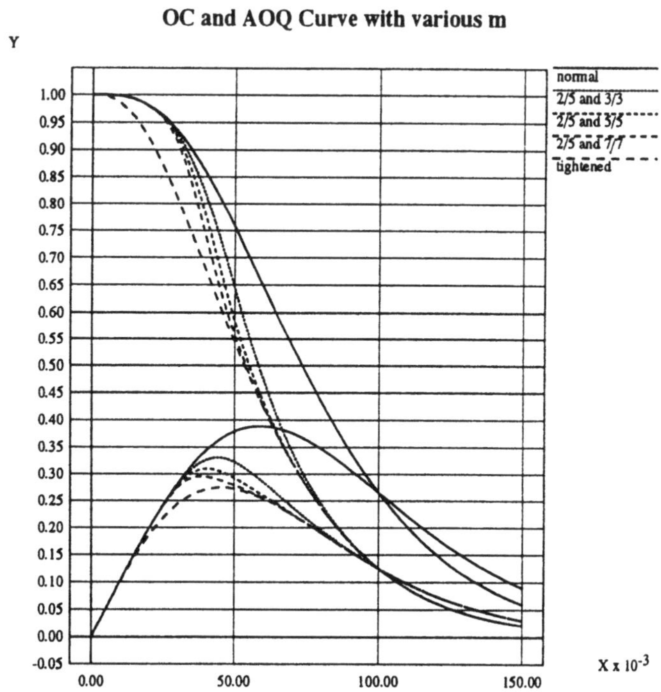

Figure 6: **Shift of Composite OC Curve with different u's and v's**

4.4. Selection of Switching Rules

As investigated so far, if we take advantage of the effect of switching rules, the composite OC curve would become more flexible. In selecting switching rules, the following criteria can and shall be considered.

1. The composite operating characteristic curve
2. The probability of being under tightened
3. The probability of shifting from normal to tightened
4. Average number of lots prior to shift
5. Average Total Inspection

5. Design of a Switching System

A customization of switching system is motivated by the fact that "ready-made" standards do not provide a flexible scheme, though it would be advantageous to make use of the basic concept of the switching in the lot-by-lot acceptance sampling system such as MIL-STD-105.

5.1. Sampling Plan with Stipulation

Numerous methods are available for determination of acceptance sampling plans that provide desired levels of protection for both producer and consumer. On the one hand, an analytic but complicated formulas to achieve the sampling plans was presented by Hald [9]. On the other hand, a simple computer program is presented by Hailey [9].

The problem can be formulated in such a way as to determine (n, c) so that $P_1 \equiv P(p_1) \geq 1 - \alpha$, $P_2 \equiv P(p_2) \leq \beta$, and c is as small as possible, where $p_1 < p_2$ and $1 - \alpha > \beta$ are as in Figure 8. In so doing, the minimum possible sample size can be determined.

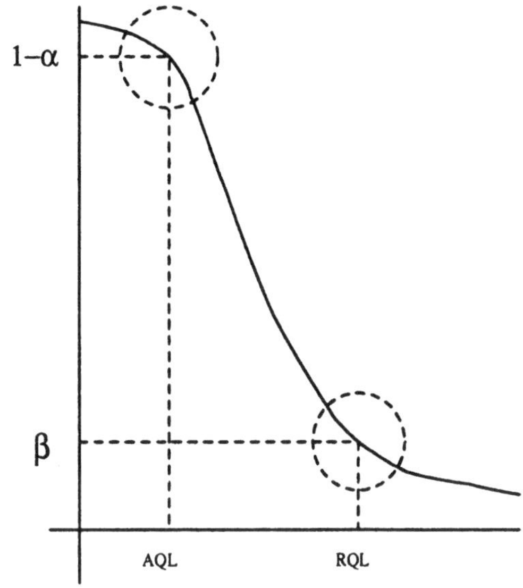

Figure 7: Sampling Plan with Stipulation

5.2. Sampling Scheme of Switching System

The switching system such as MIL-STD-105 is designed with a view to protecting the producer and thus based on the AQL concept. However, it can also protect the consumer by switching to Tightened Inspection, whenever necessary. In other words, the switching rule can add flexibility to the design of acceptance sampling system by designating a composite OC curve for stipulated producer's and consumer's risks.

The design of switching system is analogous to sampling plan with stipulation in Section 5. 1. The sampling scheme must provide, just as sampling plan, a probability of at least $1 - \alpha$ of accepting incoming lots if the proportion defective of the production process is at the AQL. The scheme must also provide a probability of acceptance of more that β if the proportion defective is at RQL.

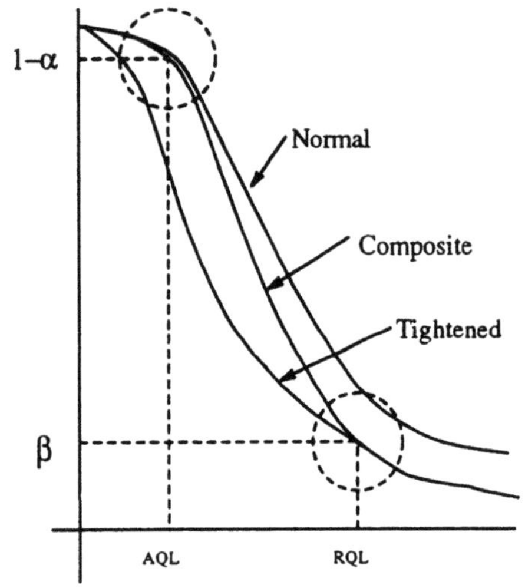

Figure 8: Design of Scheme based on Stipulation

The objective of design is to designate an effective sampling scheme, while the predetermined conditions are obeyed. The "Scheme performance" for stipulated producer's and consumer's risks are to be determined based on the following inequalities. See Fig. 8

$$1 - \alpha \leq B(c_N, s_N, p_1) \cdot \pi_{N_1} + B(c_T, s_T, p_1) \cdot \pi_{T_1} \tag{55}$$

$$\beta \geq B(c_N, s_N, p_2) \cdot \pi_{N_2} + B(c_T, s_T, p_2) \cdot \pi_{T_2} \tag{56}$$

where

$$B(c, s, p) = \sum_{i=0}^{c} f(i; s, p)$$

which is a decreasing function of s and an increasing function of c. Both π_N and π_T are function of $P_N \equiv B(c_N, s_N, p)$ and $P_T \equiv B(c_T, s_T, p)$. A necessary condition for a feasible scheme is $c_N \geq c_T$ and $s_N \leq s_T$.

5.3. Design Dilemma

Suppose consumer's and producer's risks are stipulated and switching rule is given. When it comes to designing a sampling plan or scheme, sample size and acceptance number (or critical value) are to be determined such that they are as minimized as possible. To solve the problem, these inequalities can be used recursively until the conditions of (26), (27) and $0 \leq c_T < c_N < s_N \leq s_T$ are met. As an illustration, suppose the following example is given

$$AQL = 0.02, \qquad P(p_1) = .98 \tag{57}$$

$$RQL = 0.08, \qquad P(p_2) = .10 \tag{58}$$

As seen in Figure 9, the sample size of designed scheme can be reduced drastically as compared to that of single sampling plan, while the stipulated conditions are not violated. This amazing result seems to be true at least theoretically.

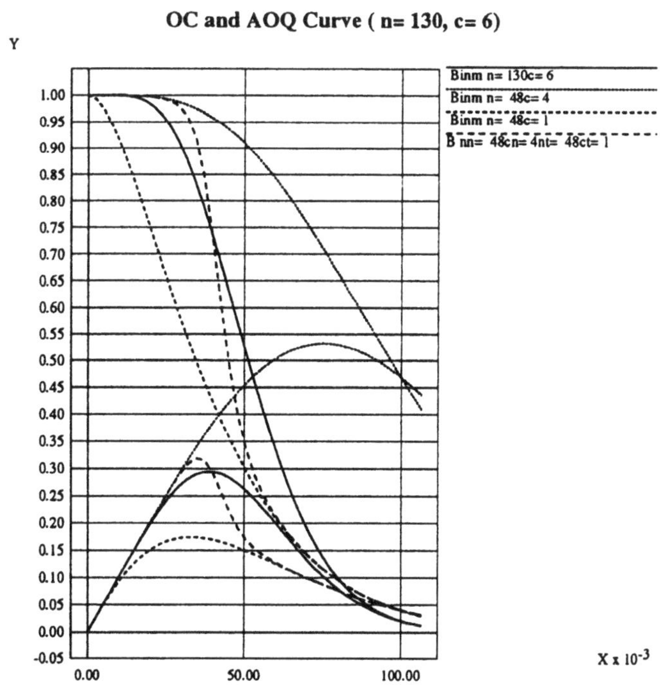

Figure 9: Design Dilemma

This type of design may not be quite practical and open to criticism. In order for the scheme performance to be realized,

(1) Number of incoming lots should be quite large. An incessant stream of incoming lots is desirable.

(2) Unless switching is mandatory, performance of this scheme may not be satisfactory. Failure of switching between Normal and Tightened may even be dangerous.

(3) AOQL is worse that that of single sampling plan.

5.4. General Strategy

In order to make the sampling scheme more practical, a general strategy is needed. The sampling strategy should be more specific than a single sampling plan.

(1) Considering the common abuse of no switching in practice of MIL-STD-105 [12], the smaller gap between OC curves of Normal and Tightened inspection would be preferred. It would be a reasonable strategy to make the acceptance probability on Tightened Inspection at AQL be no less than a predetermined level, say, 85% and/or P_{N_2} be no bigger than a specified level.

(2) In switching from Normal to Tightened Inspection, the sample size remains the same, but the acceptance number is reduced by a small number (1, 2, or 3). This general strategy is used throughout MIL-STD-105D for a transition to Tightened Inspection, which has a good reason.

(3) As pointed out by Cocca [1], minimum number of sample is not always the best policy. An optimum sample size might be obtainable via trade-off between average total sample number (ASN) and Average Total Inspection (ATI).

(4) Fixation of switching rule may cause to increase the AOQL, especially when the acceptance number is reduced. The AOQL gets higher, when k takes on large values. To offset this effect, it would be desirable to consider making the switching rule conservative. That is to say, bigger values of m and smaller values of k are

desirable. Figure 10 and Figure 11 demonstrates that AOQL can be as minimized as possible by employing conservative switching rules. As an illustration, the case of

$$AQL = 0.02, \qquad P(p_1) = .98 \tag{59}$$

$$RQL = 0.08, \qquad P(p_2) = .10 \tag{60}$$

$$P_{T_1} \geq 0.9, \qquad P_{N_2} \leq 0.5 \tag{61}$$

is under consideration. The result is shown as follows:

(i) Single Sampling Plan

$$s = 130, \quad c = 6$$

(ii) NTN Switching Sampling Scheme

$$S_N = 65, \quad C_N = 5, \quad S_T = 65, \quad C_T = 2$$

(5) The conservative switching rule may not be always desirable. It would make the probability that the procedure is run on Tightened Inspection bigger. In case of rectifying inspection of rejected lots, the ATI tend to increase.

6. Conclusion

In this study, we demonstrated that various probability measures that characterize a switching system is obtainable on the basis of reliability concept. Derivation of formulas for switching rules is closely related to the calculation of average run length on each inspection. The composite OC curve is the acceptance probability at various percent defective. The acceptance probability can be calculated as a weighted sum of average run lengths on both Normal and Tightened inspection. The composite OC curve can be diversified by varying average run lengths on each inspection. By diversifying the switching rule, the composite OC curve becomes more flexible. A customized design of switching system was attempted by taking advantage of the flexibility of the OC curve, in view of the fact that the stipulated designation with Type I and Type II errors provides fair ground for both producer and consumer. The conspicuous advantage of switching system lies in the fact that the sample size can be reduced drastically as compared with the sample size of single sampling plans. Inspite of the fact, the designation for minimum sample size would not be the best policy. Ironically, the designation for minimum sample size takes its toll. Those disadvantages are:

(1) The acceptance probability on Normal Inspection at RQL goes upwards.

(2) The acceptance probability on Tightened Inspection at AQL becomes small.

(3) The AOQL tends to move upwards and hence the AOQ deteriorates.

(4) The minimum sample size does not necessarily give the optimal ATI.

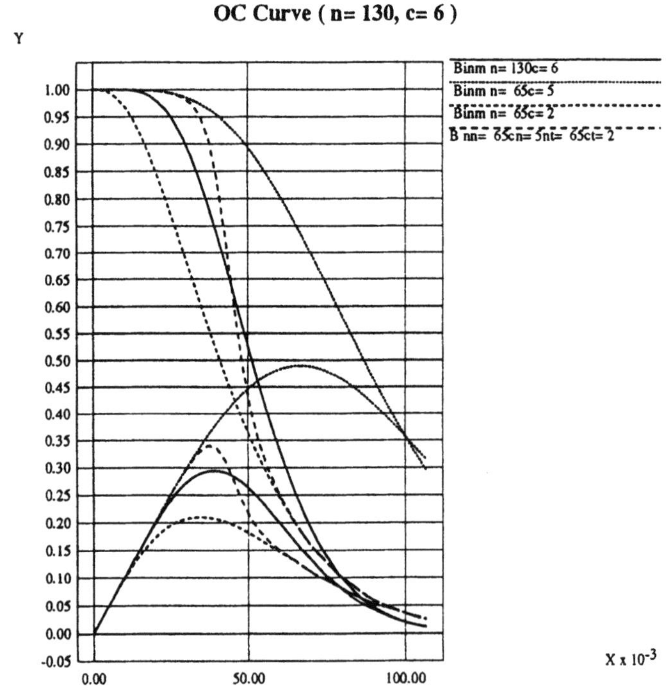

Figure 10: AOQ with switching rule of 2-out-of-5

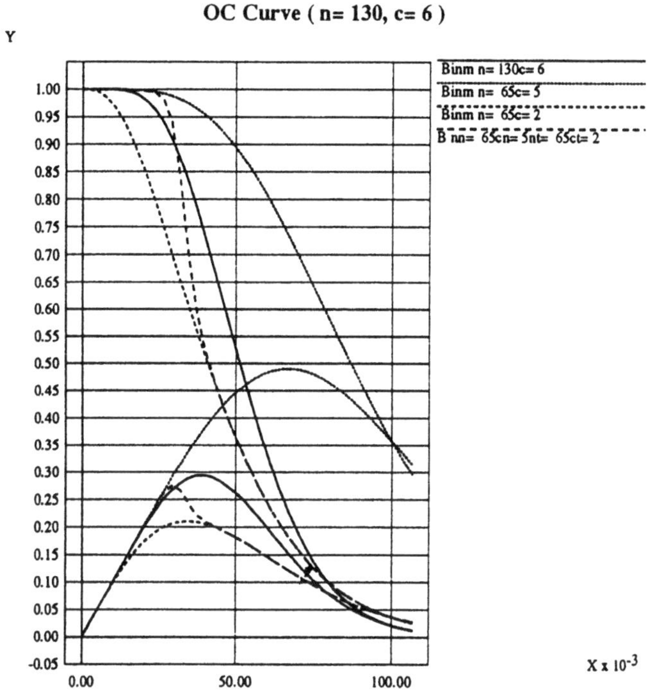

Figure 11: AOQ with switching rule of 2-out-of-9

The first and second disadvantages force us to specify additional conditions at both AQL and RQL on each inspection. The third can be offset to some extent by diversifying the switching rules. The switching rule of the standard is quite reasonable and its strategies are chosen such that the side effects from reducing sample size are moderate. However, in a customized design, predetermination of design factors may not be desirable. For example, it would be worth trying to design a sampling system which has different sample sizes on each inspection, especially when sample size is small and hence the acceptance number equals zero.

A design of a switching system is more like an art rather than a science. The optimum strategy could vary depending on the operational environment. What we learned from this study is that by diversifying the switching rules, we can get the most of customization. Although much calculation is involved in designing a switching system, the switching system based on stipulation provides more flexible sampling scheme than in sampling plans stipulated for AQL or RQL.

References

[1] Cocca, O. A., "Some Experiences Using MIL-STD-105 in the United States Air Force", *Naval Research Logistics Quarterly*, **32**, 1985, 11-16.

[2] Dodge, H. F., "Choosing Acceptance Sampling Plans", *Industrial Quality Control*, 1963, 7-12.

[3] Duncan, A. T., Mundel, A. B., Godfrey, A. B. and Patridge, V. A., "LQL indexed plan that are compatible with the structure of MIL-STD-105D", *J, Quality Technology*, **12**, No. 1, 1980, 40-46.

[4] Gibra, I. N., "Economic Design of Attribute Control Charts for Multiple Assignable Causes", *J. of Quality Technology*, **13**, No. 2, 1981, 93-99.

[5] Greenberg, I., "The First Occurrence of n Successes in N Trials", *IEEE Technometrics*, **12**, No. 3, 1970, 627-634.

[6] Griffith, W. S., "On Consecutive k-out-of-n Failure Systems and Their Generalizations", *Reliability and Quality Control*, A. P. Basu, ed., 1986, 157-165.

[7] Hahn, A. J. and Schilling, E. G., "An Introduction to the MIL-STD-105D Acceptance Sampling Scheme", *ASTM Standardization News*, **3**, No. 9, 1975, 20-30.

[8] Hailey, W. A., "Minimum Sample Size Single Sampling Plan: A Computerized Approach", *J. of Quality Technology*, **12**, No. 4, 1980, 230-235.

[9] Hald, A., "The Determination of Single Sampling Attribute Plans with Given Producer's and Consumer's Risk", *Technometrics*, **9**, No. 3, 1967, 401-415.

[10] Hill, I. D., "The Design of MIL-STD-105D Sampling Tables", *J. of Quality Technology*, **5**, No. 2, 1973, 80-83.

[11] Hwang, F. K., "Simplified reliabilities for consecutive-k-out-of-n: F system", *Siam. Alg. Disc. Math.*, **7**, 1986, 258-264.

[12] Kaplan, A. and MacDonald, E., "Instantaneous Switching Procedure for MIL-STD-105D", *J. of Quality Technology*, **1**, No. 3, 1969, 172-174.

[13] Keefe, G. J., "Attribute Sampling – MIL-STD-105", *Industrial Quality Control*, 1963, 7-12.

[14] Nelson. J. B., "Minimal-Order Models for False-Alarm Calculation on Sliding Windows", *IEEE Trans. on Aerospace and Electronics Systems*, **AES-14**, No. 2, 1978, 351-363.

[15] Pabst, W. R. Jr., "MIL-STD-105D", *Industrial Quality Control*, 1963, 4-9.

[16] Roberts, S. W., "Properties of Control Chart Zone Tests", *the Bell System Technical Journal*, 1958, 83-114.

[17] Saperstein, B., "On the Occurrence of n Successes Within N Bernoulli Trials", *Technometrics*, **15**, No. 4, 1973, 809-818.

[18] Schilling, E. G., "The Synergism of ANSI/ASQC Z1.9 and ANSI/ASQC Z1.4", *ASQC Quality Congress - Detroit*, 1982, 602-606.

[19] Schilling, E. G., "New ANSI Version of MIL-STD-414 and MIL-STD-105D", *Naval Research Logistics Quarterly*, **32**, 1985, 5-9.

[20] Schilling, E. G. and Sheesley, J. H., , "The Performance of MIL-STD-105D under the Switching Rules - Part 1", *Industrial Quality Control*, **10**, No. 2, 1978, 76-83.

[21] Schilling, E. G. and Sheesley, J. H., , "The Performance of MIL-STD-105D under the Switching Rules - Part 2", *Industrial Quality Control*, **10**, No. 3, 1978, pp.104-113.

[22] Schilling, E. G. and Sommers, D. J., "Two-point Optimal Narrow Limit Plans with Applications to MIL-STD-105D", *J. of Quality Technology*, **13**, No. 2, 1981, 85-92.

[23] Sfakianakis, M., Kounias, S. and Hillaris, A., "Consecutive k-out-of-r-from-n: F System", *IEEE Trans. Reliability*, **R-41**, 1992, 442-447.

[24] Stephens, K. S. and Larson, K. E., "An Evaluation of the MIL-STD-105D System of Sampling Plans", *Reliability and Quality Control*, 1967, 310-319.

[25] Zuo, M. and Kuo, W., "Design and Performance Analysis of Consecutive-k-out-of-n Structure", *Naval Research Logistics Quarterly*, **37**, 1990, 203-230.

[26] Kounias, S. and Sfakianakis, M., "The Reliability of linear System and its Connection with the Generalized Birthday Problem", *Statistica Applicata*, **3**, No. 4, 1991, 531-543.

EXACT AND APPROXIMATE HYPERCUBE RELIABILITIES

Anant P. Godbole
Michigan Technological University

Sanjay Khunger
Laboratory Equipment Corporation

and

Esther Ososanya
Tennessee Technological University

We consider the d-dimensional hypercube network, with a processor at each of its 2^d nodes, and study the network reliability when each of the $d \cdot 2^{d-1}$ links are assumed to be perfectly reliable. We consider several criteria for system failure and obtain exact expressions as well as Poisson approximations for the system reliability.

1. Introduction

The d-dimensional hypercube or d-cube is a highly concurrent multiprocessor architecture consisting of 2^d nodes (processing elements), one at each vertex of $\{0,1\}^d$, with each node connected to d neighbors. There are clearly $d \cdot 2^{d-1}$ (full duplex) connecting links. In large systems of this kind, it is obvious that random failures of the system components will occur. A characterization of the degraded system is therefore important to determine how many of the failures can be tolerated.

Notice that two nodes are connected iff their addresses differ in precisely one bit, and that a d-cube can be considered as being generated by connecting corresponding nodes of two $(d-1)$-dimensional cubes; see Saad and Schultz (1985) for an illustration of the latter fact when $d = 4$.

The hypercube architecture described above is important because it is capable of embedding several other architectures that are linear, ringlike, meshlike, starlike,

A. P. Godbole and S. G. Papastavridis (eds.), Runs and Patterns in Probability, 163–172.

treelike or pyramidlike in nature; consequently, our key results are easily extendible to such schemes as well. See Saad and Schultz (1985) for several more background facts on the hypercube structure and topology.

A d-cube can malfunction for several reasons; in particular, attention has focused in recent years on criteria involving some combination of node and link failures. Some authors (e.g., Yang, Wang and Lee (1987)) consider perfect nodes, with the system being operational iff the corresponding graph is connected. The advantage of this model is that the huge volume of literature on random graph theory (see e.g., Bollobás (1985)) can be invoked towards obtaining a solution; on the other hand, we will see that this model is often unrealistic. Malek and Yau, in Volume 1 of the 1988 Proceedings of the International Conference on Parallel Processing, consider a general model in which one has the following deterministic fault tolerance criterion, given an architecture graph G and a subgraph $H \subset G$: the system runs until a time t at which $G \setminus J$ does not contain a subgraph H' isomorphic to H, where J denotes the set of missing nodes and/or edges.

In Section 2, we present a basic model along the above lines; this model has been studied by Abrahams and Padmanabhan (1988), who consider the subcube reliability problem associated with the hypercube multiprocessor architecture. Following their work, we study the ability of the binary cube to embed disjoint (and fully functional) subcubes of given sizes into the damaged structure. As Abraham and Padmanabhan (1988) state, "Partitioning of a fault-free hypercube into subcubes that can be allocated to different tasks is a common practice, so that the degradation in this ability is a good measure of the effect of failures". All models presented in this paper account for node failures only. These show that the architecture is quite resilient to failures in terms of the ability to salvage functional subcubes out of a damaged hypercube.

A more specific description of the basic model of Section 2 is as follows: considering node failures only, the system functions until such a time as the faulty nodes are not contained in an i-dimension subcube ($1 \leq i \leq d-1$). In other words, the system is operational if there are $d-i$ disjoint and operational subcubes of order $d-1, d-2, \ldots, i$. A special case of importance that we focus on corresponds to $i = d-1$; in this case, the network is operational as long as there is an undamaged $(d-1)$ cube. Abrahams and Padmanabhan (1988) begin their analysis with a description of the infinitesimal generator of the process, and obtain a system of differential equations for $P_i(t)$:=the probability, at time t, of the system being in state S_i, which corresponds to the node failures being restricted to an i-cube (but not to an $(i-1)$-cube). An expression for $P_i(t)$ is obtained by induction, and $R_i(t)$, the system reliability under the above model, is obtained by *forward recursion.* In Section 2, we obtain an alternative proof of this result on using elementary combinatorial methods. An additional advantage of this technique is that (unlike the Markov chain approach) *backward recursion* can be employed to compute system reliabilities, mean times till failure (MTTF's), etc. if the fault containment criterion involves an i cube ($i \leq d-2$).

In Section 3, we consider a substantially different model, which corresponds to the case where very few node failures are tolerated: the system malfunctions if there are at least r completely inoperational s cubes ($r \geq 1, 0 \leq s \leq d-1$); the case $s = 1$ corresponds, for example, to system failure if there at least r adjacent node failures. We use the celebrated Stein-Chen method to get Poisson approximations for the system reliability, and explore possibilities for further research. Our method works equally well if the component reliabilities are different, and does not rely on the node decay model (exponential or otherwise). Finally, approximate values for the MTTF can be obtained, and limit theorems derived, as $d \to \infty$.

2. An Alternative Combinatorial Solution of the Subcube Reliability Problem

Let us denote by A_j ($1 \leq j \leq 2d$) the event that the jth $(d-1)$-cube is undamaged. We need to find $\mathbf{P}(A_1 \cup \cdots \cup A_{2d})$. By the principle of inclusion and exclusion,

$$\mathbf{P}(A_1 \cup \cdots \cup A_{2d}) = B_1 - B_2 + \cdots + B_{2d-1} - B_{2d} \tag{1}$$

where

$$B_j = \sum_{i_1 < i_2 < \cdots < i_j} \mathbf{P}(A_{i_1} \cap A_{i_2} \cap \cdots \cap A_{i_j}) \tag{2}$$

It can be shown, moreover, that all the 2^d nodes are exhausted by any choice of $(d+1)$ (or more) $(d-1)$-cubes. Thus, for $j \geq d+1$,

$$B_j = \binom{2d}{j} p^{2^d} \tag{3}$$

where p is the node reliability, assumed to be the same for each node. We need, therefore, to find B_j $(1 \leq j \leq d)$. It is evident that

$$B_1 = \sum_{i=1}^{2d} \mathbf{P}(A_i) = 2dp^{2^{d-1}} \tag{4}$$

Now, there are d ways of separating a d-cube into two non-overlapping $(d-1)$-cubes, incorporating a total of 2^d nodes. Also, the other $\binom{2d}{2} - d$ ways of arranging two $(d-1)$-subcubes in an overlapping fashion *each* incorporate a total of $2^{d-1} + 2^{d-1} - 2^{d-2} = 2^d - 2^{d-2}$ nodes. It follows that

$$B_2 = dp^{2^d} + \left(\binom{2d}{2} - d\right) p^{2^d - 2^{d-2}} \tag{5}$$

Actually (5) can be reexpressed, on using slightly different combinatorial reasoning, as

$$B_2 = \left(\binom{2d}{2} - 4\binom{d}{2}\right) p^{2^d} + 4\binom{d}{2} p^{2^d - 2^{d-2}} \tag{6}$$

In general, it is easy to check that the incorporation of $j \geq 2$ $(d-1)$-subcubes leads to the involvement of *either* 2^d *or* $2^d - 2^{d-j}$ nodes, and that

$$B_j = \left(\binom{2d}{j} - \binom{d}{j} 2^j\right) p^{2^d} + \binom{d}{j} 2^j p^{2^d - 2^{d-j}} \tag{7}$$

so that, on combining (1) through (7), we obtain

$$\mathbf{P}\left(\bigcup_{i=1}^{2d} A_i\right) = R_{d-1} = 2dp^{2^{d-1}} + \sum_{j=2}^{d}(-1)^{j+1}B_j + p^{2^d}\sum_{j=d+1}^{2d}(-1)^{j+1}\binom{2d}{j} \tag{8}$$

where B_j is given by (7) for $2 \leq j \leq d$. We have thus proved

Theorem 1. *Consider the d-dimensional hypercube interconnection network with 2^d nodes and $d2^{d-1}$ perfectly reliable links. Assume that the probability of any node being operational is p, and that the system is operational if there is at least one fully functional $(d-1)$-subcube. Then the system reliability R_{d-1} is given by (8).* □

If we now wish to calculate R_{d-i}, the system reliability when all the faulty nodes are to be embedded in an $(d-i)$-subcube, then we proceed as follows: Consider $i = 2$. We need to find $\mathbf{P}(C_1 \cup C_2 \cup \cdots \cup C_{2d})$ where C_j is the event that the jth $(d-1)$-cube is undamaged, *and* that the complementary $(d-1)$-cube has an undamaged $(d-2)$-cube embedded in it. As before,

$$R_{d-2} = \sum_{i=1}^{2d}\mathbf{P}(C_i) - \sum_{i_1<i_2}\mathbf{P}(C_{i_1} \cap C_{i_2}) + \cdots \tag{9}$$

Note now that by independence

$$\mathbf{P}(C_i) = p^{2^{d-1}}R_{d-1,d-2} \tag{10}$$

where $R_{d-1,d-2}$ is the probability, obtained by (8), that a $(d-1)$-cube has a functional $(d-2)$-cube embedded in it. Also, and more importantly,

$$\mathbf{P}(C_{i_1} \cap C_{i_2}) = \mathbf{P}(A_{i_1} \cap A_{i_2}) \tag{11}$$

where A_i is defined as before. Moreover, for any j,

$$\mathbf{P}(C_{i_1} \cap C_{i_2} \cap \cdots \cap C_{i_j}) = \mathbf{P}(A_{i_1} \cap A_{i_2} \cap \cdots \cap A_{i_j}) \tag{12}$$

It is now evident how one may use (9) through (12) set up the backward recursion equation to compute R_{d-2}. We omit the details of this calculation, and of further developments for $i \geq 3$. Table 1 below gives some values for the mean time to failure of the system given various failure rates λ.

Table 1

Dimension d	$N = 2^d$	MTTF(Hours) $\lambda = .0001$	MTTF(Hours) $\lambda = .00001$	MTTF(Hours) $\lambda = .000001$
2	4	9167	91667	916667
3	8	5179	51786	517857
4	16	2815	28155	281548
5	32	1498	14982	149818
6	64	787	7867	78669
7	128	409	4094	40944
8	256	220	2118	21176
9	512	109	1090	10901
10	1024	56	559	5591
11	2048	29	286	2861
12	4096	15	146	1460
13	8192	7	74	745
14	16384	4	37	376
15	32768	2	18	190
16	65536	1	9	92

3. Poisson Approximations for Another Subcube Reliability Model

In this section, we assume that our interconnection network will be rendered inoperational if it contains at least r *totally inoperational* s-subcubes, where $r \geq 1$ and $0 \leq s \leq d$ are arbitrary. This is clearly a very robust model that incorporates several possibilities. At the simplest level, when $s = 0$, we have an r-out-of-2^d system whose exact reliability is given by

$$\sum_{x=0}^{r-1} \binom{n}{x} q^x p^{n-x} \tag{13}$$

where $q := 1 - p$ and $n = 2^d$. If q is low, we may use the Poisson approximation

$$\sum_{x=0}^{r-1} \frac{e^{-nq}(nq)^x}{x!} \tag{14}$$

with total variation error at most $\min\{1, \frac{1}{nq}\}nq^2$. If the p's are unequal, (14) is replaced by

$$\sum_{x=0}^{r-1} \frac{e^{-Q}Q^x}{x!} \tag{15}$$

where

$$Q = \sum_{i=1}^{2^d} q_i. \tag{16}$$

In this case too, the total variation error incurred is at most

$$\min\left\{1, \frac{1}{\sum q_i}\right\} \cdot \sum q_i^2 \tag{17}$$

(see, e.g. Barbour et al (1992)); (17) is typically very small. At the next level of complexity, if $s = 1$, then we need to evaluate $\mathbf{P}(W = \sum_{j=1}^{d \cdot 2^{d-1}} I_j \leq r-1)$ where $I_j = 1$ if the jth 1-dimensional cube has both its nodes inoperational ($j = 1, 2, \ldots, d2^{d-1} = n$). Notice that

$$\mathbf{E}(I_j) = \mathbf{P}(I_j = 1) = q^2; \tag{18}$$

for a link model we would have had $\mathbf{E}(I_j) = q$. The approximate reliability is

$$\sum_{x=0}^{r-1} \frac{e^{-nq^2}(nq^2)^x}{x!} \tag{19}$$

where $n = d \cdot 2^{d-1}$, or

$$\sum_{x=0}^{r-1} \frac{e^{-Q}Q^x}{x!} \tag{20}$$

in the non-i.i.d. case, where $Q = \sum_{i=1}^{n} q_{i_1}q_{i_2}$, where the ith 1-cube has nodes $\{i_1, i_2\}$, with respective failure probabilities $\{q_{i_1}, q_{i_2}\}$. Now Corollary 2.C.5 in Barbour et al (1992) shows that the error incurred (in the i.i.d. case) is at most

$$\min\left\{1, \frac{1}{nq^2}\right\} d2^{d-1}(q^4 + 2(d-1)(q^3 + q^4)); \tag{21}$$

for example if $d = 2$, $r = 1$ and $q = .001$, (21) yields an approximation that is close to within .000000008.

We now consider the general case, starting with the following

Lemma 1. *The d-dimensional hypercube contains exactly*

$$N_{d,s} = \binom{d}{s} 2^{d-s} \tag{22}$$

subcubes of order s.

Proof. The following recursive formula is elementary:

$$N_{d,s} = 2N_{d-1,s} + N_{d-1,s-1} \tag{23}$$

(23) can be seen to be valid since a d-cube can be considered as being constructed by joining corresponding nodes of two $(d-1)$-cubes. It is now easy to check that (23) is satisfied by the function $N_{d,s} = \binom{d}{s}2^{d-s}$. □

We now set $I_j = 1$ if the jth s-cube is totally inoperational. In the i.i.d. case, $\mathbf{E}(I_j) = \mathbf{P}(I_j = 1) = q^{2^s}$, so that the approximating Poisson r.v. W has mean $\lambda = \binom{d}{s}2^{d-s}q^{2^s} = n \cdot q^{2^s}$, say. By Corollary 2.C.5 in Barbour et al (1992),

$$d_{TV}(\mathcal{L}(W), \mathrm{Po}(\lambda))$$
$$\leq \min\left(1, \frac{1}{\lambda}\right) \sum_{j=1}^{n} \left\{ q^{2^{s+1}} + \sum_{l=0}^{s-1} \binom{s}{l} 2^{s-l} \binom{d-s}{s-l} \left[q^{2^{s+1}} + q^{2^{s+1}-2^l} \right] \right\} \tag{24}$$

where d_{TV} denotes the usual total variation distance and $\mathrm{Po}(\lambda)$ the Poisson r.v. with mean λ. (24) is valid since

(i) 2 s-cubes can only intersect in an l-*cube*, where $0 \leq l \leq s-1$;

(ii) There are $\binom{s}{l}2^{s-l}\binom{d-s}{s-l}$ s-cubes that overlap a given s-cube S in an l cube. [To see this, notice that the s-cube S has $\binom{s}{l}2^{s-l}$ l-subcubes (by Lemma 1), each of which is the intersection of S with any one of $\binom{d-s}{s-l}$ s-cubes];

and

(iii) There are $2^s + 2^s - 2^l = 2^{s+1} - 2^l \geq 3 \cdot 2^{s-1}$ nodes involved in the total failure of both these s-cubes. It is easy to check now that (24) can be bounded above (somewhat crudely, but quite adequately) by

$$d_{TV}(\mathcal{L}(W), \mathrm{Po}(\lambda)) \leq$$
$$\min\left\{1, \frac{1}{\binom{d}{s}2^{d-s}q^{2^s}}\right\}\binom{d}{s}2^{d-s}\left[(\binom{d}{s}2^{d-s}+1)q^{2^{s+1}} + \binom{d}{s}2^{d-s}q^{3\cdot 2^{s-1}}\right] \quad (25)$$

For example, if $d = 10$, $s = 4$ and $q = .001$, the error bound given by (25) is at most 1.8×10^{-64}, with the approximate reliability being of order $e^{-1.3\times 10^{-44}}$ or higher. We have proved

Theorem 2. *Consider the d-dimensional hypercube interconnection network with perfectly reliable links, and for which each node fails independently with probability q. If there is system failure at the time when at least r s-dimensional cubes are totally inoperational, then the system reliability R satisfies*

$$\left|R - \sum_{j=0}^{r-1}\frac{e^{-\lambda}\lambda^j}{j!}\right| \leq$$
$$\min\left\{1, \frac{1}{\binom{d}{s}2^{d-s}q^{2^s}}\right\}\binom{d}{s}2^{d-s}\left[(\binom{d}{s}2^{d-s}+1)q^{2^{s+1}} + \binom{d}{s}2^{d-s}q^{3\cdot 2^{s-1}}\right] \quad (26)$$

where $\lambda = \binom{d}{s}2^{d-s}q^{2^s}$.

We conclude this paper by mentioning some possible avenues for further research. Firstly, we remark that the Stein-Chen method is a powerful technique that is capable of handling many more Poisson approximations, in the context of various other failure criteria. These might include criteria that combine those of Abrahams and Padmanabhan (1988) with the ones of Section 3, or which might be quite different: The cube might be inoperational if, e.g., there at at least r paths, each with a Hamming length of s or more, that are rendered useless. The general model of Malek and Yau can be considered, as can Gaussian approximations for various other models.

Acknowledgement. The authors would like to thank Chuck Cusack for finding an error in a previous version of Theorem 2, and Chuck Cusack and Qian Ye for useful discussions.

References

Abrahams, S. and K. Padmanabhan (1988). Reliability of the hypercube, in *Proceedings of the 1988 International Conference on Parallel Processing*, 90-94.

Barbour, A., L. Holst and S. Janson (1992). *Poisson Approximation.* Oxford University Press.

Bollobás, B. (1985). *Random Graphs.* Academic, New York.

Saad, Y. and M. Schultz (1985). Topological properties of hypercubes. Research Report YALEU/DCS/RR-389, Yale University.

Yang, C., J. Wang and J. Lee (1987). Reliability properties of the hypercube network. *Comput. Math. Applic.* **14**, 541-548.

RUNS AND PATTERNS WITH APPLICATIONS TO RELIABILITY

William S. Griffith

University of Kentucky

Runs and patterns are of interest in a number of areas of probability and statistics, including reliability theory. We shall examine one specialized model in reliability called the consecutive k-out-of-n failure system. Many different approaches to the analysis of such systems have been proposed. In this paper, we review the Markov chain approach of Griffith and Govindarajulu and compare it to earlier approaches. We then survey some of the extensions that can be made of this work.

1. Introduction

A consecutive k-out-of-n failure system is a system of components numbered 1 through n which fails if and only if at least k consecutively numbered components fail. Such systems arise in many different settings including telecommunications and computer networks. Since the early 1980's, a great deal of work on such systems has been published. Many different aspects and characteristics of such systems have been studied under various assumptions regarding the random variables describing the performance of the components. Early work dealt with independent and identically distributed components and considered procedures for finding the reliability of such systems. Later work undertook other aspects of system performance with more complicated assumptions.

There are two approaches which can be made in the independent and identically distributed case. These approaches can be compared to the binomial and negative binomial distributions. For example, the binomial probability of at least k successes in n independent Bernoulli trials with common success parameter p is the same as the negative binomial probability of having to wait no

A. P. Godbole and S. G. Papastavridis (eds.), Runs and Patterns in Probability, 173–181.

longer than n trials for the kth success. Likewise, when analyzing the reliability of a consecutive k-out-of-n failure system, we can view the reliability as the probability that the random number of runs of at least k consecutive failures in n independent Bernoulli trials is zero. Alternatively, we can view the reliability as the probability that the waiting time until the kth consecutive failure is greater than n. Authors prior to Griffith and Govindarajulu (1985) and Griffith (1986) adopted the former approach. Griffith and Govindarajulu took the latter approach and used as their fundamental tool, the notion of a Markov chain. We will now survey their development and review some of their extensions.

2. The Markov Chain Approach

We begin with an infinite sequence of independent Bernoulli random variables $\{Y_i\}_{i=1}^{\infty}$. We let $\{Z_i\}_{i=0}^{\infty}$ be the current length of the run of 0's at the conclusion of the ith trial. It can be shown that $\{Z_i\}$ is a Markov chain whose stationary transition probability matrix has nonzero entries of the form

$$\begin{aligned} P_{i0} &= p \\ P_{i,i+1} &= 1-p \text{ for all } i \,. \end{aligned}$$

If we then consider the first passage probabilities

$$f_{0k}^{m} = P(Z_m = k, Z_l \neq k, l = 1, 2, \ldots, m-1 | Z_0 = 0)\,,$$

we can write the reliability of our consecutive k-out-of-n failure system as

$$h(p) = 1 - \sum_{m=k}^{n} f_{0k}^{m}\,. \tag{1}$$

These first passage probabilities can be computed recursively. It is not difficult to show that

$$\begin{aligned} f_{0k}^{k} &= (1-p)^k \qquad (2) \\ f_{0k}^{m} = p(1-p)^k & \qquad \text{for } k+1 \leq m \leq 2k\,. \end{aligned}$$

and

$$f_{0k}^{m+1} = f_{0k}^{m} - f_{0k}^{k+1} f_{0k}^{m-k} \text{ if } m \geq 2k\,. \tag{3}$$

Thus, equations (1), (2), and (3) can be used to produce the system reliability.

Next, suppose one wanted to determine the component reliability that would be necessary to obtain a specified level of system reliability. One would want to be able to find the system reliability function as a function of component reliability. An equivalent problem would be to find the functional form of the system unreliability function as a function of the component unreliability. Let us denote this function by $h(q)$. This function can easily be shown to be a polynomial

$$\tilde{h}(q) = \sum_{j=k}^{n} A_j q^j\,.$$

We can show that the polynomial coefficients can be obtained by differentiation yielding

$$A_j = \frac{1}{j!} \left. \frac{d^j \tilde{h}(q)}{dq^j} \right|_{q=0} = \frac{1}{j!} \sum_{m=k}^{n} \left. \frac{d^j f_{0k}^{m}}{dq^j} \right|_{q=0}.$$

The algorithm of Section 2 for the first passage probabilities can be converted to an algorithm for the derivatives of these first passage probabilities at $q = 0$. The details can be found in Griffith and Govindarajulu. The results are fairly simple and the computation of the polynomial coefficients yield nice results.

3. Systems Requiring Multiple Runs of Consecutive Failures

In this section we will consider the first of three generalizations of consecutive k-out-of-n failure systems. In particular, we will consider systems of n independent and identically distributed components which fail if and only if there are m non-overlapping runs of k consecutive failures. We will treat a run of length lk as l runs of length k. We will adapt the Markov chain approach of Griffith and Govindarajulu to this problem.

For the sake of definiteness, we will initially consider the system requiring two runs of length k. The first passage probabilities described earlier can be used to compute the reliability of this system as shown in the following proposition.

Proposition 3.1 *For a system of n independent and identically distributed components which fails if and only if there are at least 2 runs of at least k consecutive failures the reliability may be expressed as*

$$h(p) = 1 - \sum_{i=k}^{n-k} \sum_{j=k}^{n-i} f_{0k}^{i} f_{0k}^{j} \, .$$

Proof. We again consider an infinite sequence of Bernoulli trials, a Markov chain $\{Z_i\}$ which again keeps track of the length of the current run of 0's at the end of the jth trial, and the first passage probabilities f_{0k}^{m}. If we let W_1 and W_2 be the discrete random variables describing the waiting time for the first run of k consecutive 0's and the waiting between the first and second run, respectively, then it is easy to see that W_1 and W_2 are independent and identically distributed with

$$P(W_1 = m) = P(W_2 = m) = f_{0k}^{m}$$

To analyze the system which requires 2 runs of k consecutive 0's, we observe that its reliability is

$$h(p) = 1 - P(W_1 + W_2 \leq n) \, .$$

This can be simply expressed as a convolution of distributions as

$$h(p) = 1 - \sum_{i=k}^{n-k} \sum_{j=k}^{n-i} f_{0k}^{i} f_{0k}^{j}$$

where the f_{0k}^{m}'s can be determined from the recursive relations described in section 2.

We can extend this recursive approach in a straightforward manner to the problem of computing system reliability when m runs of k consecutive failures cause system failure (of course, $mk < n$). We let W_1 be the waiting time for the first run of k consecutive 0's in our sequence of Bernoulli trials and we let W_j be the waiting time between the $(j-1)^{st}$ and j^{th} runs of k consecutive 0's.

Proposition 3.2 *The reliability of a system requiring m runs of k consecutive failures can be expressed as*

$$h(p) = 1 - P\left(\sum_{i=1}^{m} W_i \leq n\right).$$

This probability can be formulated as the convolution of the distributions of $W_1, \ldots, W_m$ and computed in terms of the first passage probabilities. It is a simple task to write a program to compute this reliability. Exact formulas were obtained by Godbole (1990) and Papastavridis (1990).

4. Consecutive k-out-of-n Failure Systems With Dependent Components

In this section we will consider consecutive k-out-of-n failure system in which components are dependent in a special sense. Our model will be that the system works by essentially having the components perform in sequence without having k components in sequence fail. We assume that the probability that a component works if there is a run of j failures immediately preceding it by p_j. If a string of failures increases the likelihood that a subsequent component will fail, then the p_j's will be decreasing.

Now to analyze the reliability of our system in this case we again consider a Markov chain $\{Z_i\}$ where Z_i can be thought of as having the interpretation of the length of the current run of consecutive failures as we have successively moved through components 1 to i. The one step transition probability matrix P for this Markov chain has nonzero entries of the form

$$\begin{aligned} P_{i,i+1} &= 1 - p_i \\ P_{i,0} &= p_i \qquad \text{for all } i \end{aligned}$$

Proposition 4.1 *The system reliability can be expressed in terms of first passage* *probabilities as*

$$h = 1 - \sum_{m=k}^{n} f_{0k}^{m} .$$

Proof. The proof is analogous to the approach of Section 2. We can obtain a recursive formula for these first passage probabilities. It is easy to verify, for example,

$$f_{0k}^{m} = P_{0k}^{m} - \sum_{j=k}^{m-1} P_{kk}^{m-j} f_{0k}^{j} \text{ for } m > k ,$$

and

$$f_{0k}^{k} = \Pi_{j=0}^{k-1}(1 - p_j) \text{ and } f_{0k}^{k+1} = p_0 \Pi_{j=0}^{k-1}(1 - p_j) .$$

This relationship can be used to generate successive first passage probabilities recursively in terms of the preceding first passage probabilities and entries in various powers of the matrix P.

5. k-Within-m-Out-Of-n Failure Systems

The last extension we will consider is the case of systems of n independent and identically distributed components which fail if and only if within any m consecutive components there are at least k failures. We shall do this using a Markov chain approach in which taboo probabilities are considered (see e.g., Chung (1967)). Chao and Lin (1983) have used taboo probabilities in their computation of the reliability of consecutive 2-out-of-n failure systems of i.i.d. components. For ease of illustration, we will consider the special case of $k = 2$ and $m = 3$. We will then comment on the straightforward extension of this to the more general case.

We begin by letting X_i be a sequence of Bernoulli trials (representing the state of our components). We will define a sequence $\{Z_j\}$ where $Z_j = (X_j, X_{j+1}, X_{j+2})$. It is not difficult to verify that $\{Z_j\}$ is a Markov chain with state space $\{(1,1,1), (1,1,0), (1,0,1), (0,1,1), (1,0,0), (0,1,0), (0,0,1), (0,0,0)\}$. We shall denote the 8 vector states by the integers 0 through 7 in the order in which they have just been displayed. We will lump the states 4,5,6, and 7 into one set T called the taboo set. Define transition probabilities among the non-taboo states by

$$ {}_TP_{ij}^{(r-1)} = P(Z_r = j, Z_l \notin T, l = 2, \ldots, r-1 | Z_1 = i) . $$

We obtain the following analogue of the standard Chapman-Kolmogorov equations:

$$ {}_TP_{ij}^{(r)} = \sum_{k \notin T} P_{ik} \; {}_TP_{kj}^{(r-1)} \text{ for } i, j \notin T . $$

If we let ${}_TP$ denote the sub-stochastic matrix describing the one-step transitions among the non-taboo states formed from the one-step transition probability matrix by deleting the rows and columns corresponding to taboo states, this relation can be shown by induction to be

$$ {}_TP^{(r)} =_T P^r $$

We summarize this development in the following proposition.

Proposition 5.1 *The reliability of the 2-within-3-out-of-n failure system can be expressed as*

$$ h(p) = \sum_{j \notin T} \sum_{i \notin T} {}_TP_{ij}^{(n-3)} P(Z_1 = i) . \tag{4} $$

The component pieces in this computation are the probabilities $P(Z_1 = i)$ for $i \notin T$ and the entries of the $(n-3)^{rd}$ power of ${}_TP$. It is easy to verify that

$$ P(Z_1 = 0) = p^3 $$

and

$$ P(Z_1 = 1) = P(Z_1 = 2) = P(Z_1 = 3) = p^2(1-p) . $$

The substochastic matrix T^p is

$$ T^p = \left\{ \begin{array}{cccc} p & 1-p & 0 & 0 \\ 0 & 0 & p & 0 \\ 0 & 0 & 0 & p \\ p & 1-p & 0 & 0 \end{array} \right\} . $$

We can now raise ${}_TP$ to the $(n-3)^{rd}$ power and compute the expression given in Proposition (5.1).

This approach can be adapted in a straightforward way for the analysis of the general k-within-m-out-of-n failure systems. There are now 2^m vectors of 0's and 1's describing the possible states of $Z_j = (X_j, X_{j+1}, \ldots, X_{j+m})$. Of these $\sum_{i=k}^{m} \binom{m}{i}$ are taboo states. These will now form the set T and ${}_TP$ is again the substochastic matrix formed by deleting the rows and columns associated with the taboo states from the one-step transition probability matrix of $\{Z_i\}$. The reliability can then be expressed in the form of equation (4) and the relevant quantities computed. See Papastavridis and Sfakianakis (1991) for a closer look at the importance of the components of a k-within-m-out-of-n failure system; see also Saperstein (1973, 1975) for generating functions and exact formulas for the reliability.

References

BARLOW, R. E. and PROSCHAN, F., Statistical Theory of Reliability and Life Testing: Probability Models, 2nd Printing (To Begin With, Silver Spring MD, 1981).

BOLLINGER, R. C. and SALVIA, A. A., Consecutive k-out-of-n: F networks, *IEEE Transactions on Reliability*, **R-31** (1982) 53-56.

BOLLINGER, R. C., Direct computation for consecutive-k-out-of-n: F systems, *IEEE Transactions on Reliability*, **R-31** (1982) 444-446.

CHAO, M. T. and LIN, G. D., Economical design of large consecutive k-out-of-n failure systems, Technical Report, Institute of Statistics, Academia Sinica, Taipai, Taiwan.

CHIANG, D. T. and NIU, S. C., Reliability of consecutive k-out-of-n: F system, *IEEE Transactions on Reliability*, **R-30** (1981) 87-89.

CHUNG, K. L, Markov Chains with Stationary Transition Probabilities, 2nd edition (Springer-Verlag, New York, 1967).

DERMAN, C., LIEBERMAN, G. J. and ROSS, S. M., On the consecutive k-out-of-n: F system, *IEEE Transactions on Reliability*, **R-31** (1982) 57-63.

GODBOLE, A. P., Specific formulæ for some success run distributions, *Statistics and Probability Letters*, **10** (1990), 119-124.

GRIFFITH, W. S., On consecutive k-out-of-n failure systems and their generalizations, *Reliability and Quality Control*, A. P. Basu, ed., Elsevier Science Publishers (1986) 157-165.

GRIFFITH, W. S. and GOVINDARAJULU, Z., Consecutive k-out-of-n failure systems: reliability and availability analysis with some multistate extensions, *American Journal of Mathematical and Management Sciences*, **5** Nos. 1-2, (1985), 125-160.

PAPASTAVRIDIS, S., m-consecutive k-out-of-n: F systems, *IEEE Transactions on Reliability*, **R-39** (1990) 386-388.

PAPASTAVRIDIS, S. and SFAKIANAKIS, M. E., Optimal arrangement and importance of components in a consecutive k-out-of-r-from-n: F system, *IEEE Transactions on Reliability*, **R-40** (1991) 277-279.

SAPERSTEIN, B., On the occurrence of n successes within N Bernoulli trials, *Technometrics*, **15** (1973) 809-818.

SAPERSTEIN, B., Note on a clustering problem, *Journal of Applied Probability*, **12** (1975) 629-632.

Acknowledgement

The work was partially supported by NSF Grant RII-8610671 and the Commonwealth of Kentucky through the Kentucky EPSCoR Program.

ROBUST PROCEDURES FOR DETECTING NON-RANDOM PATTERNS

Peter C. O'Brien

Section of Biostatistics, Mayo Clinic

Marilyn A. Agin

University of Minnesota

1. Introduction.

In tests for detecting non-random patterns, the order of occurrence of a sequence of observations is assumed to be random, and a departure from randomness is often indicative of a cycle or trend associated with the order. Violation of the initial assumptions may lead to rejection of a hypothesis of randomness and may also provide an indication of the reason for the underlying pattern. Tests for randomness are often specific to the type of pattern being considered, and frequently the application itself provides the information that indicates which test(s) should be used (Agin (1990)).

In a test for randomness, the data are assumed to be at least nominal; for our purposes each event is a Bernoulli trial. How one defines the two types of outcomes is crucial to interpreting what rejection of the null hypothesis means. Suppose M = man, W = woman, and a line in which people are buying tickets for a movie looks like M W M W M W... . This does not appear to be random, but it is up to us to interpret the reason for the pattern. As another example, much research has been done on the importance of the age at which a person contracts leukemia (Larson, Holmes, and Heath (1973), Knox (1964)). If we define S as "a person has leukemia" and F as "a person does not have leukemia," collect some data and arrange the observations according to the underlying characteristic of age, we may end up with a sequence that looks like $SSSFFFFSSS$. A rejection of a null hypothesis could

A. P. Godbole and S. G. Papastavridis (eds.), Runs and Patterns in Probability, 183–203.

lead us to conclude that it is more likely for younger or older people to contract leukemia than it is for the middle-aged. This is an example of the runs test applied to a two-sample problem (Wald and Wolfowitz (1940)). We might also conclude that $p = P(S)$ had varied from trial to trial, and that this variation produced the pattern; age may have caused the variation. In a different situation, suppose that we have a sequence of wet and dry days and we wish to test whether this sequence is random. An underlying Markovian dependency may exist between the two types of weather and a rejection of a random pattern in the weather may lead us to conclude that some type of sequential dependency exists.

Other examples include an oil pipeline with n pump stations, a telecommunication system with n relay stations (Chiang and Niu (1981)), a mobile communication system with n phototransmitting spacecraft (Chiang and Chiang (1986)), and demyelination in human and experimental neuropathy (O'Brien and Dyck (1985), Dyck et al. (1984)).

The generality of runs tests makes them easy to apply and data are usually plentiful enough to make them fairly powerful– the more data we have, the better our chances of rejecting a null hypothesis of randomness when it should be rejected (Bradley, p. 250). By a careful definition of our two types of events, we can then interpret what a rejection of randomness means.

To obtain some measure of the empirical power of various tests based on runs, ranks, and variances, simulations were performed in which sequences representative of a particular pattern were generated, and the ability of the test to detect that pattern was evaluated (O'Brien (1967), O'Brien and Dyck (1985), Agin (1990), Agin and Godbole (1992)). In this paper, we discuss the derivation of several of these tests, and give a brief historical overview of other tests for randomness.

2. A Bernoulli Process.

We consider a sequence of Bernoulli trials producing outcomes $X_1, \ldots, X_n$. The null hypothesis is that the trials are independent and identically distributed. Letting S and F represent the possible outcomes of a given trial, the alternative hypothesis is that the arrangement of F's and S's is non-random.

A relatively simple and commonly used test is the ordinary runs test (Wald and Wolfowitz (1940), and Eisenhart and Swed (1943)). Although this test is easy to apply and reasonably sensitive, power for certain alternative hypotheses (Goodman (1958)) can be improved with little loss of convenience or intuitive appeal, as will be discussed below. Power of the runs test is of particular concern when failures are rare, so that it becomes improbable that two failures will occur together. Following are several rank-based procedures for detecting trend or, more generally, unimodal clustering.

Larsen, Holmes, and Heath (1973) derived a procedure for detecting clustering where the "center" of the cluster is determined by the location of the successes.

Suppose that m successes in our sequence occur at trial numbers

$$1 \leq Y_1 \leq Y_2 \leq \ldots \leq Y_m \leq n.$$

Let $m = 2r + 1$ for m odd and $m = 2r$ for m even, and define

$$K = \sum_{i=1}^{m} |Y_i - Y_{r+1}|.$$

Small values of K will be taken as evidence of unimodal clustering, and an approximate test is based on Z, an asymptotically normal version of K. Let Z observed be

$$Z^* = \frac{K - E(K)}{\sigma_K}.$$

Let $\delta(m) = r - 2$ *for* m *odd and*

$\delta(m) = 2r - 1$ *for* m *even.*

If the m successes are distributed randomly with respect to the n trials, then

$$E(k) = \frac{(n+1)\left[\frac{m}{2}\right]\left[\frac{m+1}{2}\right]}{(m+1)}$$

$$\sigma_K^2 = \frac{r(n+1)(n-m)((m+1)^2 - 2r^2 - \delta(m))}{12\{2\left[\frac{m+1}{2}\right]+1\}^2}$$

where $[x]$ is the greatest integer in X. We will reject the null hypothesis in favor of unimodal clustering if $P(Z \leq Z^*) \leq \alpha$.

Larson, Holmes, and Heath advise that this test will not distinguish multiple clustering from randomness, and that it should only be used when unimodal clustering is the alternative hypothesis.

Haldane and Smith (1948) assumed that in a number of hereditary diseases and abnormalities, the probability of a person's contracting a disease depends on his or her birth-rank. They designed a rank-sum test based on the sum of the birth-ranks of affected cases, to try to determine if early or late births are more likely to be affected. They define a success as an affected sibling.

$$k = n = \text{ number of siblings examined}$$

$$h = \text{ number of successes}$$

$$a_i = \text{ birth} - \text{rank of } i\text{th affected sibling}$$

$$A = \sum_{i=1}^{h} a_i = \textit{ sum of the birth} - \textit{ranks of the affected siblings}$$

$$x = \frac{\left[A + \frac{1}{2} - \frac{1}{2}h(k+1)\right]}{\sqrt{(k+1)(k-h)}} + \frac{1}{2}h.$$

If A represents the sum of h independent variables each of which has a rectangular distribution, then we can find an approximation of a standardized version of A by using the test statistic

$$Z_x = \frac{x - \frac{1}{2}h}{\sqrt{\frac{h}{12}}}$$

and comparing the values to a standard normal distribution.

This test can be used to detect clustering at the beginning or end of a sequence by rejecting the null hypothesis if

$$P(Z_x \geq |Z_x^*|) \leq \frac{\alpha}{2}.$$

2.1. Runs of Varying Length.

2.1.1. One-Sample Test.

In searching for a robust test, one desires a test to be sensitive to alternatives involving multiple clusters of S's. We consider first the situation where m, the number of S's in the observed sequence, is large relative to n, the number of F's. Let Y_i represent the number of successes occurring prior to the ith failure but subsequent to the $i-1$th failure for $i = 2, \ldots, n$. Y_1 is defined as the number of successes preceding the first failure and Y_{n+1} as the number following the nth failure. We define

$$\overline{Y} = \sum_{i=1}^{n+1} \frac{Y_i}{n+1} = \frac{m}{n+1}$$

$$S^2 = \sum_{i=1}^{n+1} \frac{(Y_i - \overline{Y})^2}{n}.$$

The hypothesis of randomness is rejected for either very small or very large values of S^2, small values indicating a systematic arrangement and large values indicating clustering.

The proposed criterion is conditional on the values of m and n observed by the experimenter. (Since the error rate for type 1 error is controlled for all values of m and n, the error rate will be controlled unconditionally as well.)

In order to obtain the conditional distribution of S^2 under the hypothesis of randomness, we shall suppose that the arrangement of F's and S's has been generated by taking two independent samples $(S_1, \ldots, S_m)$ and $(F_1, \ldots, F_n)$ of size

m and n, repectively, from a standard uniform (0, 1) distribution and arranging these in ascending order of magnitude. Denote the ordered values of $F_1, \ldots, F_n$ by $F_{(1)} < \ldots < F_{(n)}$. Define

$$p_1 = F_{(1)}$$

$$p_i = F_{(i)} - F_{(i-1)},\ i = 2, \ldots, n$$

$$p_{n+1} = 1 - F_{(n)}.$$

If we take $F_1, \ldots, F_n$ as fixed and define $F_0 = 0, F_{n+1} = 1$, then for each observation in the sample $(S_1, \ldots, S_m)$, the probability of falling in the interval (F_{i-1}, F_i) is given by p_i $(i = 1, \ldots, n+1)$. Thus, the conditional density function for $(Y_1, \ldots, Y_{n+1})$ given n and $p_1, \ldots, p_{n+1}$ is given by the multinomial distribution:

$$P(y_1 \ldots, y_{n+1}|n, p_1, \ldots, p_{n+1}) = \frac{m!}{y_1! \ldots y_{n+1}!} p_1^{y_1} p_2^{y_2} \ldots p_{n+1}^{y_{n+1}}$$

and the conditional density for $(Y_1, \ldots Y_{n+1};\ p_1, \ldots, p_{n+1})$ given n is

$$P(y, p) = \frac{n!m!}{y_1! \ldots y_{n+1}!} p_1^{y_1} p_2^{y_2} \ldots p_{n+1}^{y_{n+1}}.$$

The latter distribution has been studied by Dixon (1940), who has shown that for values of m and n greater then 10, cS^2 has very nearly a chi-square distribution with v degrees of freedom (d.f.), where

$$c = \frac{n(n+2)(n+3)(n+4)}{2(m-1)(m+n+2)(n+1)} \sim \frac{n^3}{2m(m+n)},$$

and

$$v = \frac{cm(n+m+1)}{(n+2)(n+1)} \sim \frac{n}{2}.$$

(For values of m and n less than or equal to 10, Dixon has tabled the 90, 95, and 99 percentiles of ns^2/m^2.)

We next show that the true and fitted distributions converge in distribution to a normal distribution. The proof consists of showing that 1) the conditional distribution of S^2 given n and m is asymptotically identical to the distribution of the sample variance of n independent geometric random variables given their sample mean, and 2) in general, for random variables which are independently and identically distributed with first four moments finite, the conditional distribution of the sample variance given the sample mean converges to a normal distribution. The desired result then follows from the fact that the fitted chi-square distribution also converges to a normal distribution.

Lemma A: *The conditional distribution of S^2 (given n and m) is asymptotically identical to the distribution of the sample variance of n independent geometric random variables given their sample mean.*

Proof:

One may treat S^2 as having been generated by an infinite sequence of Bernoulli trials truncated at some point after nth failure but before the $n+1$th failure. Consider such a sequence of Bernoulli trials and let Z_i represent the number of trials after the $i-1$th failure but prior to the ith failure for $i = 1, 2, \ldots$.

Define

$$\bar{Z}_n = \sum_{i=1}^{n} \frac{Z_i}{n}$$

$$s_n^2 = \sum_{i=1}^{n} \frac{(Z_i - \bar{Z}_n)^2}{(n-1)}, \quad n = 1, 2, \ldots.$$

The desired result follows from noting that, asymptotically, $\bar{Z}_{n+1} \sim \bar{Z}_n$ *and* $s_{n+1}^2 \sim s_n^2$.

Lemma B: *Let $\{X_i\}_{i=1}^{\infty}$ be a sequence of independent and identically distributed random variables with first four moments finite. For $n = 1, 2, \ldots,$ define*

$$\bar{X}_n = \sum_{i=1}^{n} \frac{X_i}{n}$$

$$s_n^2 = \sum_{i=1}^{n} \frac{(X_i - \bar{X}_n)^2}{(n-1)}$$

$$E_n = E(s_n^2 | \bar{X}_n)$$

$$V_n = \mathrm{Var}(s_n^2 | \bar{X}_n)$$

$$Z_n = \frac{s_n^2 - E_n}{\sqrt{V_n}}$$

Then the conditional distribution of Z_n given $\bar{X}_n$ tends to the standard normal distribution as $n \to \infty$.

Proof: The following additional notation will facilitate the proof of the lemma:

$$Y_{1n} = \sqrt{n}(\bar{X}_x - E(X_1))$$

$$Y_{2n} = \frac{\left(\sum_{i=1}^{n} X_i^2 - nE(X_1^2)\right)}{\sqrt{n}}$$

$f_{Y_1}(y_1) =$ *the limiting probability density function (pdf) of* Y_{1n}.

$f_{Y_1,Y_2}(y_1, y_2) =$ *the limiting joint pdf of* (Y_{1n}, Y_{2n}).

$f_{Y_2|Y_1}(y_2, y_1) =$ *the limiting pdf of* Y_{2n} *conditional on* Y_{1n}.

To prove the lemma we note first the identity

$$f_{Y_2|Y_1}(y_2, y_1) = \frac{f_{Y_1,Y_2}(y_1, y_2)}{f_{Y_1}(y_1)}$$

From the central limit theorem for multivariate random variables (as stated in Feller (1966), p. 254), the numerator is a bivariate normal distribution. That $f_{Y_2|Y_1}(y_2, y_1)$ is a normal distribution follows immediately from the fact that, in general, if any two random variables have a bivariate normal distribution, then the conditional distribution of one given the other is a normal distribution (see Anderson (1958), p. 29).

2.1.2. Two-Sample Test.

The procedure described above assumes that failures are uncommon and therefore considers only the length of success runs. If two failures should occur consecutively, a success run of length 0 is defined to have occurred between the two failures. We next consider the situation where both successes and failures occur with sufficient frequency to warrant considering both types of runs. We return to the conventional definition of a run in which runs of length 0 cannot occur. For example, in considering the sequence $S-S-S-S-F-F-F-F-S-S-S-S$, we had previously viewed this as evidence of clustering of S's, with S run lengths of 4, 0, 0, 0, and 4. We now view this as evidence of a systematic arrangement with runs of length 4, 4, 4. In order to distinguish between these two types of procedures, we will refer to them as one sample (successes) and two sample (successes and failures) procedures.

The statement of the two-sample procedure requires the following notation. For successes:

$$n_S = \text{Number of successes},$$

$$r_S = \text{Number of success runs},$$

$$S_S^2 = \text{Sample variance of success run lengths},$$

$$C_S = \frac{(r^2-1)(r+2)(r+3)}{2r(n-r-1)(n+1)},$$

$$U_S = \frac{C_S n(n-r)}{r(r+1)}$$

The notation for failures is defined in similar fashion. The test statistic is

$$\chi_f^2 = C_S S_S^2 + C_F S_F^2$$

where $f = U_S + U_F$. We will reject H_0 in favor of clustering if χ_f^2 exceeds the $1-\alpha$ percentile of the chi-square distribution with f degrees of freedom (df). (In practice, it is necessary either to round off or to interpolate between tabled integer values of f.)

As alluded to previously, the two sample test is intended for situations in which both successes and failures are common. For the test to even be defined, it is required that $r_S, r_F, n_S - r_S$, and $n_F - r_F$ all exceed one, and this assumption is implicit throughout.

In order to establish the null distribution of the test statistic, it suffices to show that, conditional on n_s, n_f, r_s, and r_F, $C_S^2 S_S$ and $C_F^2 S_F$ are independent and asymptotically chi-square. Consider first the conditional distribution of S_S^2. Construct a new sequence of successes and failures as follows: (i) omit all outcomes up to and including the first success, (ii) omit all failures (if any) occurring after the last success, and (iii) replace each remaining failure run and subsequent success with a single failure. In this new sequence, let n_F^* represent the number of failures; Y_i, the number of successes occurring after the $(i-1)$th failure but prior to the ith failure $(i = 2, \ldots, n_f^*)$; Y_1, the number of successes preceding the first failure; and $Y_{n_F^*+1}$, the number of successes following the last failure. Notice that in the new sequence, $n_F^* = r_S - 1$ and the number of successes $(n_S^* = \sum Y_i)$ is given by $n_S^* = n_S - r_S$.

Let $\tilde{S}_S^2$ represent the sample variance of the Y values. Since the rule for obtaining the Y values had the effect of reducing the length of each success run by 1, the sample variance of run lengths in the new sequence is identically the same as in the old. That is, $\tilde{S}_S^2$ is identically equal to S_S^2, so that it is sufficient to consider the distribution of $\tilde{S}_S^2$.

To facilitate the correspondence between the notation here and that used by Dixon (1940) we note that, when expressed in Dixon's notation, S_S^2 has expectation

$$\frac{m(m+n+1)}{(n+1)(n+2)}$$

and variance

$$\frac{4m(m-1)(m+n+1)(m+n+2)}{n(n+2)^2(n+3)(n+4)}.$$

These expressions become, respectively,

$$E_S = \frac{(n_S - r_S)n_S}{r_S(r_S+1)}$$

and

$$V_S = \frac{4(n_S - r_S)(n_S - r_S - 1)n_S(n_S+1)}{(r_S-1)(r_S+1)^2(r_S+2)(r_S+3)}$$

when $n_S^* = n_S - r_S$ is substituted for Dixon's m and $n_F^* = r_S - 1$ is substituted for Dixon's n. (There are analogous expressions for E_F and V_F.)

Following the same approach as for the one-sample test, we have the conditional distribution of $Y_1, \ldots, Y_{n_F^*+1}$ given $\sum Y$:

$$f_Y = \int \frac{(\sum Y_i)!}{\prod Y_i!} (\prod U_I^{y_I}) \prod du_i,$$

with integration over the generalized tetrahedron defined by $u_i > 0,\ i = 1, \ldots, n_F^*$, and $\sum u_i \leq 1$, and (i) $C_S \tilde{S}_S^2$ has approximately a chi-square distribution, (ii) the approximation is quite accurate in small samples (r_s and $n_S - r_S \geq 10$), and (iii) the approximation is exact asymptotically. The same results are obtained for failure run lengths (denoted by X) using a completely analogous argument.

It should be noted that the same expression for f_Y obtains if we also condition on the individual failure run lengths, since the F's omitted in obtaining the new sequence do not enter the calculation for f_Y. Stated algebraically, $f_{Y|X} \equiv \frac{f_{YX}}{f_X} = f_Y$, so that $f_{YX} = f_X f_Y$. Thus, conditional on r_S, r_F, n_S, and n_F, the failure run lengths and success run lengths are statistically independent.

In order to assure accurate control over the size of the test, $C_S S_S^2$ and $C_F S_F^2$ have been weighted equally, thus obtaining a chi-square statistic with $V_S + V_F$ degrees freedom. Conversely, unequal weighting would result in the summation of variates having approximate gamma distributions but with unequal scale parameters, resulting in a sum which may not be well approximated by a gamma distribution.

It would be of interest to determine weights which would maximize power. It would appear that optimal weights would depend on the specific alternatives to randomness envisaged. However, we next show that equal weights are fully efficient within a large class of alternatives.

In order to evaluate the efficiency of the proposed procedure, we first establish the following results.

Theorem. *Let $X_{ij}(i = 1,2;\ j = 1,\ldots,n)$ represent a sequence of mutually independent random variables with variance σ_i^2 and distribution function F_i. To test the null hypothesis $H_0 : \sigma_i^2 = \sigma_{i0}^2$ against the alternative $H_A : \sigma_i^2 = \lambda\sigma_{i0}^2 (i = 1,2)$, consider the family of conditional tests defined by*

$$R = \{(W_1, W_2, C) : Pr(W_1 S_1^2 + W_2 S_2^2 > C | H_0, \bar{X}_1, \bar{X}_2) = \alpha\}.$$

Let r_0 represent the member of R for which

$$(W_1, W_2, C) = (\frac{n-1}{\sigma_{10}^2}, \frac{n-1}{\sigma_{20}^2}, C_{0n}),$$

where C_{0n} is chosen to maintain the size of the test.

1. If F_1 and F_2 are normal distributions, then r_0 is a uniformly most powerful (UMP) conditional test.

2. If F_1 and F_2 have finite fourth moments and r' is any other member of R, then the (Bahadur and Pitman) asymptotic relative efficiency (ARE) of r_0 relative to r' is greater than or equal to 1. Furthermore, $\lim_{n\to\infty} \frac{C_n}{\chi^2_{2n-2}} = 1$.

The proof of the first part of the theorem consists of noting that if the $\{X_{ij}\}$ are normally distributed, then the conditional distribution given the sample means is proportional to $\exp\left[\frac{-(n-1)\sum s_i^2}{(2\sigma_i^2)}\right]$. Thus, by direct application of the Neyman-Pearson lemma, the UMP test criterion is given by

$$\frac{L_0}{L_1} \propto \exp\left\{-n(n-1)\left[\sum\left(\frac{s_i^2}{2\sigma_{i0}^2}\right) - \sum\left(\frac{s_i^2}{2\lambda\sigma_{I0}^2}\right)\right]\right\}$$

$$\ln\left(\frac{L_0}{L_1}\right) \propto \left(\frac{S_1^2}{\sigma_{10}^2} = \frac{S_2^2}{\sigma_{20}^2}\right).$$

To establish the second part of the theorem we have previously shown that the conditions imposed on F_1 and F_2 are sufficient to ensure that S_1^2 and S_2^2 obey the central limit theorem. Thus, the limiting distribution of the test criterion for any member of R is independent of the underlying distributions F_1 and F_2. It follows that the first conclusion of the theorem holds asymptotically for any F_1 and F_2 with finite fourth moments, so that no other member of R can have greater asymptotic efficiency than r_0.

Corollary. *Consider conditional tests of the hypothesis of randomness as defined previously, and let σ_{SO}^2 and σ_{FO}^2 represent the null variance of run lengths for successes and failures, respectively. For alternatives in which $\sigma_{SA}^2 = \lambda\sigma_{SO}^2$ and $\sigma_{FA}^2 = \lambda\sigma_{FO}^2$, the proposed two-sample runs test has* ARE ≥ 1 *relative to any other linear combination of sample variances.*

Note first that asymptotically the run lengths $\{X_{ij}\}$ follow independent geometric distributions under H_0. Asymptotically, therefore, since the n_s successes comprise r_s runs, the expected run length is given by $1/p_S = n_S/r_S$ and the variance of the run lengths is $\frac{(1-p_S)}{p_S^2} = (n_S - r_S)\frac{n_S}{r_S^2}$. If we let $\sigma_{10}^2 = (n_S - r_S)\frac{n_S}{r_S^2}$ *and* $\sigma_{20}^2 = (n_F - r_F)\frac{n_F}{r_F^2}$, then the corresponding optimal criterion in the theorem is given by $W_S S_S^2 + W_F S_S^2$, where W_S *and* W_F are any constants such that

$$\begin{aligned}\frac{W_S}{W_F} &= \frac{\sigma_{20}^2}{\sigma_{10}^2} \\ &\approx (n_F - r_F)\frac{n_F}{(n_S - r_S)n_S}\end{aligned}$$

$$\approx \frac{C_S}{C_F}$$

The approximations are exact asymptotically and consist only of noting that r_S and r_F can differ at most by 1 and their ratio equals 1 asymptotically.

2.2. Tests based on the Binomial Distribution of Order k.

We next consider tests where our interest lies in the number of recurrent success runs of a specified length $k \geq 1$. We define a success run as a recurrent pattern with the initial situation reestablished after each occurrence of the pattern, as in Feller (1968, p. 303). We determine that a sequence of n letters S and F has as many recurrent success runs of length k as there are non-overlapping, uninterrupted blocks containing exactly k letters each. Under our null hypothesis that the Bernoulli trials are independent and identically distributed, we let $N_{n,k}$ denote the number of recurrent success runs of length $k \geq 1$. The exact probability distribution of $N_{n,k}$ was first derived independently by Hirano (1986) and by Philippou and Makri (1986) who called it the binomial distribution of order k. Let $[\cdot]$ denote the greatest integer function. Then

$$P(N_{n,k} = x) = \sum_{y=[\frac{n-kx}{k}]}^{n-kx} q^y p^{n-y} \binom{y+x}{x} \sum_{j=0}^{[\frac{n-kx-y}{k}]} (-1)^j \binom{y+1}{j} \binom{n-kx-jk}{y}$$

for $x = 0, 1, 2, \ldots, [\frac{n}{k}]$, and k a constant. The above simplified formula was derived by Godbole (1990a) and Papastavridis (1990), while Griffith (1986) published a recursive version.

2.2.1. Test based on the Conditional Distribution of Order k.

Let S_n be the total number of successes in the sequence. Then

$$P(N_{n,k} = x) = \sum_{y=[\frac{n-kx}{k}]}^{n-kx} P(N_n^k = x, S_n = n - y)$$

and by conditioning on the observed number of successes we obtain (see Agin (1990)),

$$P(N_{n,k} = x | S_n = n - y)$$
$$= \frac{\binom{y+x}{x} \sum_{j=0}^{[\frac{n-kx-y}{k}]} (-1)^j \binom{y+1}{j} \binom{n-kx-jk}{y}}{\binom{n}{y}}$$

for $x = 0, 1, 2, \ldots, [\frac{n-y}{k}]$.

This conditional distribution allows us to find the probability of observing any number of success runs of fixed but arbitrary length k in a sequence of Bernoulli trials, and it uses the same information as the tests described above.

2.2.2. Length of the Longest Run.

In the 1940's, interest in the theory of runs was quite high and began with Mood's theory of runs (1940), and Wald and Wolfowitz's application of the theory of runs to test whether two samples were from the same population (1940). In 1941, Mosteller derived a formula for the probability of getting at least one run of a given length or greater when the number of elements of each of the two kinds was the same and applied this to quality control and the occurrence of long runs above or below the median. Grant (1946) applied the concept of a longest run of at least m successes to education; he concluded that a subject had learned a concept if the subject responded correctly to at least m questions in a row. David (1947), Bateman (1948), Burr and Cane (1961), and others continued this research. Most of the interest was in obtaining a run of length k or greater; for present purposes let us consider the exact length k of the longest run of successes.

Let us define L_n as the exact length of the longest run of successes in a sequence of n Bernoulli trials, where runs are defined as for the ordinary runs test. In this context, the longest success run as defined by Feller is also the longest run of successes as defined above. Adapting Godbole's formula for $P(L_n \leq k-1, S_n = n-y)$ by conditioning on S_n, we have

$$P(L_n \leq k-1|S_n = n-y) = \frac{P(L_n \leq k-1, S_n = n-y)}{P(S_n = n-y)}$$

$$= \frac{q^y p^{n-y} \sum_{j=0}^{[\frac{n-y}{k}]} (-1)^j \binom{y+1}{j}\binom{n-jk}{y}}{\binom{n}{y} p^{(n-y)} q^y}$$

$$= \frac{\sum_{j=0}^{[\frac{n-y}{k}]} (-1)^j \binom{y+1}{j}\binom{n-jk}{y}}{\binom{n}{y}}$$

Let L_n^* be the length of the longest observed run of successes.

We could reject a null hypothesis of randomness if $P(L_n \geq L_n^*) \leq \alpha$, indicating an unusually "long" longest run of successes, or if $P(L_n \leq L_n^*) \leq \alpha$, indicating a longest run that is too "short."

With an alternative hypothesis of clustering, we are looking for the occurrence of one or more runs of unusual length, so we will use an upper tail test given by

$$P(L_n \geq L_n^*) = 1 - P(L_n \leq L_n^* - 1|S_n = n-y)$$

and reject the null hypothesis when

$$P(L_n \geq L_n^*) \leq \alpha.$$

This test makes use of the length of the longest run but does not use information about the total number of runs, the position in the sequence of the longest run, or the fact that there may be more than one run of successes of length L_n.

3. A Poisson Process.

The continuous analogue to the problem we have been considering is testing that the waiting time to occurrence of an event follows a Poisson process; that is, that the individual waiting times are independent with a common exponential distribution. Many procedures have been proposed for testing this hypothesis (see for example Cox (1955) and Bartholomew (1956)). The analogue of the one-sample procedure described previously is to consider the coefficient of variation (CV), which equals unity for an exponential distribution. The hypothesis of randomness could be rejected for small values of CV (indicative of events occurring systematically) or for large values of CV (indicative of clustering). This approach appears to have been first proposed by Moran (1951) who noted that the distribution of CV is exceedingly complex and converges only slowly to a normal distribution. However, in view of the analogy to the single sample test for a Bernoulli process, one might suppose that the distribution could be accurately approximated using a chi-square distribution.

We first approximate the distribution of CV_2 by fitting the two parameter gamma distribution

$$f(x) = \frac{x^{p-1}e^{\frac{-x}{\theta}}}{\Gamma(p)\theta^p}.$$

Moran has evaluated the moments of CV^2. The expectation is given by

$$E(CV^2) = n(n+1),$$

and the variance by

$$\text{Var}\,(CV^2) = 4n^4/(n+3)(n+2)(n+1)(n^2-1).$$

(It is of passing interest to note here that in evaluating $E(CV^2)$ we are considering a case where the expected value of a ratio equals the ratio of the expectations of the numerator and denominator.)

Thus, to fit the gamma distribution, one sets

$$p = (n+3)(n+2)(n-1)4n^2$$

$$\theta = 4n^3/(n+3)(n+2)(n^2-1).$$

Equivalently, $k\ CV^2$ has approximately a chi-square distribution with v d.f., where

$$k = (n+3)(n+2)(n^2-1)/2n^3 \sim \frac{n}{2}$$

$$v = (n+3)(n+2)(n-1)/2n^2 \sim \frac{n}{2}.$$

Moran has shown that CV^2 is asymptotically normally distributed. This result is also readily apparent from Lemma B: since the conditional distribution of s^2 (and therefore of CV^2) given $\bar{X}$ is asymptotically normal, the convergence of the unconditional distribution of CV^2 is assured by the independence of CV^2 and $\bar{X}$. Since the fitted distribution also tends to a normal distribution, the true and fitted distributions are asymptotically identical.

References

Agin, Marilyn A. *A New Exact Test for Randomness Based on the Binomial Distribution of Order k*, M.S. thesis, Michigan Technological University, 1990.

Agin, Marilyn A. and Anant P. Godbole. "A New Exact Runs Test for Randomness," *Computing Science and Statistics: Statistics of Many Parameters: Curves, Images, Spatial Models*, C. Page and R. LePage, eds., New York, 1992, 281-285.

Barbour, A.D. and G.K. Eagleson. "Poisson Approximation for Some Statistics Based on Exchangeable Trials," *Advances in Applied Probability* **15**, 585-600, 1983.

Barton, D.E. and F.N. David. "Nonrandomness in a Sequence of Two Alternatives (II. Runs Tests)," *Biometrika* **45**, 253-256, 1958.

Barton, D.E. and F.N. David. "A Test for Birth Order Effect," *Annals of Human Genetics* **22** (Vol. 1, No. 3), 250-257, 1958.

Bateman, G. "On the Power Function of the Longest Run as a Test for Randomness in a Sequence of Alternatives," *Biometrika* **35**, 97-111, 1948.

Burr, E.J. and Gwenda Cane. "Longest run of consecutive observations having a specified attribute." *Biometrika* **48**, 461-265, 1961.

Chiang, D.T. and S.C. Niu. "Reliability of Consecutive-k-out of-n: F System," *IEEE Transactions on Reliability* **R-30**, 87-89, 1981.

Chiang, D.T. and Ru-fang Chiang. "Relayed Communication via Consecutive k-out-of-n: F System," *IEEE Transactions on Reliability* **R-35**, 65-67, 1986.

David, F.N. "A Power Function for Tests of Randomness in a Sequence of Alternatives," *Biometrika* **34**, 335-339, 1947.

Dixon, W.J. "A Criterion for Testing the Hypothesis that Two Samples are from the Same Population," *Annals of Mathematical Statistics* **11**, 199-204, 1940.

Dyck, P.J., P.K. Thomas, E.H. Lambert and R. Bunge. *Peripheral Neuropathy, Second Edition, Volumes I and II,* Philadelphia, 1984.

Feller, William. *An Introduction to Probability Theory and Its Applications, Volume I,* New York, 1968.

[1] Gibbons, Jean Dickinson. *Nonparametric Statistical Inference.* New York 1971.

[2] Gibbons, Jean Dickinson. *Nonparametric Methods for Quantitative Analysis, Second Edition.* Columbus, 1985.

[1] Godbole, Anant P. "Specific Formulae for Some Success Run Distributions." *Statistics and Probability Letters* **10**, 119-124, 1990.

[2] Godbole, Anant P. "Degenerate and Poisson Convergence Criteria for Success Runs." *Statistics and Probability Letters* **10**, 247-255, 1990.

Goodman, Leo A. "Simplified Runs Tests and Likelihood Ratio Tests for Markoff Chains," *Biometrika* **45**, 181-197, 1958.

Grant, David A. "New Statistical Criteria for Learning and Problem Solution in Experiment Involving Repeated Trials," *Psychological Bulletin* **43**, 273-281, 1946.

Griffith, W.S. "On Reliability of Consecutive-k-out-of-n Failure Systems and Their Generalizations," *Reliability and Quality Control,* 157-165, 1986.

Haldane, J.B.S. and C.A.B. Smith. "A Simple Exact Test for Birth-Order Effect," *Annals of Eugenics* **14(2)**, 117-124, 1948.

Hirano, K. "Some Properties of the Distributions of Order k," in A.N. Philippou et al (eds.) *Fibonacci Numbers and Their Applications.* Dordrecht, 1986.

Larsen, R.J., C.L. Holmes and C.W. Heath, Jr. "A Statistical Test for Measuring Unimodal Clustering: A Description of the Test and of its Application to Cases of Acute Leukemia in Metropolitan Atlanta, Georgia," *Biometrics* **29**, 301-309, 1973.

Mantel, Nathan. "The Detection of Disease Clustering and a Generalized Regression Approach," *Cancer Research* **27**, (Part 1, No. 2), 209-220, 1967.

Mood, A.M. "The Distribution Theory of Runs," *Annals of Mathematical Statistics* **11**, 367-392, 1940.

Mosteller, F. "Note on an Application of Runs to Quality Control Charts," *Annals of Mathematical Statistics* **12**, 228-232, 1941.

O'Brien, P.C. "A Test for Randomness," *Biometrics* **32**, 391-401, 1976.

O'Brien, P.C. and P.J. Dyck. "A Runs Test Based on Run Lengths," *Biometrics* **41**, 237-244, 1985.

[1] Papastavridis, S. "The Number of Failed Components in a Consecutive-k-out-of-$n : F$ System," *IEEE Transactions on Reliability* **R-38**, 338-340, 1989.

[2] Papastavridis, S. "m-Consecutive-k-out-of-n : F Systems," *IEEE Transactions on Reliability* **R-39**, 386-388, 1990.

Philippou, A.N. and F.S. Makri, "Successes, Runs and Longest Runs," *Statistics and Probability Letters* **4**, 211-215, 1986.

Swed, F.S. and C. Eisenhart. "Tables for Testing Randomness of Grouping in a Sequence of Alternatives," *Annals of Mathematical Statistics* **14**, 66-87, 1943.

Wald, A. and J. Wolfowitz. "On a Test Whether Two Samples are From the Same Population," *Annals of Mathematical Statistics* **11**, 147-162, 1940.

RELIABILITY OF CONSECUTIVE-WEIGHTED-K-OUT-of-N:F SYSTEM

Jer-Shyan Wu and Rong-Jaye Chen
National Chiao-Tung University, Taiwan

In this paper, we propose a new model that is more general than previously considered ones: the consecutive-weighted-k-out-of-n:F system. Suppose that there are n components, and each component has its own positive integer weight. The consecutive-weighted-k-out-of-n:F system fails if and only if the total weight of a sequence of failed consecutive components is at least k. Assume that the weight of each component is 1. It is obvious that the consecutive-k-out-of-n:F system is a special case of the consecutive-weighted-k-out-of-n:F system. An algorithm is proposed to compute the reliability of the consecutive-weighted-k-out-of-n:F system in $O(n)$ time.

CONSECUTIVE-WEIGHTED-K-OUT-OF-n:F SYSTEM, SYSTEM RELIABILITY, ALGORITHM.

1. Introduction

Computing the reliability of a consecutive-k-out-of-n:F system is widely discussed from 1981, and the optimal algorithm run is $O(n)$. A consecutive-k-out-of-n:F system consists of a sequence of n ordered components such that the system fails if and only if at least k consecutive components fail. The reliability of this system was first studied by Kontoleon [6] and Chiang and Niu [2], and later by Derman, Lieberman, Ross [3], Hwang [4], Shanthikumar [5], Antonopoulou and Papastavridis [1] and Papastavridis [7].

Now, we propose a new model that is more general than the consecutive-k-out-of-n:F system: the consecutive-*weighted*-k-out-of-n:F system. Assume that each component has its own positive integer weight: we have system failure if and only

A. P. Godbole and S. G. Papastavridis (eds.), Runs and Patterns in Probability, 205–211.

if the total weight of a sequence of failed consecutive components is at least k. For example, a consecutive-weighted-5-out-of-7:F system with weights: 3,2,2,1,3,4,1, has failed consecutive components (1-2,2-4,3-5,5-6, and 6-7) that yield the system failure. Consider the special case where each component has weight 1. It is obvious that we get the consecutive-k-out-of-n:F system, which is thereby a special case of our new model.

In this paper, we propose an algorithm to evaluate the reliability of a consecutive-weighted-k-out-of-n:F system: this algorithm needs only $O(n)$ computing time. Because there are n components, it is likely that the algorithm is optimal.

2. The Optimal Algorithm

In this section, we present our algorithm for computing the reliability of a consecutive-weighted-k-out-of-n:F system. The notation to be used consistently throughout this paper is described as follows.

Notation:

n: number of components in a system;

k: minimum total weight of failed consecutive components which cause the system failure;

w_i: the weight of component i;

p_i, q_i: probability that component i functions, fails; $p_i + q_i = 1$;

S_i: event that the minimum failed consecutive components whose total weight is at least k;

m: total number of all possible S_i;

$Beg(i)$: the first component of S_i;

$End(i)$: the last component of S_i;

$Wet(i)$: the total weight of S_i;

$Q(i)$: $\prod_{j=Beg(i)}^{End(i)} q_j$;

$R(i)$: the reliability of the consecutive-weighted-k-out-of-i:F system;

$F(i)$: $Pr\{\text{system fails}\}; R(i) + F(i) = 1$.

Before deriving the algorithm, we need the following lemmas.

Lemma 1: *Consider a consecutive-weighted-k-out-of-n:F system. It needs $O(n)$ computing time to derive each S_i,* for $i = 1, \ldots, m$.

Proof: Initially, we scan from the first component and compute the total weight until the total weight is greater than or equal to k. Then we check whether this event is minimum. We try to remove components from the beginning of this event until the total weight is less than k. Hence we get the S_1. Furthermore, let the beginning component of the second event be the one following $Beg(1)$, and continue to scan from the component following $End(1)$. By the above method, we can get the other S_i, for $i = 2, \ldots, m$. The details of this method are presented as follows.

begin

$m = 0; event = 1;$ //Set initial value for variables m, $event$, $Wet(1)$,

$Wet(event) = 0; Q(event) = 1;$ $Q(1)$ and$Beg(1)$.

$Beg(event) = 1$

for $component = 1$ **to** n //Scan components from the first one.

begin

$Wet(event) = Wet(event) + W_{component};$ //Compute the values of

$Q(event) = Q(event)q_{component};$ $Wet(event)$and$Q(event)$.

if $Wet(event) \geq k$ **then** //New event appears.

begin

$m = m + 1;$

$End(event) = component;$ //Set the value of $End(event)$.

while $(Wet(event) - W_{Beg(event)} \geq k)$ **then** //Check whether the event is minimum.

begin

$Wet(event) = Wet(event) - W_{Beg(event)}$;

$Q(event) = Q(event)/q_{Beg(event)}$;

$Beg(event) = Beg(event) + 1$;

end

$Beg(event+1) = Beg(event) + 1$; //Set values for the next event: Beg(event+1),Wet(event+1) and Q(event+1).

$Wet(event+1) = Wet(event) - W_{Beg(event)}$;

$Q(event+1) = Q(event)/q_{Beg(event)}$;

$event = event + 1$;

end

end

end.

The time that the worst case of the while loop takes is $Beg(1)+[Beg(2)-Beg(1)]+[Beg(3)-Beg(2)]+\ldots+[Beg(m)-Beg(m-1)] = Beg(m) = O(n)$. So it needs $O(n)$ computing time to derive each S_i, and get $Beg(i), End(i)$ and $Q(i)$, for $i = 1, 2, \ldots, m$.

Now, we want to derive the value of $F(i)$, for $i = 0, 1, \ldots, n$.

Lemma 2: *Consider a consecutive-weighted-k-out-of-n:F system; the value of* $F(i)$ *for* $i = 0, 1, \ldots, n$, *is as follows.*

$$F(i) = 0, \quad (i = 0, 1, \ldots, End(1) - 1), \tag{1}$$

$$F(i) = Pr(S_1 \cup S_2 \cup \ldots \cup S_j) = F(End(j)),$$
$$(i = End(j), End(j)+1, \ldots, End(j+1)-1; \;\; j = 1, 2, \ldots, m-1) \tag{2}$$

$$F(i) = Pr(S_1 \cup S_2 \cup \ldots \cup S_m) = F(End(m)), \quad (i = End(m), End(m)+1, \ldots, n) \tag{3}$$

Proof: First, it is trivial that $F(i) = 0$, for $i = 0, 1, \ldots, End(1)-1$, and it is obvious that $F(i) = Pr(S_1 \cup S_2 \cup \ldots \cup S_j)$, for $i = End(j)$, and $j = 1, 2, \ldots, m$. Now, consider a consecutive-weighted-k-out-of-i:F system, for $i = End(j), End(j)+1, \ldots End(j+1)-1$, and $j = 1, 2, \ldots, m-1$, all the events that the minimum failed consecutive components whose total weight are at least k, are still $S_1, S_2, \ldots,$ *and* S_j. So Eq. (2) is shown. If one considers only components from $End(m)$ to n, Eq. (3) is similar to Eq (2).

The above two lemmas show that it needs only $O(n)$ computing time to derive the value of $F(n)$. We now prove

Theorem 3: *Consider a consecutive-weighted-k-out-of-n:F system. The value of $F(End(j))$, for $j = 2, 3, \ldots, m$, is as follows:*

$$F(End(j)) = F(End(j-1)) +$$
$$\sum_{i=0}^{\Lambda}(1 - F(Beg(j-1)+i-1))p_{Beg(j-1)+i}q_{Beg(j-1)+i+1}\cdots q_{Beg(j)-1}Q(j), \quad (4)$$

where $\Lambda = Beg(j) - Beg(j-1) - 1$; the time to get the value of $R(n)$ is $O(n)$.

Proof: By the union-of-disjoint-sets method, the value of $F(End(j))$, for $j = 2, 3, \ldots, m$, is

$$\begin{aligned}
F(End(j)) =& Pr(S_1 \cup S_2 \cup \ldots \cup S_j) \\
=& Pr(S_1 \cup S_2 \cup \ldots \cup S_{j-1}) + Pr(\overline{S_1 \cup S_2 \cup \ldots \cup S_{j-1}} \cap S_j) \\
=& F(End(j-1)) + R(Beg(j)-2)p_{Beg(j)-1}Q(j) \\
& + R(Beg(j)-3)p_{Beg(j)-2}q_{Beg(j)-1}Q(j) + \ldots + \\
& R(Beg(j-1)-1)p_{Beg(j-1)}q_{Beg(j-1)+1}\cdots q_{Beg(j)-1}Q(j) \quad (5) \\
=& F(End(j-1)) + \sum_{i=0}^{Beg(j)-Beg(j-1)-1} \Bigg\{
\end{aligned}$$

$$(1 - F(Beg(j-1) + i - 1))p_{Beg(j-1)+i}q_{Beg(j-1)+i+1} \cdots q_{Beg(j)-1}Q(j)\Big\}$$

Applying Lemma 1, it needs only $O(n)$ computing time to derive each S_i, and get $Beg(i), End(i)$ and $Q(i)$, for $i = 1, 2, \ldots, m$. And by Lemma 2, we know each $F(i)$, for $i = 0, 1, \ldots, n$, is zero or any one of the $F(End(j))$, for $j = 1, 2, 3, \ldots, m$. By (5), the time for the worst case of deriving all individual values of $F(End(j))$, for $j = 1, 2, 3, \ldots, m$, is: $Beg(1) + (Beg(2) - Beg(1)) + (Beg(3) - Beg(2)) + \ldots + (Beg(m) - Beg(m-1)) = Beg(m) = O(n)$. Because $R(n) = 1 - F(n)$, it takes $O(n)$ computing time to derive $R(n)$.

An example. In order to explain our algorithm, we present an example. For instance, a consecutive-weighted-10-out-of-10:F system with weights 2,1,3,5,1,1,2, 2,3,4, has failed consecutive components ($S_1 : 1-4, S_2 : 2-5, S_3 : 3-6, S_4 : 4-8$, and $S_5 : 7-10$) that yield the system failure. So the reliability of the consecutive-weighted-10-out-of-10:F system is $R(10) = 1 - Pr(S_1 \cup S_2 \cup S_3 \cup S_4 \cup S_5)$.

First, it is trivial that $F(0), F(1), F(2)$ and $F(3)$ are 0. Secondly, considering the last component for each S_i, we get $F(4) = Pr(S_1), F(5) = Pr(S_1 \cup S_2), F(6) = Pr(S_1 \cup S_2 \cup S_3), F(8) = Pr(S_1 \cup S_2 \cup S_3 \cup S_4)$ and $F(10) = Pr(S_1 \cup S_2 \cup S_3 \cup S_4 \cup S_5)$. Finally, $F(7) = Pr(S_1 \cup S_2 \cup S_3) = F(6)$, and we get $F(9) = F(8)$, similarly.

Therefore, $F(10) = Pr(S_1 \cup S_2 \cup S_3 \cup S_4 \cup S_5) = Pr(S_1 \cup S_2 \cup S_3 \cup S_4) + Pr(\overline{S_1 \cup S_2 \cup S_3 \cup S_4} \cap S_5) = F(8) + R(5)p_6Q(5) + R(4)p_5q_6Q(5) + R(3)p_4q_5q_6Q(5)$, $F(8) = Pr(S_1 \cup S_2 \cup S_3 \cup S_4) = Pr(S_1 \cup S_2 \cup S_3) + Pr(\overline{S_1 \cup S_2 \cup S_3} \cap S_4) = F(6) + R(2)p_3Q(4)$, $F(6) = Pr(S_1 \cup S_2 \cup S_3) = Pr(S_1 \cup S_2) + Pr(\overline{S_1 \cup S_2} \cap S_3) = F(5) + R(1)p_2Q(3)$, $F(5) = Pr(S_1 \cup S_2) = Pr(S_1) + Pr(\overline{S_1} \cap S_2) = F(4) + R(0)p_1Q(2)$ and $F(4) = q_1q_2q_3q_4$. Hence, we get $R(10) = 1 - F(10)$.

References

[1] I. Antonopoulou, S. Papastavridis, "Fast recursive algorithm to evaluate the reliability of a circular consecutive-k-out-of-n:F system," *IEEE Trans. Reliability*, **R-36**, 1987, 83-84.

[2] D.T. Chiang, S.C. Niu, "Reliability of consecutive-k-out-of-n:F system," *IEEE Trans. Reliability,* **R-30**, 1981, 87-89.

[3] C. Derman, G. Lieberman, S. Ross, "On the consecutive-k-out-of-n:F system," *IEEE Trans. Reliability,* **R-31**, 1982, 57-63.

[4] F.K. Hwang, "Fast solutions for consecutive-k-out-of-n:F system," *IEEE Trans. Reliability,* **R-31**, 1982, 447-448.

[5] J.G. Shanthikumar, "Recursive algorithm to evaluate the reliability of a consecutive-k-out-of-n:F system," *IEEE Trans. Reliability,* **R-31**, 1982, 442-443.

[6] J. M. Kontoleon, "Reliability determination of a r-successive-out-of-n:F system," *IEEE Trans. Reliability,* **R-29**, 1980, 437.

[7] S. G. Papastavridis, "Upper and lower bounds for the reliability of consecutive-k-out-of-n : F system," *IEEE Trans. Reliability,* **R-35**, 1986, 607-610.

PATTERN RECOGNITION IN MARKOV CHAINS

Dragan Banjević
University of Belgrade

We study the first recognition of the occurrence of any pattern from a set of patterns in a Markov chain. We recognize patterns with given probabilities. A class of statistics connected with the sequence up to the first recognition is considered.

PATTERN RECOGNITION; WAITING TIME; OCCURRENCE OF PATTERNS

1. Introduction and background

Recently, there has been a growing interest in the investigation of the occurrence of sequence patterns in a sequence of random experiments. Let the experiment have countably many outcomes, and suppose we are given the set of finite sequence patterns of outcomes. We are mainly interested in the waiting time for the first appearance of any given pattern. For instance, we toss the coin till the sequence of three heads or three tails appears. Problems of this kind were discussed earlier, as mentioned in Blom and Thorburn (1982). From a modern point of view, this problem is considered by Feller (1968), who treats it as a basic example for his theory of recurrent events. We have to note that the first edition of Feller's book appeared in 1950, where he said in the Preface: "In particular, the theory of recurrent events has been developed for the purpose of this book."

Feller considers a sequence of Bernoulli trials with probability p of success (S), and probability q of failure (F). In ch. XIII.7, he obtains the generating function of the waiting time for a run of r successes, and derives the related mean waiting time (m.w.t.) $\mu = (1-p^r)/(qp^r)$. He notes that his method is applicable to patterns of any kind (say $SSFFSS$). Moreover, in ch. XIII.8 he illustrates the power of his method by obtaining the generating function of the waiting time for "either a success run of length r or a failure run of length ρ". Additionally, Feller finds the probability

A. P. Godbole and S. G. Papastavridis (eds.), Runs and Patterns in Probability, 213–230.

that a success run occurs before a failure run. So, we can say that almost all later results related to i.i.d. random sequences are simple applications and generalizations of Feller's method.

We mention three problems considered by Feller: the waiting time till a given pattern occurs, the waiting time till any of given patterns occurs and the probability that some of patterns will occur first. These are the main problems which are discussed in almost all papers related to this area.

Solov'ev (1966) gives very general results for the case with one stopping pattern, and obtains the m.w.t. for the case of i.i.d. random sequence. Let $X_1, X_2, \ldots$ be the sequence of an i.i.d. random experiments, and let N be the waiting time till the given pattern of outcomes $a_1 a_2 \ldots a_n$ occurs in the sequence $X_1, X_2, \ldots$. Then by Feller (1968), Solov'ev (1966) and later authors, the m.w.t. for N is

(*)
$$\mathbf{E}(N) = \sum_{i=1}^{n} \frac{\mathbf{P}(X_i = a_1, X_{i+1} = a_2, \ldots, X_{i+n-1} = a_n \mid X_1 = a_1, \ldots, X_n = a_n)}{\mathbf{P}(X_1 = a_1, \ldots, X_n = a_n)}.$$

From eq. (*) we see that $\mathbf{E}(N)$ essentially depends on the overlaps of prefixes and suffixes in the sequence $a_1, a_2, \ldots, a_n$, i.e., of the positions i for which $a_1 a_2 \cdots a_{n-i+1} = a_i a_{i+1} \cdots a_n$. This fact was pointed out by many authors. Johnson (1968) in his note considers the waiting time for a repetition of the first n Bernoulli trials with outcomes S and F. Referring to the approach of Bizley (1962), he gives the m.w.t. for a given pattern in terms of "critical points" depending on overlaps in that pattern. It is of interest to note Johnson's final result: If a trial has m outcomes with probabilities $p_1, p_2, \ldots p_m$, then the m.w.t. for a repetition is $m^n + n - 1$, and so does not depend on $p_1, p_2, \ldots p_m$ (the variance depends on the distribution)! Note that Blom (1983) considers Johnson's problem in a slightly different way. In Solov'ev (1966) "periods of string" are equivalent to the "critical points" of Johnson (1968), the "leading numbers" of Conway (see Li (1980) and Blom and Thorburn (1982)), and to the "bifix indicators" of Nielsen (1973), where applications in information theory are mentioned. Nielsen considers the m.w.t. of a search to find a given sequence in stream of random data. In digital communication systems it is of interest to insert a fixed pattern at regular intervals into the transmitted sequence, so that it is important to find an example of such a pattern that has not been inserted by periodic insertions.

Let us now return to formula ($*$). We see that more overlaps imply a longer m.w.t for given pattern. It can give surprising results. For instance, if $p = q = 0.5$, the m.w.t. for $SSSSSS$ is $2^7 - 2 = 126$, while the m.w.t. for $FSSSSS$ is $2^6 = 64$! An interesting mathematical coin-tossing "game" was introduced by Penney in 1969. For history and more information on it see Gardner (1974), Li (1980), Guibas and Odlyzko (1981) and Chen and Zame (1979). This game with two players starts with choosing an integer $n \geq 2$. Player A then selects a pattern A composed of n heads (H) and tails (T), and after that Player B, knowing what is the selection of Player A, selects his own pattern B of length n. Then the coin is tossed until one of two selected patterns is obtained, and the corresponding player wins. Surprising theoretical results are obtained. Let us denote $A < B$ if $\mathbf{P}(B \text{ beats } A) > 0.5$ (i.e., if the game is favourable for B). Let $n \geq 3$. Then for each A there exists an B (which can be explicitly determined) such that $A < B$. The game is "nontransitive" because $A < B$ and $B < C$ does not implies that $A < C$. Moreover, a "tremendous" paradox may occur: if the m.w.t. of B is less than the m.w.t. of A, it does not imply that $B > A$! Gardner gives an example: let $A = THTH$ and $B = HTHH$. Then the m.w.t. of $A = 20$, the m.w.t. of $B = 18$, but $\mathbf{P}(B \text{ beats } A) = 5/14 < 0.5$. Confused readers may refer to Blom (1984) where the more "proper" variant of the game is discussed, i.e. players toss their own coins simultaneously (in the case that patterns A and B are obtained at the same time, tossing starts again). In that game

$$\text{m.w.t. of } A > \text{m.w.t. of } B \Leftrightarrow \mathbf{P}(B \text{ beats } A) > \mathbf{P}(A \text{ beats } B).$$

In recent years, applications to probabilistic models of DNA sequences motivated new investigations. Following Breen et al. (1985) and Biggins and Cannings (1987), we can understand a DNA sequence as a long word composed of letters A,T,G,C.

We are given one (or more) restriction enzyme which cuts the DNA sequence when a corresponding (short) pattern appears. After the first cut, the enzyme works on the rest of the sequence and so it goes on. Breen et al. (1985) treat DNA sequences as sequences of independent trials and generalize some of Feller's (1968) results. A more realistic model for a DNA sequence is a Markov chain, which is considered by Bishop et al. (1983). This model is pointed out in Schwager (1983) too, where high-order and nonhomogeneous Markov chains are considered and other possible applications in psychology, sociology, ecology, radar astronomy and nonparametric statistics are mentioned.

Introduction of Markov chains produces a situation in which intervals between renewals (cuts in DNA sequences) are not i.i.d. random variables. Anyway, their distribution depends only on the preceding and ending patterns (cuts), so that it is natural to use the theory of Markov renewal processes. In another context, this fact was mentioned by Blom and Thorburn (1982). Biggins and Cannings (1987) and Biggins (1987) exploit these facts to obtain and generalize former results on waiting times in Markov chains. We will refer to some of their results later. Finally, Chryssaphinou and Papastavridis (1990) considered a first-order Markov chain, and used a combinatorial argument to study the waiting time for the first appearance of a specific pattern, as well the waiting time for the first appearance of a member of a finite set of patterns.

In this paper we use only elementary facts from the theory of Markov chains, and ideas from Feller's technique of recurrent events. It is of interest to note that some of the basic results (for i.i.d. random sequences) were alternatively obtained by the martingale approach in Li (1980) and Gerber and Li (1981). Note, too, that a review of some other papers which are not mentioned here is given in Banjević (1988).

2. Notation and formulation of the problem

Let $\Omega = \{a, b, \ldots\}$ be the finite or infinite set of outcomes of an experiment. Let $W, W_0, W_1, W_2, \ldots$ be strings or sequence patterns composed of elements of Ω. Let

AB be the pattern composed of patterns A and B, by concatenation. It is useful to introduce an empty pattern Δ, so that $\Delta A = A\Delta = A$, for all A. Let $l(W)$ be the length of pattern W, so that $l(\Delta) = 0$. We shall frequently use the notation $d_i = l(W_i)$.

Let $A = a_1 a_2 \dots a_m$, $a_i \in \Omega$. We introduce two operations φ_n and ψ_n defined on the set of all finite patterns, so that $\varphi_n(A)$ is composed of the last n elements of A, and $\psi_n(A)$ is composed of the first n elements of A, i.e.,

$$\varphi_n(a_1 a_2 \dots a_m) = a_{m-n+1} \dots a_m, \quad 1 \le n \le m,$$

$$\psi_n(a_1 a_2 \dots a_m) = a_1 \dots a_n, \quad 1 \le n \le m,$$

$$\varphi_n(A) = \psi_n(A) = \Delta, \quad n = 0 \quad \text{or} \quad n > m.$$

Let us use the notation $A \subset B$ if $l(A) < l(B)$ and A is a sub pattern of B, i.e., if there exist patterns C, D such that $CD \neq \Delta$, and $CAD = B$. If $S = \{A, B, C, \dots\}$ is a set of patterns, we shall say that S is reduced to S^0 if $S^0 = S \backslash \{W \in S : \exists A \in S, A \subset W\}$. The set S is reduced if $S = S^0$ (terminology of Guibas and Odlyzko (1981)).

We are mainly interested in the first occurrence of any pattern from a reduced set S in an infinite sequence of elements from Ω. This problem is primarily discussed for a finite set S, and finite or infinite Ω. The case that S is reduced and infinite is not impossible and unnatural. For example $\Omega = \{1, 2, 3, \dots\}$ and $S = \{2, 4, 6, \dots\}$, or $\Omega = \{0, 1\}$ and $S = \{11, 101, 1001, \dots\}$. We think that it is of interest to consider the problem for infinite sets S. On the other hand if S is reduced and finite, we can ask ourselves if there exists an infinite and reduced set S', such that $S \subseteq S'$. If there exists such a set S', S is called consistent in the terminology of Guibas and Odlyzko (1981).We shall say that the set S is complete if for all V such that $S \subseteq V$, $V^0 = S$. For example if $\Omega = \{0, 1\}$, $S_1 = \{1, 00\}$, $S_2 = \{11, 101, 00, 010\}$, $S_3 = \{01, 11, 000\}$, are complete (S_3 is given in Guibas and Odlyzko (1981)).

Let us now consider a homogeneous Markov chain $\dots, X_{-2}, X_{-1}, X_0, X_1, X_2, \dots$ with state space Ω. Let $S = \{W_1, W_2, \dots\}$, $W_i \neq \Delta$, be a reduced and finite set of patterns, and W_0 be a pattern such that $l(W_0) = d_0 \geq 0$. Let $a \in \Omega$. We shall refer to a as initial state and W_0 as initial pattern such that $X_{-d_0} = a$ and $X_{-d_0}X_{-d_0+1}\dots X_0 = aW_0$. Let

$$(1) \qquad N_{W_0} = N_{W_0}(S) = \min\{n \geq 1 : \exists j, \varphi_{d_j}(W_0X_1\dots X_n) = W_j\}$$

be the moment of the first occurrence of any W_j in $W_0X_1X_2\dots$ not before $n = 1$. Initial state a has the influence on the distribution of N_{W_0} if $W_0 = \Delta$. Definition of N_{W_0} is known. Let us consider this waiting time problem in the following way: We are given the sequence $X_{-d_0+1}\dots X_0X_1X_n, n \geq 1$, and we check if $X_n \in S$. If it is not, we check if $X_{n-1}X_n \in S$, and so on up to the beginning of the sequence. If there is no i such that $X_iX_{i+1}\dots X_n \in S$, we repeat the procedure on the sequence $X_{-d_0+1}\dots X_nX_{n+1}$. We shall assume that sometimes we cannot decide if $X_iX_{i+1}\dots X_n \in S$, or not, which can be understood in the following sense: let $B_{ij}(X_iX_{i+1}\dots X_j)$, $i \leq j$, be a sequence of binary random variables defined on $\{X_n\}$, such that B_{ij} are independent conditionally $\{X_n\}$, and $\mathbf{P}(B_{ij}(X_iX_{i+1}\dots X_j) = 1 \mid \{X_n\}) = p(X_iX_{i+1}\dots X_j)$, where the function $p(x_ix_{i+1}\dots x_j)$ is given. Then we can define

$$(2) \qquad \tilde{N}_{W_0} = \min\{j \geq 1 : \exists i, B_{ij}(X_iX_{i+1}\dots X_j) = 1\}.$$

We get the definition in eq. (1), if we define $p(W)$ such that $p(W) = 1$, $W \in S$ and $p(W) = 0$, $W \notin S$. In this paper we shall consider the situation such that we are given the reduced and finite set $S = \{W_1, W_2, \dots\}$, and the function $p(W)$ such that $p_i = p(W_i)$, $i = 1, 2, \dots, 0 < p_i \leq 1$, and $p(W) = 0$, $W \notin S$. It means that if $X_{n-d_j+1}\dots X_n = W_j$, we "recognize" W_j if $B_{n-d_j+1,n}(W_j) = 1$, with probability p_j, and "recognize" nothing, with probability $1 - p_j$. (For brevity, we denote $B_{n-d_j+1,n}(W_j)$ by $B(W_j)$.) Then we rewrite the definition of waiting time

$\tilde{N}_{W_0}$ given in eq. (2) by

(3) $$\tilde{N}_{W_0} = \tilde{N}_{W_0}(S) = \min\{n \geq 1 : \exists j, \varphi_{d_j}(W_0 X_1 \ldots X_n) = W_j, B(W_j) = 1\}.$$

3. Distribution of waiting time

Let $p_{ij}(n)$ be the transition probabilities in n steps, $n = 1, 2, \ldots$, for a given Markov chain $\{X_n\}$, $p_{ij}(1) = p_{ij}$, $i, j \in \Omega$. Let, for a given finite and reduced set of patterns $S = \{W_1, W_2, \ldots\}$, $l(W_i) = d_i \geq 1$, recognizing probabilities $p(W_i) = p_i$ and initial state a and initial sequence $W_0, l(W_0) = d_0 \geq 0$, be

(4) $$\begin{aligned} {}_a\tilde{P}_{W_0,j}(n) =& \mathbf{P}(X_{n-d_j+1} \ldots X_n = W_j, \\ & B(W_j) = 1 \mid X_{-d_0} X_{-d_0+1} \ldots X_0 = aW_0), \quad n \geq 1 \end{aligned}$$

(5) $${}_aP_{W_0,j}(n) = \mathbf{P}(X_{n-d_j+1} \ldots X_n = W_j \mid X_{-d_0} X_{-d_0+1} \ldots X_0 = aW_0), n \geq 1.$$

If $n - d_j + 1 < -d_0 + 1$, we have ${}_a\tilde{P}_{W_0,j}(n) = {}_aP_{W_0,j}(n) = 0, 1 \leq n < \max\{1, d_j - d_0\}$, so that

(6) $${}_a\tilde{P}_{W_0,j}(n) = p_j \times {}_aP_{W_0,j}(n), \quad n \geq 1.$$

We have to consider ${}_aP_{W_0,j}(n)$ in two separate situations: either $X_{n-d_j+1} \ldots X_n$ and $X_{-d_0+1} \ldots X_0$ do overlap or do not overlap. In the overlapping case, i.e., if $n - d_j + 1 \leq 0$, or $n < d_j$, we have

(7)

$${}_aP_{W_0,j}(n) = \mathbf{P}\big(X_1 \ldots X_n = \varphi_n(W_j) \mid X_0 = \varphi_1(aW_0)\big)\mathbf{I}(\varphi_{d_j-n}(W_0) = \psi_{d_j-n}(W_j)),$$
$$\max\{1, d_j - d_0\} \leq n < d_j,$$

where $\mathbf{I}(E)$ is the indicator of event E .

If $W = w_1 w_2 \ldots w_k$, let us denote $P(W \mid b) = \mathbf{P}(X_1 \ldots X_k = w_1 \ldots w_k \mid X_0 = b) = p_{bw_1} p_{w_1 w_2} \cdots p_{w_{k-1} w_k}$. By definition of $\varphi, \varphi_m(W) = \Delta$, if $m > l(W)$, so that it is convenient to define the function

(8) $${}_aS_{W_0,j}(n) = P\big(\varphi_n(W_j) \mid \varphi_1(aW_0)\big)\mathbf{I}(\varphi_{d_j-n}(W_0) = \psi_{d_j-n}(W_j)), 1 \leq n \leq d_j - 1.$$

Note that ${}_aS_{W_0,j}(n) = 0$ if $1 \leq n < \max\{1, d_j - d_0\}$, and that ${}_aS_{W_0,j}(n) = 0$ if $W_0 = \Delta$, or $d_j = 1$. Then, from (7), we have

$$(9) \qquad {}_aP_{W_0,j}(n) = {}_aS_{W_0,j}(n), 1 \leq n \leq d_j - 1$$

If there is no overlapping, i.e., if $n - d_j + 1 \geq 1$, we have

$$(10) \qquad {}_aP_{W_0,j}(n) = \mathbf{P}\big(X_{n-d_j+1} \ldots X_n = W_j \mid X_0 = \varphi_1(aW_0)\big) =$$

$$= p_{\varphi_1(aW_0),A(j)}(n - d_j + 1)P(W_j' \mid A(j)), \quad n \geq d_j,$$

where we denote

$$W_j = A(j)B(j)\ldots E(j)F(j),$$

$$W_j' = \varphi_{d_j-1}(W_j) = B(j)\ldots E(j)F(j),$$

i.e., $A(j)W_j' = W_j$, and $P(\Delta \mid A(j)) = 1$.

Let us introduce the relevant generating functions

$$(11) \qquad {}_ar_{W_0,j}(z) = \sum_{n=1}^{\infty} z^n {}_aP_{W_0,j}(n),$$

$${}_aq_{W_0,j}(z) = \sum_{n=1}^{d_j-1} z^n {}_aS_{W_0,j}(n),$$

$${}_ah_{W_0,j}(z) = \sum_{n=d_j}^{\infty} z^n {}_aP_{W_0,j}(n),$$

$$h_{ab}(z) = \sum_{n=1}^{\infty} z^n p_{ab}(n), \quad 0 \leq z \leq 1.$$

Note that ${}_aq_{W_0,j}(z) = 0$, if $W_0 = \Delta$, and that ${}_aq_{W_0,j}(z)$ is the extension of matching polynomials of Breen et al. (1985), as pointed out in Biggins and Cannings (1987). From (10) we have

$$(12) \qquad {}_ah_{W_0,j}(z) = P(W_j' \mid A(j))z^{d_j-1}h_{\varphi_1(aW_0),A(j)}(z).$$

Let $f_{ab}(z)$ be the generating function of the first-passage time from a to b, in the chain $\{X_n\}$. It is well known from the theory of Markov chains that

$$h_{ab}(z) = \frac{f_{ab}(z)}{1 - f_{bb}(z)}, \tag{13}$$

so that

$$_a h_{W_0,j}(z) = P(W_j' \mid A(j)) z^{d_j - 1} \frac{f_{\varphi_1(aW_0),A(j)}(z)}{1 - f_{A(j),A(j)}(z)}, \tag{14}$$

which could be more convenient in calculations then (12). Note that eq. (14) is equivalent to eq. (6.11) in Biggins and Cannings (1987). Finally, we have

$$_a r_{W_0,j}(z) = {_a q_{W_0,j}(z)} + {_a h_{W_0,j}(z)}, \tag{15}$$

where we can determine $_a h_{W_0,j}(z)$ from (12) or (14).

We are now ready to determine the distribution of $\tilde{N}_{W_0} = \tilde{N}_{W_0}(S)$, defined in (3). Let $\tilde{W}^f = W_j$, if W_j is the first recognized pattern. Let

(16)

$$_a \tilde{Q}_{W_0,k}(m) = \mathbf{P}(\tilde{N}_{W_0} = m, \tilde{W}^f = W_k \mid X_{-d_0} = a, X_{-d_0+1} \dots X_0 = W_0), \quad m \geq 1,$$

and

$$_a \tilde{H}_{W_0,j}(z) = \sum_{m=1}^{\infty} z^m {_a \tilde{Q}_{W_0,j}(m)}, \quad 0 \leq z \leq 1. \tag{17}$$

Theorem 1. *The generating functions* $_a \tilde{H}_{W_0,j}(z), j = 1, 2, \dots$ *satisfy the system of linear equations*

$$_a r_{W_0,j}(z) = \sum_k {_a \tilde{H}_{W_0,k}(z)} \times {_{A(k)} r_{W_k',j}(z)} + \frac{1}{p_j} {_a \tilde{H}_{W_0,j}(z)}. \tag{18}$$

Proof. Using the formula of total probability, we have

$$_a \tilde{P}_{W_0,j}(n) = \sum_{m=1}^{n-1} \sum_k {_a \tilde{Q}_{W_0,k}(m)} \times {_{A(k)} \tilde{P}_{W_k',j}(n-m)} + {_a \tilde{Q}_{W_0,j}(n)}, \quad n \geq 1. \tag{19}$$

Using eq. (6) and the definitions of ${}_a r_{W_0,j}(z), {}_a\tilde{H}_{W_0,k}(z)$, simple algebra gives the result. □

If we denote $a_{0j}(z) = {}_a\tilde{H}_{W_0,j}(z), j \geq 1, \rho_{kj}(z) = {}_{A(k)}r_{W'_k,j}(z) + \mathbf{I}(k = j), k, j \geq 1, \rho_{0j}(z) = {}_a r_{W_0,j}(z), j \geq 1$, we get the system

$$\rho_{0j}(z) = \sum_k a_{0k}(z)\rho_{kj}(z) + \frac{1-p_j}{p_j} a_{0j}(z), \quad j = 1, 2, \ldots, \tag{20}$$

which is direct generalization of the system given in Biggins and Cannings (1987), p.539.

Remark. To avoid confusion, we have to make a distinction between whether some pattern is initial or not. It means that if we want to find $a_{ij}(z) = {}_{A(i)}\tilde{H}_{W'_i,j}(z), \quad j = 1, 2, \ldots,$ we have that $\rho_{0j}(z) = {}_{A(i)}r_{W'_i,j}(z) = \rho_{ij}(z) - \mathbf{I}(i = j)$. We have the same problem if we consider the notation $q_{ij}(z)$ in eq. (6.4) of Biggins and Cannings (1987). If we set $q_{ij}(z) = {}_{A(i)}q_{W'_i,j}(z) + \mathbf{I}(i = j)$, and $W_0 = W'_i$, then $q_{0j}(z) = q_{ij}(z) - \mathbf{I}(i = j)$. Let us illustrate the possible solution to avoid confusion in the case when $S = \{W_1\}$ and W_0 are given. Let $\hat{q}_{01}(z) = {}_a q_{W_0,1}(z)$, $\hat{q}_{11}(z) = {}_{A(1)}q_{W'_1,1}(z)$, $\tilde{H}_{01}(z) = {}_a\tilde{H}_{W_0,1}(z)$, $\tilde{H}_{11}(z) = {}_{A(1)}\tilde{H}_{W'_1,1}(z)$, $r_{01}(z) = {}_a r_{W_0,1}(z)$, $r_{11}(z) = {}_{A(1)}r_{W'_1,1}(z)$, $f_{01}(z) = f_{\varphi_1(aW_0),A(1)}(z)$, $f_{11}(z) = f_{A(1),A(1)}(z)$. Then, from (18), we have

$$\tilde{H}_{i1}(z) = \frac{r_{i1}(z)}{r_{11}(z) + 1/p_1} = \frac{\hat{q}_{i1}(z) + P(W'_1 \mid A(1))z^{d_1-1} f_{i1}(z)/(1 - f_{11}(z))}{\hat{q}_{11}(z) + 1/p_1 + P(W'_1 \mid A(1))z^{d_1-1} f_{11}(z)/(1 - f_{11}(z))}$$
$$= \frac{\tilde{q}_{i0}(z)(1 - f_{11}(z)) + f_{i1}(z)}{\tilde{q}_{i1}(z)(1 - f_{11}(z)) + f_{11}(z)}, \quad i = 0, 1, \tag{21}$$

where

$$\tilde{q}_{i0}(z) = \frac{\hat{q}_{i1}(z)}{P(W'_1 \mid A(1))z^{d_1-1}}, \quad \tilde{q}_{i1}(z) = \frac{\hat{q}_{11}(z) + 1/p_1}{P(W'_1 \mid A(1))z^{d_1-1}}, \quad i = 0, 1. \tag{22}$$

which agrees with formula for $a_{01}(z)$ in Biggins and Cannings (1987), p.539.

4. Relation between the distributions of N_{W_0} and $\tilde{N}_{W_0}$

Let $N_{W_0} = N_{W_0}(S)$ be defined as in (1) (or in (3), if $p_j = 1, j = 1, 2, \dots$), and let W^f be the first pattern from S which appears in $\{X_n\}$. Set

$$(23)\quad {}_aQ_{W_0,j}(m) = \mathbf{P}(N_{W_0} = m, W^f = W_j \mid X_{-d_0}X_{-d_0+1}\dots X_0 = aW_0), \quad m \geq 1,$$

$$(24)\qquad {}_aH_{W_0,j}(z) = \sum_{m=1}^{\infty} z^m\, {}_aQ_{W_0,j}(m), \quad 0 \leq z \leq 1.$$

Theorem 2. *Let* ${}_aH_{W_0,j}(z), \quad j = 1, 2, \dots$ *be given. Then*

$$(25)\qquad {}_a\tilde{H}_{W_0,j}(z) = \sum_k (1 - p_k)\, {}_aH_{W_0,k}(z) \times {}_{A(k)}\tilde{H}_{W'_k,j}(z) + p_j \times {}_aH_{W_0,j}(z).$$

Proof. We have that

$${}_a\tilde{Q}_{W_0,j}(m) = \sum_{n=1}^{m-1} \sum_k {}_aQ_{W_0,k}(n)(1 - p_k) \times {}_{A(k)}\tilde{Q}_{W'_k,j}(m - n) + {}_aQ_{W_0,j}(m)p_j.$$

Using (17) and (24), simple algebra gives the result. □

Suppose we are given $H_{0j}(z) = {}_aH_{W_0,j}(z), j \geq 1$, $H_{kj}(z) = {}_{A(k)}H_{W'_k,j}(z), k, j \geq 1$, and set $\tilde{H}_{0,j}(z) = {}_a\tilde{H}_{W_0,j}(z), j \geq 1$, $\tilde{H}_{ij}(z) = {}_{A(i)}\tilde{H}_{W'_i,j}(z), \quad i, j \geq 1$.

We have first to solve the system

$$(26)\qquad \tilde{H}_{ij}(z) = \sum_k (1 - p_k) H_{ik}(z) \tilde{H}_{kj}(z) + p_j H_{ij}(z),$$

with respect to $\tilde{H}_{ij}(z)$, $i = 1, 2, \dots$, and then

$$(27)\qquad \tilde{H}_{0j}(z) = \sum_k (1 - p_k) H_{0k}(z) \tilde{H}_{kj}(z) + p_j H_{0j}(z).$$

When we have $S = \{W_1\}$ and W_0, from (26) and (27), we have

$$(28)\qquad \tilde{H}_{i1}(z) = H_{i1}(z) \frac{p_1}{1 - (1 - p_1) H_{11}(z)}, \quad i = 0, 1,$$

which agrees with (21).

Set $\mathbf{E}_a(\tilde{N}_{W_0}) = \mathbf{E}(\tilde{N}_{W_0} \mid X_{-d_0}X_{-d_0+1}\dots X_0 = aW_0)$, and suppose that the chain $\{X_n\}$ is persistent, irreducible and aperiodic. Then $\sum_j \tilde{H}_{kj}(1) = 1$, $k \geq 0$, $j \geq 1$, and $\mathbf{E}_a(\tilde{N}_{W_0}) = (\sum_j \tilde{H}_{0j}(1))'$. Differentiating both sides of (27), and setting $z = 1$, we have

$$\mathbf{E}_a(\tilde{N}_{W_0}) = \mathbf{E}_a(N_{W_0}) + \sum_k (1-p_k)P_a(W^f = W_k \mid W_0)\mathbf{E}_{A(k)}(\tilde{N}_{W_k'}), \tag{29}$$

and

$$\mathbf{E}_{A(i)}(\tilde{N}_{W_i'}) = \mathbf{E}_{A(i)}(N_{W_i'}) + \sum_k (1-p_k)P_{A(i)}(W^f = W_k \mid W_i')\mathbf{E}_{A(k)}(\tilde{N}_{W_k'}), \tag{30}$$

where $P_a(W^f = W_k \mid W_0) = H_{0k}(1)$, $P_{A(i)}(W^f = W_k \mid W_i') = H_{ik}(1)$. Eq (29) and (30) can be obtained directly.

When we have one pattern W_1, we have, from (30), that

$$\mathbf{E}_{A(1)}(\tilde{N}_{W_1'}) = \frac{1}{p_j}\mathbf{E}_{A(1)}(N_{W_1'}), \tag{31}$$

$$\mathbf{E}_a(\tilde{N}_{W_0}) = \mathbf{E}_a(N_{W_0}) + \frac{1-p_1}{p_1}\mathbf{E}_{A(1)}(N_{W_1'}), \tag{32}$$

which can be obtained by differentiation from (28). Setting $p_1 = 1$ in (21), we get

$$\mathbf{E}_{A(1)}(N_{W_1'}) = H_{11}'(1) = \frac{f_{11}'(1)}{P(W_1' \mid A(1))} = \frac{1}{\pi_1},$$

where $\pi_1 = \nu_1 P(W_1' \mid A(1))$, and ν_1 is the stationary probability for state $A(1)$, as pointed out in Biggins and Cannings (1987). Then, we have

$$\mathbf{E}_a(\tilde{N}_{W_0}) = \mathbf{E}_a(N_{W_0}) + \frac{1-p_1}{p_1}\frac{1}{\pi_1}. \tag{33}$$

In the case of independent r.v. X_i, Lemma 2.4. from Li (1980), gives (using notation from Li (1980))

$$\mathbf{E}_a(\tilde{N}_{W_0}) = \mathbf{E}(N_{W_0W_1}) + \frac{1-p_1}{p_1}\mathbf{E}(N_{W_1'W_1})$$

$$= W_1 * W_1 - W_0 * W_1 + \frac{1-p_1}{p_1}\frac{1}{p(A(1))\dots p(F(1))},$$

where $p(x)$ is the probability of state x.

5. Distribution of some statistics defined on Markov chain

Set $\tilde{N} = \tilde{N}_{W_0}$. It could be of interest to ask how many times state b appears in the sequence $X_1, X_2, \dots, X_{\tilde{N}}$, or how many pure runs appears in that sequence. We want to give results on the distributions of statistics of that type. In the case in which S is the set of pure runs, Banjević (1988, 1990) defines a class of statistics and gives some results.

Let S be a reduced and finite set of patterns, and let a and W_0 be given. We shall introduce the class of integer valued functions defined on $aW_0X_1X_2\dots$, with the "additive" property such that for all j

(34) $$T^n(aW_0X_1\dots X_n)\mathbf{I}(X_{m-d_j+1}\dots X_m = W_j) =$$

$$= T^m(aW_0X_1\dots X_m)\mathbf{I}(X_{m-d_j+1}\dots X_m = W_j)$$

$$+T^{n-m}(W_jX_{m+1}\dots X_n)\mathbf{I}(X_{m-d_j+1}\dots X_m = W_j), \quad 1 \le m < n.$$

Set

$${}_a\tilde{Q}_{W_0,j}(m,i) = \mathbf{P}(\tilde{N}_{W_0} = m, \tilde{W}^f = W_j, T^m = i \mid X_{-d_0}X_{-d_0+1}\dots X_0 = aW_0),$$

$$m \ge 1, i \ge 0,$$

$${}_a\tilde{P}_{W_0,j}(n,i) =$$

$$= \mathbf{P}(T^n = i, X_{n-d_j+1}\dots X_n = W_j, B(W_j) = 1 \mid X_{-d_0}X_{-d_0+1}\dots X_0 = aW_0),$$

(35) $${}_aP_{W_0,j}(n,i) =$$

$$= \mathbf{P}(T^n = i, X_{n-d_j+1}\dots X_n = W_j, \mid X_{-d_0}X_{-d_0+1}\dots X_0 = aW_0), n \ge 1, i \ge 0,$$

$$ {}_a r_{W_0,j}(z,v) = \sum_{n\geq 1, i\geq 0} z^n v^i {}_a P_{W_0,j}(n,i), $$

$$ {}_a \tilde{H}_{W_0,k}(z,v) = \sum_{m\geq 1, i\geq 0} z^m v^i {}_a \tilde{Q}_{W_0,k}(m,i). $$

Then we have ${}_a\tilde{P}_{W_0,j}(n,i) = {}_a P_{W_0,j}(n,i) \times p_j$, $n \geq 1, i \geq 0$, and

$$ (36)\quad {}_a\tilde{P}_{W_0,j}(n,i) = \sum_{m=1}^{n-1}\sum_{l=0}^{i}\sum_{k} {}_a\tilde{Q}_{W_0,k}(m,l) \times {}_{A(k)}\tilde{P}_{W'_k,j}(n-m,i-l) + {}_a\tilde{Q}_{W_0,j}(n,i), $$

which gives the main formula

$$ (37)\qquad {}_a r_{W_0,j}(z,v) = \sum_k {}_a\tilde{H}_{W_0,k}(z,v) \times {}_{A(k)} r_{W'_k,j}(z,v) + \frac{1}{p_j} {}_a\tilde{H}_{W_0,j}(z,v). $$

We have to determine ${}_a r_{W_0,j}(z,v)$. If $1 \leq n \leq d_j - 1$, we have

$$ {}_a P_{W_0,j}(n,i) = {}_a S_{W_0,j}(n,i) = P(\varphi_n(W_j) \mid \varphi_1(aW_0)) $$

$$ \times \mathbf{I}(\varphi_{d_j-n}(W_0) = \psi_{d_j-n}(W_j))\mathbf{I}(T^n(aW_0\varphi_n(W_j)) = i), \quad 1 \leq n \leq d_j - 1. $$

If $n \geq d_j$, in the general case we cannot simplify ${}_a P_{W_0,j}(n,i)$, so that if we define

$$ {}_a q_{W_0,j}(z,v) = \sum_{n=1}^{d_j-1}\sum_{i\geq 0} z^n v^i {}_a S_{W_0,j}(n,i), $$

$$ (38)\qquad {}_a h_{W_0,j}(z,v) = \sum_{n\geq d_j}\sum_{i\geq 0} z^n v^i {}_a P_{W_0,j}(n,i), $$

we have

$$ {}_a r_{W_0,j}(z,v) = {}_a q_{W_0,j}(z,v) + {}_a h_{W_0,j}(z,v). $$

For example, let us consider the number of occurrences of patterns $W_{i_1}, W_{i_2}, \ldots$ from S , in the sequence $aW_0X_1X_2\ldots X_{\tilde{N}}$, starting from X_1. Let $\alpha(k) = 1$, $k \in \{i_1, i_2, \ldots\}$, $\alpha(k) = 0$, $k \notin \{i_1, i_2, \ldots\}$, e.g., if we are interested only in W_{k_0}, then

$\alpha(k_0) = 1$, $\alpha(k) = 0$, $k \neq k_0$. It means that T^n is the statistic of the type defined in (34), but with the special property

$$T^n(aW_0X_1 \ldots X_n)\mathbf{I}(X_{n-d_j+1} \ldots X_n = W_j) =$$

$$= (T^{n-1}(aW_0X_1 \ldots X_{n-1}) + \alpha(j))\mathbf{I}(X_{n-d_j+1} \ldots X_n = W_j).$$

Then, we have

$${}_aP_{W_0,j}(n,i) = \sum_{m=1}^{n-1} \sum_k {}_aQ_{W_0,k}(m) \times {}_{A(k)}P_{W'_k,j}(n-m, i-\alpha(k)) +$$

$$+ {}_aQ_{W_0,j}(n)\mathbf{I}(i = \alpha(j)),$$

so that

$$(39) \qquad {}_ar_{W_0,j}(z,v) = \sum_k {}_aH_{W_0,k}(z) \times {}_{A(k)}r_{W'_k,j}(z,v)v^{\alpha(k)} + {}_aH_{W_0,j}(z)v^{\alpha(j)}.$$

First, we have to determine ${}_aH_{W_0,k}(z)$, ${}_{A(i)}H_{W'_i,k}(z)$, as was done in Section 3, and then to solve the system

$$(40) \quad {}_{A(i)}r_{W'_i,j}(z,v) = \sum_k {}_{A(i)}H_{W'_i,k}(z) \times {}_{A(k)}r_{W'_k,j}(z,v)v^{\alpha(k)} + {}_{A(i)}H_{W'_i,j}(z)v^{\alpha(j)},$$

with respect to ${}_{A(i)}r_{W'_i,j}$. Then, from (39), we get ${}_ar_{W_0,j}$, and from (37), we get ${}_a\tilde{H}_{W_0,k}$.

For example, if $S = \{W_1\}$, so that $\alpha(1) = 1$, and $r_{01} = {}_ar_{W_0,1}$, $r_{11} = {}_{A(1)}r_{W'_1,1}$, we get from (40) $r_{11} = vH_{11}/(1 - vH_{11})$, and from (39) $r_{01} = vH_{01}/(1 - vH_{11})$, so that from (37), we have

$$\tilde{H}_{01}(z,v) = \frac{r_{01}(z,v)}{r_{11}(z,v) + 1/p_1} = \frac{p_1 v H_{01}(z)}{1 - v(1-p_1)H_{11}(z)}.$$

This agrees with (28), if we set $v = 1$. If we set $z = 1$, we have $H_{01}(1) = H_{11}(1) = 1$, and $\tilde{H}_{01}(1,v) = p_1 v/(1 - v(1-p_1))$, which implies the evident conclusion that the number of occurrences of W_1, has a geometric distribution $(1-p_1)^{i-1}p_1, i = 1, 2, \ldots$.

Let us now consider more simple function T^n, such that $T^n(aW_0X_1\dots X_n) = T^n(X_0X_1\dots X_n)$, $X_0 = \varphi_1(aW_0)$, and

(41)
$$T^n(X_0X_1\dots X_n) = T^m(X_0X_1\dots X_m) + T^{n-m}(X_mX_{m+1}\dots X_n), \quad 1 \le m < n,$$

so that

$$T^n(X_0X_1\dots X_n) = T^1(X_0X_1) + T^1(X_1X_2) + T^1(X_{n-1}X_n).$$

It means that the integer valued function $T^1(xy)$ characterizes the function T^n. We shall give some examples.

(a) The number of outcomes b in a sequence $x_1x_2\dots x_n$ is defined by $T^1(xy) = \mathbf{I}(y = b)$.

(b) The number of pure runs in $x_0x_1\dots x_n$, excluding last run, is defined by $T^1(xy) = \mathbf{I}(x \ne y)$. We can understand that T^n counts the number of alternations of states in $x_0x_1\dots x_n$.

(c) The function $T^1(xy) = \mathbf{I}(x \ne b, y = b)$ counts the number of transitions in runs of type b. If we define $T^1(xy) = \mathbf{I}(x = b, y \ne b)$, we get the counter of transitions from runs of type b. The number of pure runs of type b is in both cases less than T^n at most for 1.

From (41) and (35), we have that, for $n \ge d_j$,

(42)
$$_aP_{W_0,j}(n,i) =$$

$$= \mathbf{P}(T^{n-d_j+1} = i - T^{d_j-1}(W_j), X_{n-d_j+1} = A(j) \mid X_0 = \varphi_1(aW_0))P(W_j' \mid A(j)),$$

so that we reduce our problem on evaluating

(43)
$$p_{ac}(n,i) = \mathbf{P}(T^n = i, X_n = c \mid X_0 = a), \quad n \ge 1, i \ge 0,$$

or

(44)
$$h_{ac}(z,v) = \sum_{n\ge1,i\ge0} z^n v^i p_{ac}(n,i).$$

From (38) and (42), we have

$$_a h_{W_0,j}(z,v) = z^{d_j-1} v^{T^{d_j-1}(W_j)} P(W_j' \mid A(j)) h_{\varphi_1(aW_0),A(j)}(z,v).$$

Let N_{ab} be the first passage time from a to b in the chain $\{X_n\}$, and $F_{ab}(z,v)$ be the joint generating function of $T^{N_{ab}}$ and N_{ab}, i.e.,

$$F_{ab}(z,v) = \sum_{n\geq 1, i\geq 0} z^n v^i \mathbf{P}(T^n = i, N_{ab} = n \mid X_0 = a).$$

Then from

$$p_{ac}(n,i) = \sum_{m=1}^{n-1} \sum_{l=0}^{i} \mathbf{P}(T^m = l, N_{ac} = m \mid X_0 = a) p_{cc}(n-m, i-l)$$
$$+\mathbf{P}(T^n = i, N_{ac} = n \mid X_0 = a),$$

we have that

$$h_{ac}(z,v) = \frac{F_{ac}(z,v)}{1 - F_{cc}(z,v)},$$

so that we have

$$_a r_{W_0,j}(z,v) = {}_a q_{W_0,j}(z,v) + z^{d_j-1} v^{T^{d_j-1}(W_j)} P(W_j' \mid A(j)) \frac{F_{\varphi_1(aW_0),A(j)}(z,v)}{1 - F_{A(j),A(j)}(z,v)},$$

which is the generalization of the formula in (14).

There is an alternative method for determining $h_{ac}(z,v)$: From (43), we have

(45) $$p_{ac}(1,i) = \mathbf{P}(T^1(X_0X_1) = i, X_1 = c \mid X_0 = a) = p_{ac}\mathbf{I}(T^1(ac) = i),$$

(46) $$p_{ac}(n,i) = \sum_f p_{af}(n-1, i - T^1(fc)) p_{fc}$$
$$= \sum_f p_{af} p_{fc}(n-1, i - T^1(af)), \quad n \geq 2.$$

Using (44), (45), (46), simple algebra gives the system of equations

$$h_{ac}(z,v) = z(p_{ac} v^{T^1(ac)} + \sum_f v^{T^1(fc)} h_{af}(z,v) p_{fc}),$$

or

$$h_{ac}(z,v) = z(p_{ac} v^{T^1(ac)} + \sum_f v^{T^1(af)} p_{af} h_{fc}(z,v)),$$

which is convenient for calculation if the number of elements in Ω is small.

References

Banjević, D. (1988), *On some statistics connected with runs in Markov chains*, J. Appl. Prob. **25**, 815–821.

Banjević, D. (1990), *On order statistics in waiting time for runs in Markov chains*, Stat. Prob. Letters. **9**, 125–127.

Biggins, J. D. (1987), *A note on repeated sequences in Markov chains*, Adv. Appl. Prob. **19**, 739–742.

Biggins, J. D. and Cannings, C. (1987), *Markov renewal processes, counters and repeated sequences in Markov chains*, Adv. Appl. Prob. **19**, 521–545.

Bishop, D. T. , Williamson, J. A. and Skolnick, M. H. (1983), *A model for restriction fragment length distributions*, Amer. J. Hum. Genet. **35**, 795–815.

Bizley, M. T. L. (1962), *Patterns in repeated trials*, J. Inst. Actu. **88**, 360–366.

Blom, G. (1983), *The mean waiting time to a repetition*, Adv. Appl. Prob. **15**, 216–218.

Blom, G. (1984), *On the stochastic ordering of waiting times for patterns in sequences of random digits*, J. Appl. Prob. **21**, 730–737.

Blom, G. and Thorburn, D. (1982), *How many random digits are required until given sequences are obtained?*, J. Appl. Prob. **19**, 518–531.

Breen, S., Waterman, M. S. and Zhang, N. (1985), *Renewal theory for several patterns*, J.Appl.Prob. **22**, 228–234.

Chen, R. and Zame, A. (1979), *On fair coin-tossing games*, J. Multivariate Anal. **9**, 150–156.

Chryssaphinou, O. and Papastavridis, S. (1990), *The occurrence of a sequence of patterns in repeated dependent experiments*, Theory Probab. Appl. **35**, 167–173.

Feller, W. (1968), *An Introduction to Probability Theory and its Applications.* I, Wiley, New York.

Gardner, M. (1974), *Mathematical games*, Sci. Amer. **10**, 120–125.

Gerber, H. U. and Li, S. R. (1981), *The occurence of sequence patterns in repeated experiments and hitting times in a Markov chain*, Stoch. Proc. Appl. **11**, 101–108.

Guibas, L. J. and Odlyzko, A. M. (1981), *String overlaps, pattern matching, and nontransitive games*, J. Combin. Theory A **30**, 183–208.

Johnson, N. L. (1968), *Repetitions*, Amer. Math. Monthly **75**, 382–383.

Li, S. R. (1980), *A martingale approach to the study of occurrence of sequence patterns in repeated experiments*, Ann. Prob. **8**, 1171–1176.

Nielsen, P. T. (1973), *On the expected duration of a search for a fixed pattern in random data*, IEEE Trans. Inform. Theory **19**, 702–704.

Solov'ev, A. D. (1966), *A combinatorial identity and its application to the problem concerning the first occurrence of a rare event*, Theory Prob. Appl. **11**, 276–282.

Schwager, S. J. (1983), *Run probabilities in sequences of Markov-dependent trials*, J. Amer. Statist. Assoc. **78**, 168–175.

ON THE WAITING TIME OF APPEARANCE OF GIVEN PATTERNS

Ourania Chryssaphinou, Stavros Papastavridis, Theothosios Tsapelas
University of Athens

Let $X_1, X_2, \cdots$ be a sequence of independent and identically distributed r.v.'s, which take values in an alphabet $\Omega = \{\omega_1, \cdots, \omega_v\}, v \geq 2$. A pattern is a finite sequence of letters from the alphabet Ω produced under the above scheme. Considering a finite set of k patterns $A = \{A_i, i = 1, \cdots, k\}$, which are non-overlapping pairwise, we study the waiting time in order to observe A_1 or A_2 or $\cdots$ or A_k whichever comes first and the waiting time in order to observe m different patterns from A whichever they may be. The above results generalize results by Ebneshahrashoob and Sobel (1990) and Ling (1990). Finally we give a limiting result concerning the waiting time for whichever pattern comes first.

1. Introduction

Let $X_1, X_2, \cdots$ be a sequence of independent and identically distributed r.v.'s which take values in an alphabet $\Omega = \{\omega_1, \cdots, \omega_v\}, v \geq 2$ with probabilities $p_j = P(X_i = \omega_j)$ $i = 1, 2, \cdots,$ $j = 1, \cdots, v,$ $p_1 + \cdots + p_v = 1$. A pattern is a finite sequence of letters from the alphabet Ω. We consider a finite set of k different patterns $A = \{A_i, i = 1, \cdots, k\}$. We assume that if $i \neq j$, then the patterns A_i and A_j do not overlap and denote by $| A_i |$ the length of pattern A_i $i = 1, \cdots, k$.

In this paper we study problems concerning the following type of questions: how long do we have to wait to observe m different patterns from $A_1, A_2, \cdots, A_k$?

The case of Bernoulli trials and having patterns A_1 and A_2 to be a run of s successes and a run of r failures respectively has been studied by Feller (1968) and Ebneshahrashoob and Sobel (1990). Ling (1990) has examined the problem of the first coming pattern for the case $p_1 = \cdots = p_v = p$ and $| A_1 | = \cdots = | A_k |$. For the

A. P. Godbole and S. G. Papastavridis (eds.), Runs and Patterns in Probability, 231–241.

same model Ivanov and Novicov (1977) provided interesting limiting results.

In this paper we treat these problems under very general conditions, assuming all kinds of patterns which do not overlap pairwise. We compute the relevant generating function and we prove a limiting result. We rely on methods developed in Solov'ev (1966), Guibas and Odlyzko (1981) and Blom and Thorburn (1982).

2. Waiting time for the appearance of the first pattern

Let us consider a finite set of patterns $A = \{A_i, i = 1, \cdots, k\}$, which are nonoverlapping pairwise. Our notation is slightly different from that of Guibas and Odlyzko (1981), but it comes down to the same thing. We define $s(n)$ to be the probability that none of $A_1, \cdots, A_k$ will occur in the first n trials with $s(0) = 1$, and $w_{A_i}(n)$ to be the probability that A_i comes up on the nth trial but that none of $A_1, \cdots, A_k$ come up on any of the preceding trials. We consider the generating functions

$$S(z) = \sum_{n=0}^{\infty} s(n) z^n \qquad \text{and} \qquad W_{A_i}(z) = \sum_{n=0}^{\infty} w_{A_i}(n) z^n. \tag{2.1}$$

$i = 1, \cdots, k$, with $s(0) = 1$ and $w_{A_i}(0) = 0$.

According to combinatorial arguments of §3 of Guibas and Odlyzko (1981) the following basic relations are valid

$$s(n) = s(n+1) + w_{A_1}(n+1) + \cdots + w_{A_k}(n+1)$$

and

$$P(A_i)s(n) = \sum_{r=1}^{|A_i|} w_{A_i}(n+r) P(a_{i_{r+1}} \cdots a_{i_{|A_i|}}) \tag{2.2}$$

where $A_i = (a_{i_1}, \cdots, a_{i_{|A_i|}})$. Multiplying (2.2) by z^n and summing over n we obtain the relations

$$(1-z)S(z) + W_{A_1}(z) + \cdots + W_{A_k}(z) = 1$$

and

$$S(z) = W_{A_i}(z) C_{A_i}(z) \qquad i = 1, \cdots, k \tag{2.3}$$

where

$$C_{A_i}(z) = \sum_{r=1}^{|A_i|} z^{-r}[P(a_{i_1}, \cdots, a_{i_r})]^{-1}$$

Hence, the generating functions $S(z)$, $W_{A_1}(z), \cdots, W_{A_k}(z)$ satisfy the system (2.3) of linear equations, which is a special case of Theorem 3.3 of Guibas and Odlyzko (1981). Solving the system we get

$$S(z) = [(1-z) + C(z)]^{-1} \tag{2.4}$$

where

$$C(z) = \sum_{A_i \epsilon A} C_{A_i}^{-1}(z) \tag{2.5}$$

Now, let us define W_1 be the waiting time in order to observe A_1 or A_2 or $\cdots$ or A_k whichever comes first. It is clear that

$$P(W_1 = n) = \sum_{A_i \epsilon A} w_{A_i}(n) \tag{2.6}$$

so that

$$W_1(z) = \sum_{n=0}^{\infty} z^n P(W_1 = n) = \sum_{A_i \epsilon A} W_{A_i}(z) \tag{2.7}$$

Using (2.3), (2.4) and (2.5) we obtain the following result:

Theorem 2.1. *With notation as above, the generating function $W_1(z)$ is given by*

$$W_1(z) = 1 - (1-z)[(1-z) + C(z)]^{-1} \tag{2.8}$$

Furthermore, we have

$$S(z) = \frac{1 - W_1(z)}{1-z} \tag{2.9}$$

Corollary 2.2. *We have*

$$E(W_1) = C^{-1}(1) \tag{2.10}$$

and

$$\mathrm{Var}(W_1) = \frac{1 + C(1) - 2C'(1)}{C^2(1)} \tag{2.11}$$

where C(1) is given by (2.5) for z=1 and $C'(1)$ denotes the first derivative of C(z) with respect to z for z=1.

Corollary 2.3. *With the same notation as above and if $P(X_i = w_j) = 1/v = p$, $i = 1, 2, \cdots;$ $j = 1, \cdots, v$, and $A_i = (\omega_i, \cdots, \omega_i)$, $i = 1, \cdots, v$, then*

a)

$$W_1(z) = 1 - (1-z)\Big[1 - z + (1-pz)\big(\sum_{A_i \epsilon A} \frac{(pz)^{|A_i|}}{1-(pz)^{|A_i|}}\big)\Big]^{-1} \tag{2.12}$$

where $A = \{A_1, \cdots, A_v\}$.

b)

If $|A_i| = m$ *for* $i = 1, \cdots, v$, *then*

$$W_1(z) = \frac{v(pz)^m(1-pz)}{(1-z)[1-(pz)^m] + v(1-pz)(pz)^m} \tag{2.13}$$

The following recursive formula is valid

$$\begin{aligned} P(W_1 = n) &= 0, \qquad n = 1, \cdots, m-1 \\ &= vp^m \qquad n = m \\ &= v(p^m - p^{m+1}) \qquad n = m+1, \cdots, 2m-1 \\ &= P(W_1 = n-1) - (v-1)p^m P(W_1 = n-m) \\ &\quad + (vp-1)p^m P(W_1 = n-m-1), \qquad n = 2m, 2m+1, \cdots \end{aligned} \tag{2.14}$$

Proof. The polynomial $C_{A_i}(z)$ is written as

$$\begin{aligned} C_A(z) &= \sum_{r=1}^{|A_i|} z^{-r}[P(a_{i_1}, \cdots, a_{i_r})]^{-1} \\ &= \sum_{r=1}^{|A_i|} z^{-r}p^{-r} \\ &= \frac{1}{(pz)^{|A_i|}} \frac{1-(pz)^{|A_i|}}{1-pz} \end{aligned}$$

so that

$$C(z) = \sum_{A_i \epsilon A} \frac{(pz)^{|A_i|}(1-pz)}{1-(pz)^{|A_i|}}$$

which together with (2.8) gives us (2.12). Relation (2.13), (2.14) can be easily derived.

Furthermore, considering the alphabet $\Omega = \{1, \cdots, m\}$ with $P(X_i = j) = 1/m$, $j = 1, \cdots, m$, $i = 1, 2, \cdots$ and the patterns $A_i = (1, \cdots, 1)$, $\mid A_1 \mid = k$, $\cdots$, $A_m = (m, \cdots, m)$, $\mid A_m \mid = k$ i.e., the very special case studied by Ling (1990), we get immediately the following corollary, which improves, in terms of complexity, Theorems 4.4 and 4.6 of Ling (1990).

Corollary 2.4. *With notation as above we have*

$$W_1(z) = \frac{(\frac{z}{m})^k (m-z)}{(1-z) + (\frac{z}{m})^k (m-1)}$$

and the following recursive formula is valid

$$\begin{aligned} P(W_i = n) &= 0 \qquad n = 0, \cdots, k-1 \\ &= m^{-k+1} \qquad n = k \\ &= (m-1)m^{-k} \qquad n = k+1, \cdots, 2k-1 \\ &= P(W_1 = n-1) - \frac{m-1}{m^k} P(W_1 = n-k), \qquad n = 2k, 2k+1, \cdots. \end{aligned}$$

Remark 1. In the case of having the alphabet $\Omega = \{S, F\}$ and the patterns $A_1 = (S, \cdots, S)$, $\mid A_1 \mid = s$ and $A_2 = (F, \cdots, F)$, $\mid A_2 \mid = r$ relations (2.10), (2.11) and (2.12) gives us relations (2), (3) and (8) of Ebneshahrashoob and Sobel (1990).

Finally, using Corollary 2.4 we derive

$$E(W_1) = \frac{m^k - 1}{m - 1}$$

which agrees with that obtained by Blom and Thorburn (1982) p.527.

3. Waiting time for the appearance of the mth pattern

Let us consider a finite set of patterns $A = \{A_i, i = 1, \cdots, k\}$ which are nonoverlapping pairwise, and the events $A_i(n)$: the pattern A_i will appear in the n first trials

$i = 1, \cdots, k$. Let us define the random variable W_m to be the waiting time for the appearance of m different patterns from the collection $A_1, \cdots, A_k$ and its generating function

$$W_m(z) = \sum_{n=0}^{\infty} z^n P(W_n = n), \qquad |\, z \,| \leq 1$$

We notice that if a pattern appears many times, it is counted only once. We consider the probabilities $P(W_m > n) \quad m = 1, \cdots, k, \quad n \geq 0$, and its generating function

$$S_m(z) = \sum_{n=0}^{\infty} z^n P(W_m > n), \qquad |\, z \,| < 1$$

It is known, see Feller (1968) p.265, that the generating functions $W_m(z), S_m(z)$ satisfy the relation

$$W_m(z) = 1 - (1 - z) S_m(z) \tag{3.1}$$

But

$$\begin{aligned} S_m(z) &= \sum_{n=0}^{\infty} z^n P(\text{at most } (m-1) \text{ patterns will appear till the } n\text{th trial}) \\ &= \sum_{n=0}^{\infty} z^n P(\text{at least } k - m + 1 \text{ patterns will not appear till the } n\text{th trial}) \\ &= \sum_{n=0}^{\infty} z^n \sum_{r=1}^{m} (-1)^{r-1} \binom{k - m + r - 1}{k - m} S_{r+k-m} \end{aligned} \tag{3.2}$$

where

$$S_{r+k-m} = \sum_{1 < i_1 < \cdots < i_{r+k-m} < k} P(\bar{A}_{i_1}(n) \cdots \bar{A}_{i_{r+k-m}}(n))$$

and $\bar{A}_{i_s}(n)$ be the complement of $A_{i_s}(n) \quad s = 1, \cdots, k$.

The relation (3.2) is written as

$$S_m(z) = \sum_{r=1}^{m} (-1)^{r-1} \binom{k - m + r - 1}{r - m} \sum_{1 \leq i_1 \leq \cdots \leq i_{r+k-m} \leq k} S(A_{i_1}, \cdots, A_{i_{r+k-m}}; z) \tag{3.3}$$

where

$$S(A_{i_1},\cdots,A_{i_{r+k-m}};z)=\sum_{n=0}^{\infty} z^n P(\bar{A}_{i_1}(n)\cdots\bar{A}_{i_{r+k-m}}(n))$$

Since $P(\bar{A}_{i_1}(n)\cdots\bar{A}_{i_{r+k-m}}(n))$ is the probability that in first n trials none of the patterns $A_{i_1},\cdots,A_{i_{r+k-m}}$ will appear, it is clear that $S(A_{i_1},\cdots,A_{i_{r+k-m}};z)$ is a generating function of the same kind as that defined in (2.1), i.e., $S(A_{i_1},\cdots,A_{i_{r+k-m}};z)=S(z)$ and so that it is given by (2.4). Hence, combining (3.1) and (3.3) we get the following result:

Theorem 3.1. *With notation as above we have*

$$W_m(z)=1-(1-z)\sum_{r=1}^{m}(-1)^{r-1}\binom{k-m+r-1}{k-m}\sum_{1\le i_1\le\cdots\le i_{r+k-m}\le k} S(A_{i_1},\cdots,A_{i_{r+k-m}};z) \tag{3.4}$$

The following are immediate consequences of Theorem 3.1.

Corollary 3.2. *The generating function for the waiting time for the last coming is given by*

$$W_k(z)=1-(1-z)\sum_{r=1}^{k}(-1)^{r-1}\sum_{1\le i_1\le\cdots\le i_r\le k} S(A_{i_1},\cdots,A_{i_r};z) \tag{3.5}$$

Corollary 3.3 *If* $\Omega=\{1,2,\cdots,v\}$ *with* $P(X_j=i)=1/v,\quad i=1,\cdots,v,\quad j=1,2,\cdots$ *and* $A_1=(1,\cdots,1),\ |A_1|=l,\cdots,\ A_v=(v,\cdots,v),\ |A_v|=l$, *then*

$$W_m(z)$$
$$=1-(1-z)\sum_{r=1}^{m}(-1)^{r-1}\binom{v-m+r-1}{v-m}\binom{v}{v-m+r}\frac{1-(\frac{z}{v})^l}{(1-z)[1-(\frac{z}{v})^l]+(v-m+r)(1-\frac{z}{v})(\frac{z}{v})^l} \tag{3.6}$$

Remark 2. It is clear that using (3.4) we can obtain the expectation and the variance of the r.v. W_m. Obviously, relation (3.4) for $m=1$ gives us relation (2.8).

Furthermore, with the notation as in Corollary 3.3 we get the generating function of the waiting time for last coming for $m = v$

$$W_v(z) = 1 - (1-z)\sum_{r=1}^{v}(-1)^{r-1}\binom{v}{r}\frac{1-(\frac{z}{v})^l}{(1-z)[1-(\frac{z}{v})^l]+r(1-\frac{z}{v})(\frac{z}{v})^l} \tag{3.7}$$

which leads us to

$$E(W_v) = \sum_{r=1}^{v}(-1)^{r-1}\binom{v}{r}\frac{1-(\frac{1}{v})^l}{r(1-\frac{1}{v})(\frac{1}{v})^l} \tag{3.8}$$

In the case of having the alphabet $\Omega = \{S, F\}$ and the patterns $A_1 = (S, \cdots, S)$, $\mid A_1 \mid = s$ and $A_2 = (F, \cdots, F)$, $\mid A_2 \mid = r$ relation (3.5) gives us relation 11 of Ebneshahrashoob and Sobel (1990) for $x = y = 1$. We notice that (3.8) is the same as that obtained by Blom and Thorburn (1982), Theorem 4.1.

4. Limit theorem

Let us consider an infinite sequence of sets of patterns

$$\mathcal{A}_n = \{A_{n_i} : \quad i = 1, \cdots, k(n)\}, \quad n = 1, 2, \cdots$$

where $\{k(n); n \geq 1\}$ is bounded and A_{n_i}, A_{n_j} do not overlap for $i \neq j$ $n = 1, 2, \cdots$. Throughout this section let $a_n = \min\{\mid A_{n_i} \mid, i = 1, \cdots, k(n)\}$ $n = 1, 2, \cdots$, and $p = \max\{p_i, i = 1, \cdots, v\} < 1$.

We are going to prove the following result concerning the r.v. W_{1_n} i.e. the waiting time for whichever comes first from the set of the patterns $\mathcal{A}_n$ $n = 1, 2, \cdots$:

Theorem 4.1. *If $a_n \to \infty$ as $n \to \infty$, then the r.v. $W_{1_n}/E(W_{1_n})$ converges in distribution to the exponential with parameter 1.*

Proof. We consider the Laplace transform for the r.v. W_{1_n}/μ_n where $\mu_n = E(W_{1_n})$, which is according to relation (2.5) and (2.8) as follows

$$\begin{aligned} W_1(e^{-z/\mu_n}) &= E(e^{-zW_{1_n}/\mu_n}) \\ &= \frac{1}{1+(1-e^{-z/\mu_n})C_n{}^{-1}(e^{-z/\mu_n})}, \quad z \geq 0 \end{aligned} \tag{4.1}$$

where

$$C_n(e^{-z/\mu_n}) = \sum_{A_{n_i}\epsilon A_n} \frac{e^{-z|A_{n_i}|/\mu_n} P(A_{n_i})}{\phi_{A_{n_i}}(e^{-z/\mu_n})} \tag{4.2}$$

and

$$\phi_{A_{n_i}}(e^{-z/\mu_n}) = \sum_{r=1}^{|A_{n_i}|} e^{-z(|A_{n_i}|-r)/\mu_n} P(a_{n_{i_{r+1}}}, \cdots, a_{n_{i_{|A_{n_i}|}}})$$

considering the pattern $A_{n_i} = (a_{n_{i_1}}, \cdots, a_{n_{i_{|A_{n_i}|}}})$.

Rewriting (4.1) as

$$W_{1_n}(e^{-z/\mu_n}) = \frac{1}{1 + \frac{1-e^{-z/\mu_n}}{1/\mu_n} \frac{1/\mu_n}{C_n(e^{-z/\mu_n})}} \tag{4.3}$$

we observe that

$$\lim_{n\to\infty} \frac{1 - e^{-z/\mu_n}}{1/\mu_n} = z$$

since according to (2.10) we have

$$\frac{1}{\mu_n} = C_n(1) = \sum_{A_{n_i}\epsilon A_n} \frac{P(A_{n_i})}{\phi_{A_{n_i}}(1)} \le \sum_{A_{n_i}\epsilon A_n} P(A_{n_i}) \le \sum_{A_{n_i}\epsilon A_n} p^{|A_{n_i}|} \to 0$$

as $n \to \infty$. So it remains to prove that

$$\lim_{n\to\infty} \frac{1/\mu_n}{C_n(e^{-z/\mu_n})} = 1$$

or

$$\lim_{n\to\infty} \frac{C_n(e^{-z/\mu_n})}{C_n(1)} = 1$$

Using (4.2) we have

$$\begin{aligned} &\left| \sum_{A_{n_i}\epsilon A_n} \frac{e^{-z|A_{n_i}|/\mu_n} P(A_{n_i})}{\phi_{A_{n_i}}(e^{-z/\mu_n})} - \sum_{A_{n_i}\epsilon A_n} \frac{P(A_{n_i})}{\phi_{A_{n_i}}(1)} \right| \Big/ C_n(1) \\ &\le \sum_{A_{n_i}\epsilon A_n} \left| \frac{e^{-z|A_{n_i}|/\mu_n} P(A_{n_i})}{\phi_{A_{n_i}}(e^{-z/\mu_n})} - \frac{P(A_{n_i})}{\phi_{A_{n_i}}(1)} \right| \Big/ C_n(1) \end{aligned} \tag{4.4}$$

But

$$\left|\frac{e^{-z|A_{n_i}|/\mu_n}P(A_{n_i})\phi_{A_{n_i}}(1)-P(A_{n_i})\phi_{A_{n_i}}(e^{-z/\mu_n})}{\phi_{A_{n_i}}(1)\phi_{A_{n_i}}(e^{-z/\mu_n})\sum_{A_{n_j}\epsilon A_n}\frac{P(A_{n_j})}{\phi_{A_{n_j}}(1)}}\right|$$

$$\leq \left|\frac{e^{-z|A_{n_i}|/\mu_n}\phi_{A_{n_i}}(1)-\phi_{A_{n_i}}(e^{-z/\mu_n})}{\phi_{A_{n_i}}(e^{-z/\mu_n})}\right|$$

$$\leq \left|e^{-z|A_{n_i}|/\mu_n}\phi_{A_{n_i}}(1)-\phi_{A_{n_i}}(e^{-z/\mu_n})\right|$$

$$= \left|e^{-z|A_{n_i}|/\mu_n}\sum_{r=1}^{|A_{n_i}|}P(a_{n_{i_{r+1}}},\cdots,a_{n_{i_{|A_{n_i}|}}})\right.$$

$$\left.-\sum_{r=1}^{|A_{n_i}|}e^{-z(|A_{n_i}|-r)/\mu_n}P(a_{n_{i_{r+1}}},\cdots,a_{n_{i_{|A_{n_i}|}}})\right|$$

$$= \left|\sum_{r=1}^{|A_{n_i}|}(e^{-z|A_{n_i}|/\mu_n}-e^{-z(|A_{n_i}|-r)/\mu_n})P(a_{n_{i_{r+1}}},\cdots,a_{n_{i_{|A_{n_i}|}}})\right|$$

$$\leq \sum_{r=1}^{|A_{n_i}|}(e^{-z|A_{n_i}|/\mu_n}-e^{-z(|A_{n_i}|-r)/\mu_n})p^{|A_{n_i}|-r}$$

$$= \sum_{m=0}^{|A_{n_i}|}(e^{-z|A_{n_i}|/\mu_n}-e^{-zm/\mu_n})p^{m}$$

$$\leq e^{-z|A_{n_i}|/\mu_n}\frac{1}{1-p}-\frac{1}{1-pe^{-z/\mu_n}}\to 0 \qquad \text{as} \quad n\to\infty$$

since

$$\frac{|A_{n_i}|}{\mu_n}=|A_{n_i}|\sum_{A_{n_j}\epsilon A_n}\frac{P(A_{n_j})}{\phi_{A_{n_j}}(1)}$$

$$\leq \sum_{A_{n_j}\epsilon A_n}|A_{n_i}|P(A_{n_j})$$

$$\leq \sum_{A_{n_j}\epsilon A_n}|A_{n_i}|p^{|A_{n_j}|}\to 0 \qquad \text{as} \quad n\to\infty$$

and the proof is complete.

References

Blom G. and Thorburn D. (1982). How many random digits are required until given sequences are obtained? *J. Appl. Prob.* **19**, 518-531.

Ebneshahrashoob M. and Sobel M. (1990). Sooner and later waiting time problems for Bernoulli trials: Frequency and run quotas. *Statistics and Probability Letters* **9**, 5-11.

Feller W. (1968). *An introduction to probability theory and its applications, Vol. 1*, Wiley.

Guibas L.J. and Odlyzko A. M. (1981). String overlaps, pattern matching and nontransitive games. *J. Comb. Theory Series A* **30**, 183-208.

Ivanov V. A. and Novicov A. E. (1977). On the distribution of the time up to the first occurrence of a given number of different l-tuple series. *Theory Prob. Appl.* **22**, 533-542.

Ling, K.D. (1990). On Geometric distributions of order $(k_1, \cdots, k_m)$. *Statistics and Probability Letters* **9**, 163-171.

Solov'ev, A. D. (1966). A combinatorial identity and its application to the problem concerning the first occurrence of a rare event. *Theory Prob. Appl.* **11**, 276-282.

RUNS AND EXCURSIONS

Antonia Földes*

City University of New York

In this short note some repetitive patterns of the excursions of a simple symmetric random walk are discussed in comparison to the similar properties of a coin tossing sequence. Some related open problems are also addressed.

Let $X_1, X_2, \cdots$ be a sequence of i.i.d. random variables with $P(X_1 = +1) = P(X_1 = -1) = 1/2$. The repetitive patterns, runs of the sequence $\{X_n\}_{n=1}^{\infty}$, were extensively studied in the literature.

Consider now $S_0 = 0$, $S_n = X_1 + X_2 + \cdots + X_n$, $n = 1, 2, \cdots$, S_n is the so-called simple symmetric random walk. Define

$$\rho_0 = 0$$

$$\rho_1 = \inf\{n; n > 0, S_n = 0\}$$

$$\vdots$$

$$\rho_k = \inf\{n; n > \rho_{k-1}, S_n = 0\}$$

Then $\{S_n; \rho_k < n \leq \rho_{k+1}\}$ $k = 0, 1, 2, \cdots$ is a sequence of i.i.d. random processes, the so-called *excursions* of the walk S_n. These excursions share many properties with the original $\{X_n\}$ sequence, and this similarity is the topic of this paper. Namely, the repetitive patterns of $\{X_n\}$ and the sequence of excursions will be compared.

These results are simple consequences of an interesting general result of Csáki, Földes, Komlós (1987). Some open questions will also be discussed. To set the stage for further discussion, we present a special case of the above mentioned result.

---

* Research Supported by CUNY Research Grant No. 662349

A. P. Godbole and S. G. Papastavridis (eds.), Runs and Patterns in Probability, 243–251.

Let $X_1, X_2, \cdots$ be a sequence of i.i.d. r.v.-s. and let $\mathcal{F}_{n,m}$ denote the σ-algebra generated by the block of variables $X_n, X_{n+1}, \cdots X_{n+m-1}$. We will denote by A_n^m an event which is $\mathcal{F}_{n,m}$ measurable. In what follows the m will be suppressed, thus A_n stands for A_n^m.

Lemma A. (Csáki, Földes, Komlós [1]). *Assume that there is a number p, $0 < p \leq 1$, such that the following three conditions hold for some $k < m, \epsilon > 0$.*

(i) $|P(\bar{A}_2\bar{A}_3\cdots\bar{A}_k|A_1) - p| < \epsilon$

(ii) $\sum_{k\leq i\leq 2m} P(A_i|A_1) < \epsilon$

(iii) $P(A_1) < \epsilon/m$

Then, for all $N > 1$

$$\left|\frac{P(\bar{A}_2\cdots\bar{A}_N|A_1)}{P(\bar{A}_2\cdots\bar{A}_N)} - p\right| < 7\epsilon$$

and

$$e^{-(p+10\epsilon)NP(A_1)-2mP(A_1)} < P(\bar{A}_1\bar{A}_2\cdots\bar{A}_N) < e^{-(p-10\epsilon)NP(A_1)+2mP(A_1)}$$

Since $A_n(= A_n^m)$ is an event determined by $X_n, X_{n+1}, \cdots, X_{n+m-1}, P(\bar{A}_1\bar{A}_2\cdots\bar{A}_N)$ is the probability that in *none* of the first N blocks of length m occurred the event represented by $A_1(= A_1^m)$, or in other words, the waiting time for a pattern of the length m described by A_1, is longer than N. By verifying the conditions of this lemma, with an appropriate choice of k and ϵ, one can get the limit distribution of the waiting time $\tau_m = \{$first n; such that A_n occurs$\}$ (thus the waiting time for a pattern of length m represented by A_1). Hence, the above lemma gives us a unified approach to prove results in diverse situations, where the strong dependence between nearby blocks is controlled by the conditions of the lemma. To mention first a very simple example, the properties of the length of the longest head runs are well known (Erdős-Rényi [2], Erdős-Révész [3], Földes [4], Komlós-Tusnády [6], Guibas-Odlyzko [5], Samarova [11]). However, many of these results can be recovered from the above mentioned result as we pointed out in [1]. Indeed, with $P(X_n = 1) = \alpha(=$

$1/2), P(A_n) = \alpha^m, P(\bar{A}_2 \cdots \bar{A}_k | A_1) = P(X_m = -1) = 1 - \alpha(= 1/2)$, for $2 \leq k \leq m$, hence $p = 1 - \alpha(= 1/2)$.

The second and third conditions follow from the observation that

$$P(A_i|A_1) = \begin{cases} \alpha^{i-1} & \text{for } 2 \leq i \leq m \\ \alpha^m & \text{for } i > m \end{cases}$$

leading to the following limit theorem

$$\lim_{m \to \infty} P\left(\tau_m > x\alpha^{-m} \frac{1}{1-\alpha}\right) = e^{-x}$$

where τ_m is the waiting time for the first head run of length m.

Remark: It would have been much simpler to write the above result as

$$\lim_{m \to \infty} P(\tau_m > x2^{m+1}) = e^{-x};$$

however, the above particular form is the one which always appears, where only α changes from problem to problem. A standard Borel Cantelli argument leads to the following well-known results. If $\mu(n)$ denotes the longest head run up to n, then we have, with

$$A_n = \sum_{n=1}^{\infty} e^{-(\log 2) a_n},$$

that

$$a_n \epsilon UUC(\mu(n)) \qquad \text{if } A_n < \infty$$

and

$$a_n \epsilon ULC(\mu(n)) \qquad \text{if } A_n = \infty$$

Morever, for any $\epsilon > 0$

$$\mu(n) \geq \lfloor f(n) - \epsilon \rfloor \qquad \text{for } n > n_0(w)$$

and

$$\mu(n) < \lceil f(n) + \epsilon \rceil \qquad \text{for infinitely many } n$$

where

$$f(n) = \frac{\log n - \log\log\log n - 1}{\log 2}$$

and $\lfloor . \rfloor, \lceil . \rceil$ stand for the floor and ceiling functions respectively (rounding down, rounding up).

This question can be translated to the excursion sequences without any further effort. Namely, let's call the k-th excursion positive if $S_{\rho_{k-1}+1} > 0, S_{\rho_{k-1}+2} > 0, \cdots, S_{\rho_k - 1} > 0$ or which is the same as $S_{\rho_{k-1}+1} > 0$. Then we might ask, considering the first n excursions of the walk, what is the longest sequence of consecutive positive excursions? If τ_m^* is the waiting time for a consecutive positive excursion sequence of length m, and $\mu^*(n)$ is the length of the longest positive sequence of excursions, then all the previously listed results for τ_m and $\mu(n)$ hold for τ_m^* and $\mu^*(n)$. (To see this, it is enough to observe that the positivity of the k-th excursion is determined by its first step $X_{\rho_{k-1}+1}, k = 1, 2, \cdots, n$, which is a sequence of i.i.d. r.v.-s. with the same distribution as X_1.)

Consider now the following question: Let M_i be the absolute height of the i-th excursion, thus

$$M_i = \sup_{\rho_{i-1} < j < \rho_i} |S_j|$$

The distribution of M_i is well-known:

$$P(M_i = k) = \frac{1}{k(k+1)} \quad k = 1, 2, \cdots; \quad P(M_i \le k) = \frac{k}{k+1}$$

This result is a simple consequence of the following important lemma (see e.g. Révész [10], p.23):

Lemma B. *Let* $0 \le i \le k$. *Then for* $\ell \ge i$, *we have*

$$p(0, i, k) = P(\min\{j, j \ge \ell, S_j = k\} < \min\{j, j \ge \ell, S_j = 0\} | S_\ell = i) = \frac{i}{k}$$

The above lemma immediately implies that $P(M_i \ge k) = 1/k$.

We might ask, what can we say about $\tau_m^{(k)}$, the waiting time for a consecutive sequence of m excursions having heights less than or equal to a fixed k? A straightforward application of our Lemma A gives the limit distribution of $\tau_m^{(k)}$, namely,

$$\lim_{m\to\infty} P\Big(\tau_m^{(k)} > x\big(\frac{k+1}{k}\big)^m(k+1)\Big) = e^{-x} \quad [\alpha = \frac{k}{k+1}]$$

Similarly, if $\mu^{(k)}(n)$ is the length of the longest block of excursions with heights at most k, we get with the usual Borel-Cantelli argument, that if

$$A_n^{(k)} = \sum_{k=1}^{\infty} e^{-(\log \frac{k+1}{k})a_n}$$

then

$$a_n \epsilon UUC(\mu^{(k)}(n)) \qquad \text{if } A_n^{(k)} < \infty$$

and

$$a_n \epsilon ULC(\mu^{(k)}(n)) \qquad \text{if } A_n^{(k)} = \infty$$

Moreover, for any $\epsilon > 0$

$$\mu^{(k)}(n) \geq \lfloor f^{(k)}(n) - \epsilon \rfloor \qquad \text{for } n > n_0(\omega)$$

and

$$\mu^{(k)}(n) < \lceil f^{(k)}(n) - \epsilon \rceil \qquad \text{for infinitely many } n$$

where

$$f^{(k)}(n) = \frac{\log n - \log\log\log n + \log \frac{1}{k+1}}{\log(\frac{k+1}{k})}$$

Remark: Observe that the k=1 case gives the same results that we got in the pure head run case. This coincidence is readily explained by the following observation: $M_i \leq 1$ means that we have an excursion, which consists of exactly one +1 step and exactly one −1 step. However, the probability that an excursion is of the above type (having exactly two steps) is 1/2.

The same method can be applied to the question of sequences of excursions having a given absolute height k.

Then we arrive at the following result:

If $\tau_m^{*(k)}$ denotes the waiting time for a sequence of m consecutive excursions with absolute heights k, then

$$\lim_{m\to\infty} P(\tau_m^{*(k)} > x(k(k+1))^m \frac{k(k+1)}{k^2+k-1}) = e^{-x} \qquad [\text{as } \alpha = \frac{1}{k(k+1)}]$$

For $\mu^{*(k)}(n)$ (= the length of the longest block of excursions with exact absolute heights k) we get a result parallel to $\mu^{(k)}(n)$ but with

$$A_n^{*(k)} = \sum_{k=1}^{\infty} e^{-\log(k(k+1))a_n}$$

and

$$f^{*(k)}(n) = \frac{\log n - \log\log\log n + \log(1 - \frac{1}{k(k+1)})}{\log(k(k+1))}$$

Another interesting well-known problem is the question of the longest monotone block. More precisely, let $\{Y_i\}_{i=1}$ be a sequence of i.i.d. r.v.-s with a *continuous* distribution function. Then we might ask the length of the longest monotone block in a sample of size n. (Clearly $A_k^m = \{Y_k \le Y_{k+1} \le \cdots \le Y_{k+m-1}\}$ is a monotone block of size m.) This problem was first investigated by Pittel [7], [8] and Révész [9]. Let V_n denote the length of the longest monotone block up to n. Révész [9], proved (besides many interesting results) that V_n has only three of four possible values with probability 1, if n is big enough, and V_n is around

$$s(n) = \frac{\log n}{\log\log n - \log\log\log n - 1 + \sigma(1)}$$

Again the above result can be recovered from our Lemma A. However, this question can be raised for our i.i.d. sequence $\{X_i\}_{i=1}^{\infty}$, too. However, as X_i takes on only the values $+1$ or -1, this means that a monotone sequence is barely a block of (-1)

followed by a block of (+1). This can be treated similarly to the problem of the longest head run. Our Lemma A leads to the following limit theorem:

Theorem. *Let τ_m^M be the waiting time for the first monotone block of length m, then*

$$\lim_{m\to\infty} P\Big(\tau_m^M > x\frac{2^{m+1}}{m+1}\Big) = e^{-x}$$

which of course gives us again a possibility to get the upper and lower class type results for the longest monotone block of the sequence $\{X_i\}_{i=1}^{\infty}$.

However, it is clear that as the $\{X_i\}_{i=1}^{\infty}$ variables are *discrete*, we do not have a general method to work with, as we had in the continuous distribution function case. If we consider again the sequence of excursions, we might formulate the following interesting question: What can we say about the length of the longest monotone "increasing" sequence of excursions? To be more precise, let M_i (as before) be the absolute height of the i-th excursion. We have a *monotone* block of excursions of length k if $M_i \leq M_{i+1} \leq \cdots \leq M_{i+k-1}$ Moreover, we have a *strongly monotone* sequence of excursions of length k if $M_i < M_{i+1} < \cdots < M_{i+k-1}$. As to the first case, the *monotone* sequence for the excursions, consider the following special case. Take the excursions with absolute height 1. As we have seen earlier, the probability that a given excursion is of this type is 1/2. Again, it is easy to see that to have a sequence of consecutive excursions of this type is equivalent with the event of having a pure head run of the same length. Hence, the length of a monotone sequence of excursions is at least as much as the length of a pure head run.

However, considering the *strongly monotone* sequences of excursions, it is easy to see that the length of the longest strongly monotone block is (much) shorter. Namely, if we have a sequence of k excursions with k different heights, then their $k!$ possible orderings are equally probable. Thus the probability that their heights form a strongly monotone sequence is (much) less than $1/k!$ This implies that the length of the longest monotone block will be less than or equal to $s(n)$. It seems to be an interesting question to find the limit distribution of the waiting time for a monotone, and for a strongly monotone sequence of excursions of length m. However, for this purpose one has to evaluate the asymptotic probability of A_m and A_m^s when $m \to \infty$, where $A_m = \{M_i \le M_{i+1} \le \cdots \le M_{i+k-1}\}$ and $A_m^s = \{M_i < M_{i+1} < \cdots < M_{i+k-1}\}$. These questions are open. To give the flavor of them let us consider A_m. Define

$$P_1(j) = P(M_1 \le j) = \frac{j}{j+1}$$
$$P_k(j) = P(M_1 \le M_2 \le \cdots \le M_{k-1} \le M_k \le j)$$

It would be enough to determine $P_k(j)$ as

$$\sum_{j=1}^{\infty} P(M_k = j) P_{k-1}(j) = P(A_k)$$

It is an easy computation to show that

$$P_2(j) = \sum_{\ell=1}^{j} \frac{1}{(\ell+1)^2}$$

Moreover,

$$P_m(j) = \sum_{\ell=1}^{j} P(M_m = \ell) P(M_1 \le M_2 \le \cdots \le M_{m-1} \le \ell)$$

thus

$$P_m(j) = \sum_{\ell=1}^{j} \frac{1}{\ell(\ell+1)} P_{m-1}(\ell)$$

The problem is to get a closed form for $P_m(j)$, and/or get $\lim_{j\to\infty} P_m(j)$. Besides the investigation of the longest (strongly) monotone sequence of excursions, it also seems to be important to raise the question of the longest monotone sequence for an i.i.d. $\{Y_j\}_{j=1}^{\infty}$ where $P(Y_j < x) = F(x)$ where $F(x)$ is arbitrary. As it was pointed out above, the only known case is when $F(x)$ is continuous. The above discussed special cases indicate that the limit distribution of the waiting time for a monotone or strongly monotone block of length m highly depends on $F(x)$. The author intends to return to these questions later on.

References

[1] CSÁKI, E., FÖLDES, A., KOMLÓS, J., Limit Theorems for Erdős-Rényi type problems, *Studia Sci. Math. Hung.* **22** (1987), 321-332.

[2] ERDŐS, P., and RÉNYI, A., On a new law of large numbers, *J. Analyse Math.* **23** (1970), 103-111.

[3] ERDŐS, P., and RÉVÉSZ, P., On the length of the longest head run, Topics Information Theory, *Colloq. Math. Soc. János Bolyai.* **16**, 219-228, North-Holland, Amsterdam, 1977.

[4] FÖLDES, A., The limit distribution of the length of the longest head run, *Period. Math. Hungar.* **10** (1979), 301-310.

[5] GUIBAS, L. J. and ODLYZKO, A. M., Long repetitive patterns in random sequences, *Z. Wahrsch. Verw. Gebiete* **53** (1980), 241-262.

[6] KOMLÓS, J. and TUSNÁDY, G., On sequences of "pure heads", *Ann. Probab.* **3** (1975), 608-617.

[7] PITTEL, B. G., A process of runs and its convergence to the Brownian motion, *Stoch. Process. Appl.* **10** (1980), 33-48.

[8] PITTEL, B. G., Limiting behaviour of a process of runs, *Ann. Probab.* **9** (1981), 119-129.

[9] RÉVÉSZ, P., Three problems on the length of increasing runs, *Stochastic Process. Appl.* **15** (1983), 169-179.

[10] RÉVÉSZ, P., Random Walk in Random and Non-Random Environments, World Scientific, Singapore, 1990.

[11] SAMAROVA, S. S., On the length of the longest head-run for the Markov chain with two states, *Teor. Verojatnost. i Primenen* **26** (1981), 510-520 (in Russian).

APPLICATIONS OF THE STEIN-CHEN METHOD TO THE THEORY OF PATTERNS AND RUNS:
An Annotated Bibliography

Anant P. Godbole
Michigan Technological University

We provide a (fairly) complete list of references, with accompanying notes, describing recent progress made, using the Stein-Chen method, towards understanding the probabilistic basis of randomly generated patterns. Some suggestions for further research are offered.

1. Introduction

The Stein-Chen method consists of a group of versatile, powerful and (ultimately) simple techniques that can be used in various situations to establish the validity of a Poisson approximation for sums of dependent indicator random variables. Since its introduction by Stein in 1972 [see Stein (1986)] and Chen (1974), the technique has grown in scope and importance, primarily through the work of Andrew Barbour and his co-workers.

The Method, as we shall refer to it in the sequel, has been applied to problems in fields as diverse as random graph theory; Ramsey theory; number theory; reliability theory; molecular genetics; insurance; the statistics of extremes; classical combinatorial probability; spacings; random mappings, partitions, permutations and other random combinatorial structures; randomness testing, etc. The above list can go on and on, but we shall resist the temptation to give too many references and details, primarily due to the fact that the full story, up till 1992 at any rate, has already been told in the wonderful monograph of Barbour, Holst and Janson (1992) - which is the definitive reference on the *coupling* approach to Stein-Chen approximation. The *local* approach, which is easily implementable if a local dependence structure exists for the indicator variables in question, is the subject of two recent papers by Arratia,

A. P. Godbole and S. G. Papastavridis (eds.), Runs and Patterns in Probability, 253–261.

Goldstein and Gordon [(1989), (1990)]. The primary advantage of the Method over other approaches - such as Bonferroni-type inequalities or the method of moments - is that it yields, in a telling way, a total variation error bound for the deviation between the distributions of the counting variable under consideration, and that of a Poisson r.v. with the same (or occasionally different) mean. Readers wishing to gain an understanding of the Stein-Chen "magic formula" and of the Method in general are urged to study Barbour, Holst and Janson (1992) and Arratia, Goldstein and Gordon (1990); the breadth and depth of the examples covered therein is indeed impressive.

Our main goal is far more modest: We shall provide a (fairly) complete list of papers written on the use of the Method in the probabilistic analysis of (primarily linear or circular) patterns and runs, the main subject of this volume. We hope that readers will find this list useful, and that it will stimulate further research. We apologize in advance for any inadvertent omissions. After listing a paper, we offer a short description of its contents. We end by stating some problems for further research.

2. Use of the Stein-Chen Method in the Theory of Patterns and Runs: An Annotated Bibliography.

[1] Arratia, R., Goldstein, L. and Gordon, L. (1989). Two moments suffice for Poisson approximations: the Chen-Stein method, *Ann. Probab.* **17**, 9-25.

A definitive study of runs in independent Bernoulli trials. The authors use general theorems on the local approach to Stein-Chen approximation to study the properties of the length of the longest run; the Erdős-Rényi law in distribution and the Erdős-Rényi law for matching two random sequences.

[2] Arratia, R., Goldstein, L. and Gordon, L. (1990). Poisson approximation and the Chen-Stein method, *Statist. Sci.* **5**, 403-440.

A survey paper in which the agenda of Arratia, Goldstein and Gordon (1989) is carried further. In particular, process versions of the Method are employed to provide a compound Poisson approximation for the number of overlapping success runs of a given length.

[3] Arratia, R., Gordon, L. and Waterman, M. (1992). The Erdős-Rényi law in distribution, for coin tossing and sequence matching, *Ann. Statist.* **18**, 539-570.

A basic paper in which the Erdős-Rényi law in distribution and the the Erdős-Rényi law for matching two random sequences are studied using the Stein-Chen method. The material in this paper is summarized in Arratia, Goldstein and Gordon (1989).

[4] Barbour, A., Holst, L. and Janson, S. (1992). *Poisson Approximation.* Oxford University Press.

The basic reference source. Chapter 8 is devoted to a study of exceedances and extremes, of which our problems are special cases. Of special relevance is the material on pp. 162-169, where Theorem 8.H [on the number of visits to a rare set by a stationary ergodic Markov chain] is used to provide an improved Poisson approximation for the number of non-overlapping success runs of a given length. Compound Poisson limits are treated in Chapter 10, and the definitive improvements on pp. 244-249 show how the "magic factors", that are absent in several multivariate situations, can be reintroduced.

[5] Barbour, A., Chryssaphinou, O. and Roos, M. (1993a). Compound Poisson approximation in systems reliability. Preprint.

The results of Roos' thesis [Roos (1993a)] are employed in the context of two-dimensional consecutive-k-out-of-n:F systems, and consecutive-2 and connected-s systems. A double pipeline model is studied. The local approach to compound Poisson approximation is employed.

[6] Barbour, A., Chryssaphinou, O. and Roos, M. (1993b). A compound Poisson approximation for the consecutive-k-out-of-n:F systems. Preprint.

Similar in spirit to Barbour, Chryssaphinou and Roos (1993a), except that the systems under discussion are the consecutive-k-out-of-n:F systems, for which significant improvements are developed using Roos' (1993a) CP-local approach.

[7] Burghardt, P., Godbole, A. and Prengaman, A. (1994). A Poisson approximation for the number of k-matches. *Statist. Probab. Lett.* To appear.

A different kind of problem is studied here; a Poisson approximation is developed, using both monotone and non-monotone couplings, for the number of times that the same letter appears within a memory window of length k.

[8] Campbell, S., Godbole, A. and Schaller, S. (1993). Discriminating between sequences of Bernoulli and Markov-Bernoulli trials. Preprint.

The simple case of $k = 1$ *is studied in this paper, but in the context of both Bernoulli and Markov trials. An unconditional runs test is proposed, and, using the Method, is shown to be almost consistent.*

[9] Chryssaphinou, O. and Papastavridis, S. (1988). A limit theorem for the number of non-overlapping occurrences of a pattern in a sequence of independent trials, *J. Appl. Probab.* **25**, 428-431.

This paper was the source of the author's initial interest in applying the Method to the theory of patterns and runs. The Method is used in the form developed by Barbour and Eagleson (1984).

[10] Chryssaphinou, O. and Papastavridis, S. (1990). Limit distribution for a consecutive-k-out-of-n:F system, *Adv. Appl. Probab.* **22**, 491-493.

Barbour and Eagleson's (1984) formula is applied to derive an approximate expression for the reliability of a consecutive-k-out-of-n:F system.

[11] Dembo, A. and Karlin, S. (1992). Poisson approximations for r-scan processes, *Ann. Appl. Probab.* **2**, 329-357.

Several general problems are considered, and Godbole's (1991) result on runs of successes is derived as a corollary of a theorem on the Poisson approximation of scan statistics.

[12] Fu, J. and Koutras, M. (1993). Poisson approximations for 2 dimensional patterns, *Ann. Instit. Statist. Math.* To appear.

The Method is not used; correlation inequalities are employed to derive the reliability of two-dimensional consecutive-k-out-of-n:F systems. Comparisons are made to bounds obtained on using the Stein-Chen technique; the latter usually perform poorer.

[13] Geske, M., Godbole, A., Schaffner, A., Skolnick, A., and Wallstrom, G. (1993). Compound Poisson approximations for word patterns under Markovian hypotheses. Preprint.

Pattern generation from a stationary Markov chain is studied. Process versions of the Method (in its local form) are used to derive compound Poisson approximations. It would be interesting to see whether Roos' (1993a) ideas could be extended to reintroduce the "magic factor".

[14] Godbole, A. (1991). Poisson approximations for runs and patterns of rare events, *Adv. Appl. Probab.* **23**, 851-865.

Both overlapping and non-overlapping runs and patterns are studied. The long range dependence in the Markov case is analyzed. Most of the results here have been superceded by those in Godbole and Schaffner (1993).

[15] Godbole, A. (1993). Approximate reliabilities of m-consecutive-k-out-of-n: Failure systems. *Statist. Sinica* **3**, 321-327.

The reliability of m-consecutive-k-out-of-n:F systems is studied. A novel feature is the study of the higher-order Markov situation.

[16] Godbole, A. and Deegan, L. (1993). Strict consecutive-k-out-of-n:F systems: Poisson approximations and asymptotic infeasibility. Preprint.

The so-called strict-consecutive-k-out-of-n: failure system is studied; a Poisson approximation is derived, and the system is shown to be asymptotically infeasible.

[17] Godbole, A. and Gornowicz, M. (1992). Exact and approximate runs distributions. *Comm. Statist.* **A21**, 2151-2167.

Normal and Poisson approximations are derived for the total number of runs in a sequence of Bernoulli trials. The latter approximant is supported on the odd integers. This work led to the developments in Campbell, Godbole and Schaller (1993).

[18] Godbole, A., Khunger, S. and Ososanya, E. (1993). Exact and approximate hypercube reliabilities. In this volume.

A reliability model for the binary d-cube is offered, and the Method employed to get a Poisson approximation. Improvements and extensions seem possible.

[19] Godbole, A., Papastavridis, S. and Weishaar, R. (1993). Formulæ, recursions and approximations for the joint distribution of success runs of several lengths. Preprint.

Motivated by statistical considerations, the authors use the Method to derive a multivariate Poisson approximation for the numbers of non-overlapping runs of several different lengths.

[20] Godbole, A. and Schaffner, A. (1993). Improved Poisson approximations for word patterns, *Adv. Appl. Probab.* **25**, 334-347.

General word patterns are considered, and both the independent and Markov cases are studied; the former is tackled using methodology in Barbour, Holst and Janson's text (1992), while extensive use is made, in the Markov case, of eigenvalue bounds on convergence to stationarity.

[21] Goldstein, L. (1990). Poisson approximation and DNA sequence matching, *Comm. Statist.* **A19**, 4167-4179.

Similar in spirit to Arratia, Goldstein and Gordon (1989); new examples are considered.

[22] Koutras, M., Papadopoulos, G. and Papastavridis, S. (1993a). Runs on a circle. Preprint.

The local version of the Stein-Chen method is employed to prove Poisson convergence results for the numbers of non-overlapping success runs, given a circular sequence of independent but not necessarily i.i.d. trials. Applications to randomness testing are given.

[23] Koutras, M., Papadopoulos, G. and Papastavridis, S. (1993b). Reliability of 2-dimensional consecutive-k-out-of-n:F systems, *IEEE Trans. Reliab.* To appear.

The Method is employed to obtain a Weibull limit for the time to failure of a two-dimensional consecutive-k-out-of-n: failure system. Both the i.i.d. and non-i.i.d. cases are treated.

[24] Koutras, M., Papadopoulos, G. and Papastavridis, S. (1993c). Circular overlapping success runs. In this volume.

Analogous treatment as in Koutras, Papadopoulos, and Papastavridis (1993a), with the focus being on overlapping runs on a circle. The local method is employed, and the non-i.i.d. case is studied. Limit theorems are derived.

[25] Koutras, M. and Papastavridis, S. (1993). Application of the Stein-Chen method for bounds and limit theorems in the reliability of coherent structures, *Naval Research Logistics* **40**, 617-631.

The Method is used to study Poisson approximations for various coherent systems. Minimal cut and minimal path bounds are provided.

[26] Papastavridis, S. (1988). A Weibull limit for the reliability of a consecutive k-out-of-n system, *Adv. Appl. Probab.* **20**, 690-692.

The suitably normalized time to failure of consecutive k-out-of-n systems is shown to follow a Weibull distribution.

[27] Roos, M. (1993a). Stein-Chen method for compound Poisson approximation. Ph. D. dissertation, Universität Zürich.

The local approach of Chen (1974) is extended to the case of compound Poisson approximation, using the compound Poisson Stein-Chen equation introduced by Barbour, Chen and Loh (1992). A fundamental consequence is the reintroduction of the "magic factor". Applications to a vast variety of situations are given.

We conclude this section by mentioning two general problems that we would like to investigate further. Firstly, the case of several patterns is a vast topic that might be vulnerable to attack using the Stein-Chen method. In many cases, remarkable results have been obtained on using ingenious methods such as the independence principle; see, e.g., the vast body of results of Móri [references may be found in Móri (1993)]. The paper of Chryssaphinou, Papastavridis and Tsapelas (1993) is a little more restricted (the patterns are disjoint), but sharp results are obtained. How does one use the Method to bypass the problem of the arbitrary overlaps between various patterns?

The problem of designing optimal statistical tests for randomness based on the material in the above papers is a vast one. The optimality of these procedures needs to be addressed; in Campbell, Godbole and Schaller (1993), the Method was employed in one such instance, but much more needs to be done, and many more test criteria and alternatives to randomness need to be considered, as in Agin and Godbole (1992) or O'Brien and Agin (1993).

Additional References

Agin, M. and Godbole, A. (1992). A new exact runs test for randomness, in *Computing Science and Statistics*, Proceedings of the 22nd Symposium on the Interface, C. Page and R. LePage, eds., Springer Verlag, New York, 281-285.

Barbour, A., Chen, L. H-Y and Loh, W-L (1992). Compound Poisson approximation for nonnegative random variables via Stein's method, *Ann. Probab.* **20**, 1843-1866.

Barbour, A. and Eagleson, G. (1984). Poisson convergence for dissociated statistics, *J. Royal Statist. Soc. B* **46**, 397-402.

Chen, L. H-Y (1974). Poisson approximation for dependent trials, *Ann. Probab.* **3**, 534-545.

Chryssaphinou, O. Papastavridis, S. and Tsapelas, T. (1993). On the waiting time of appearance of given patterns. In this volume.

Móri, T. (1993). Asymptotic joint distribution of cover times. In this volume.

O'Brien, P. and Agin, M. (1993). Robust procedures for detecting non-random patterns. In this volume.

Roos, M. (1993b). Stein's method for compound Poisson approximation: the local approach. Preprint.

Stein, C. (1986). *Approximate Computation of Expectations*, Institute of Mathematical Statistics, Hayward, California.

RUNS OF SUPERIMPOSED POISSON PROCESSES

Anant P. Godbole
Michigan Technological University

Hao Zhang
University of Pennsylvania

Consider two independent Poisson processes $\{X_t\}$ and $\{Y_t\}$ with intensities α and β respectively, and let $N_{k,t}$ denote the number of recurrent runs of length k, of events associated with the first process, in the interval $[0,t]$. We derive the exact probability distribution of $N_{k,t}$, and of the random variables E_k and $\Gamma_{k,r}$, defined, respectively, as the waiting times for the first and rth such runs ($r \geq 2$). Applications to statistical tests for randomness are presented. Next, let R_t denote the total number of runs, of events associated with either process, in $[0,t]$. The exact distribution of R_t in the sequence of time-based observations is found. Of special importance is the case $\alpha = \beta$, for which the moments and generating functions of R_t can be calculated, leading to a central limit theorem for R_t. Several exponential inequalities are proved that lend credibility to the validity of a law of the iterated logarithm for R_t. More applications are presented.

1. Introduction.

Let $\{X_t\}$ and $\{Y_t\}$, $t \geq 0$, be two independent and simultaneously observable Poisson processes with parameters α and β respectively. Alternately, one may think of a single thinned Poisson process $\{Z_t\}$ for which the arrivals may not be (or choose not to be) registered (certain counter models and queues with balking are specific examples) or for which each successive occurrence independently causes a shock with probability p (component lifetime models are often based on such an assumption). We shall refer to the two types of occurrences as A's and B's, or, alternately, as successes (S) and failures (F). In this paper, we shall start, in Section 2, with a

A. P. Godbole and S. G. Papastavridis (eds.), Runs and Patterns in Probability, 263–286.

study of the intertwining of the two arrival processes, as manifested in a specific characteristic, namely the number of times in the interval $[0, t]$ that one observes a run of k consecutive successes, where k is a fixed but arbitrary integer. This quantity will be denoted by $N = N_{k,t}$; related random variables of interest are the waiting times for the first and the rth ($r \geq 2$) run of A's; these will be denoted by $E = E_k$ and $\Gamma = \Gamma_{k,r}$ respectively. The exact distributions of N, E and Γ are derived (Theorem 1 and Corollary 1). Based on our computation of the one-dimensional distributions of $N_{k,t}$ and our analysis of $N_{k,t}$ as a simple point process (Theorem 2), we shall call N the *Poisson process of order k*, and E and Γ the *exponential and gamma random variables of order k* respectively. The kth-order random variables introduced above will be seen to specialize to the usual Poisson process, and to the exponential and gamma random variables respectively, on setting $k = 1$, as, indeed, they must. Throughout this article, runs of length k will be counted as in Feller (1968). As a specific example, we note that the sequence $FSSSSSF$ will be considered to have two success runs of length 2 (and one of length 3, etc.) rather than as having (i) no runs of length 2, but one of length 5 or (ii) four runs of length 2, viz. $SS---$, $-SS--$, $--SS-$ and $---SS$. The 'order k' terminology was introduced by Philippou, Georghiou and Philippou (1983), and by Philippou and Makri (1986), who derived the exact form of several (discrete) success run distributions, all defined in a natural way, and under the recurrent scheme of Feller (1968). Our derivation of the exact probability distribution of $N_{k,t}$ is based on the same technique as in Godbole (1990a), (1990b), where the formulæ of Philippou et al (1983), (1986) and of Panaretos and Xekalaki (1986) were reformulated in a simpler format that is computationally and algorithmically efficient. The distributions of E_k, $\Gamma_{k,r}$ and L_t, (the length of the longest A-run in $[0, t]$) follow immediately on noting that $P(E_k > t) = P(L_t \leq k - 1) = P(N_{k,t} = 0)$ and that $P(\Gamma_{k,r} > t) = P(N_{k,t} \leq r - 1)$. Incidentally, E_k and $\Gamma_{k,r}$ appear to be the first continuous random variables of order k to have been considered in the literature, and $N_{k,t}$ seems to be the first such random process.

The formulæ of Theorem 1 are interesting in their own right - they enable one to quickly solve textbook-style exercises such the following "find the probability that there are four consecutive arrivals of imported cars at a tollbooth on at least three occasions between noon and 3 p.m."; less frivolous examples of applications will be discussed in sub-section 2.2, where, after briefly discussing a shock model, a statistical test for the independence of the two superimposed Poisson processes will be proposed and its merits considered. These tests can be based on $N_{k,t}$ or, in the case of rare events, on L_t or $\Gamma_{k,r}$, and can, in either case, assume a conditional or an unconditional form. The nature of these test statistics allow one, moreover, to treat both the synchronous (i.e. observations made until the nth arrival) as well as asynchronous (i.e. observations made until time t) cases. See Karr (1986), pp. 123 and 131, for a discussion on the comparison of two independent Poisson processes. Several computational aspects of our formulæ that are relevant to the abovementioned testing procedures will be discussed in Section 2.1; in particular, Laplace transform methods and exponential probability inequalities will be employed to get a better handle on the various expressions we derive in Section 2.

Let $R = R_t$ denote the total number of runs, of events associated with either process, in the interval $[0, t]$. In Section 3, we begin by deriving the exact distribution of R. A further simplification is obtained on using Laplace transform methods. When $\alpha = \beta$, the above distribution will be seen to assume a particularly convenient form that allows for an easy computation of moments and generating functions; these calculations lead to a central limit theorem for R. In addition, several exponential inequalities are obtained that lend credibility to the validity of a law of the iterated logarithm for R_t. More applications to statistical tests are presented.

2. The Generalized Poisson Process.

To begin with, we observe, as in Godbole (1990a), Lemma 1, that the probability of the event $\{N_{k,t} = x, T_t = n, F_t = y\}$ is positive if and only if $n \geq kx$ and

$[(n-kx)/k] \leq y \leq n-kx$, where T_t and F_t represent, respectively, the total number of arrivals and the number of failures in $[0,t]$. This fact enables us to place exact bounds on many of the sums that follow. We have

$$\begin{aligned}
&\mathbf{P}(N_{k,t}=x)\\
&=\sum_n\sum_y \mathbf{P}(N_{k,t}=x, T_t=n, F_t=y)\\
&=\sum_n\sum_y \mathbf{P}(T_t=n)\mathbf{P}(F_t=y|T_t=n)\mathbf{P}(N_{k,t}=x|T_t=n,F_t=y)\\
&=\sum_n\sum_y e^{-(\alpha+\beta)t}\frac{[(\alpha+\beta)t]^n}{n!}\binom{n}{y}(\frac{\alpha}{\alpha+\beta})^{n-y}(\frac{\beta}{\alpha+\beta})^y\mathbf{P}(N_{k,t}=x|T_t=n,F_t=y),
\end{aligned} \tag{1}$$

where n varies between kx and ∞, and y satisfies $[(n-kx)/k] \leq y \leq n-kx$; (1) holds because T_t has a Poisson distribution with parameter $\alpha+\beta$, and the failures in a superimposed Poisson scheme have a binomial distribution (Karlin and Taylor (1975), p. 128).

We need, next, to compute the conditional probability in (1); this can be accomplished as in Godbole (1990a), Theorem 1. This yields

$$\begin{aligned}
&P(N_{k,t}=x)\\
&=\sum_n e^{-(\alpha+\beta)t}\frac{[(\alpha+\beta)t]^n}{n!}\bullet\sum_{y=[(n-kx)/k]}^{n-kx}\binom{y+x}{x}(\frac{\alpha}{\alpha+\beta})^{n-y}(\frac{\beta}{\alpha+\beta})^y\\
&\bullet\sum_{j=0}^{[(n-kx-y)/k]}(-1)^j\binom{y+1}{j}\binom{n-kx-jk}{y}
\end{aligned} \tag{2}$$

We have thus proved

Theorem 1. *Let $\{X_t\}$ and $\{Y_t\}$ be two independent Poisson processes with arrival rates α and β respectively, and let $N_{k,t}$ represent the number of non-overlapping runs of length $k \geq 1$, of particles associated with the X process, in the interval $[0,t]$. Then $\mathbf{P}(N_{k,t}=x)$ is given by (2) for $x=0,1,2,\ldots$.* □

The next corollary follows immediately from the observations made in Section 1:

Corollary 1. *Let L_t and E_k denote, respectively, the length of the longest run of A-particles in the period $[0,t]$, and the waiting time for the first run of length k. Then*

$$\begin{aligned}
&\mathbf{P}(L_t \le k-1) = \mathbf{P}(E_k > t) \\
&= \sum_{n=0}^{\infty} e^{-(\alpha+\beta)t} \frac{[(\alpha+\beta)t]^n}{n!} \bullet \sum_{y=[n/k]}^{n} (\frac{\alpha}{\alpha+\beta})^{n-y} (\frac{\beta}{\alpha+\beta})^{y} \\
&\bullet \sum_{j=0}^{[(n-y)/k]} (-1)^j \binom{y+1}{j} \binom{n-jk}{y}
\end{aligned} \tag{3}$$

It is easy to verify that the expression for $P(N_{k,t} = x)$ in (2) reduces to $\exp(-\alpha t)$ $(\alpha t)^x / x!$ on setting $k = 1$, with all the terms involving β dropping out. Note also that the tail probability of the distribution of E_k can be written in the form

$$\begin{aligned}
&\mathbf{P}(E_k > t) \\
&= e^{-(\alpha+\beta)t} \sum_{n=0}^{\infty} \frac{(\alpha t)^n}{n!} \sum_{y=[n/k]}^{n} (\frac{\beta}{\alpha})^y \sum_{j=0}^{[(n-y)/k]} (-1)^j \binom{y+1}{j} \binom{n-jk}{y} \\
&= e^{-(\alpha+\beta)t} \sum_{n=0}^{\infty} a_{n,\alpha,\beta} \frac{(\alpha t)^n}{n!}, \quad \text{say,}
\end{aligned} \tag{4}$$

The second component of (4) represents a power series that clearly converges for each real t (to $\exp[(a+b)t] \bullet \mathbf{P}(E_k > t)$). It follows that $\mathbf{P}(E_k > t)$ is differentiable for each t, and that the required differentiation may be carried out term by term. We conclude, from (4), that the following result holds:

Corollary 2. *The probability density function of E_k is given by*

$$f(t) = \frac{e^{-(\alpha+\beta)t}}{t} \sum_{n=0}^{\infty} a_{n,\alpha,\beta} \frac{(\alpha t)^n}{n!} \{(\alpha+\beta)t - n)\} \tag{5}$$

If we set $k = 1$ in (5), then it can easily be verified that $a_{n,\alpha,\beta}$ equals $(\beta/\alpha)^n$, that the terms involving β drop out, and that (5) reduces to $\alpha \exp(-\alpha t)$, as, of course, it must. Note also that the distribution of the waiting time Γ for the rth run is obtained from the relation $\mathbf{P}(\Gamma_{k,r} > t) = \mathbf{P}(N_{k,t} \leq r-1)$. Its density may then be obtained easily, as was the case with E_k. The moments, moment generating function etc. of E_k (and $\Gamma_{k,r}$) can also be readily expressed as an infinite sum: for example, either (4) or (5) can be used to show that

$$\mathbf{E}(E_k) = \frac{1}{\alpha+\beta}\sum_{n=0}^{\infty} a_{n,\alpha,\beta}\left(\frac{\alpha}{\alpha+\beta}\right)^n \tag{6}$$

which may be verified to specialize to the familiar mean $1/\alpha$ of the exponential r.v. on setting $k = 1$; the series in (4) converges uniformly on $\mathbf{R}^+$, allowing us to integrate it term by term.

We have seen above that the *one-dimensional* distributions of the process $\{N_{k,t} : t \geq 0\}$ coincide with those of the ordinary Poisson process on setting $k = 1$; this is not sufficient, however, for us to be able to characterize the former as a generalized Poisson process. Our next result is based on an analysis of N as a simple point process, and justifies the terminology *"Poisson process of order k"* that we shall be employing, henceforth, for $\{N_{k,t} : t \geq 0\}$; E_k and $\Gamma_{k,r}$ will, likewise, be called the *exponential and gamma distributions of order k* respectively:

Theorem 2. *The right continuous, non-decreasing, simple point process $\{N_{k,t} : t \geq 0\}$ is a generalization of the simple Poisson process $\{P_t : t \geq 0\}$ with parameter α in the sense that the distribution of the former reduces to that of the latter on setting $k = 1$.*

Proof. It is well-known that two simple point processes R and S are identical if and only if their *zero-probability functionals* (ZPF's) $z_R(\bullet)$ and $z_S(\bullet)$ coincide; i.e., if for each $A \in \mathcal{B}(\mathbf{R})$, we have $z_R(A)(= \mathbf{P}(R(A) = 0)) = z_S(A)(= \mathbf{P}(S(A) = 0))$ - see, e.g. Karr (1986), Theorem 1.12. In other words, a simple point process is fully characterized by its ZPF, and we accordingly define $N_k(A)$ to be the number of non-overlapping k-long runs of the X process in the time period determined by the set A. We need to show that the ZPF of N_k reduces to that of the simple Poisson process P on setting $k=1$. Suppose we are able to show that $\mathbf{P}(N_k(A)=0)|_{k=1} = \mathbf{P}(P(A)=0)$ for each set A that is a finite union of disjoint left-half-open intervals, then the monotone class theorem would lead us to the required conclusion, i.e. that $\mathbf{P}(N_k(A)=0)|_{k=1} = \mathbf{P}(P(A)=0)$ for each $A \in \mathcal{B}(\mathbf{R})$. Now if A is a union of intervals of the specified type, one of which is of infinite length, then it is evident that both $\mathbf{P}(N_k(A)=0)$ and $\mathbf{P}(P(A)=0)$ equal 0. It may be assumed without loss of generality, therefore, that A is the union of finitely many disjoint, left-half-open and bounded intervals. If $A = \cup_{j=1}^{n}(a_j, b_j]$, then

$$\mathbf{P}(N_k(A)=0)|_{k=1} = \prod_{j=1}^{n} \mathbf{P}(N_k(a_j,b_j]=0|N_k(a_j,b_j]=0, 1 \le i \le j-1)|_{k=1} \quad (7)$$

Now the process $\{N_{1,t}\}$ has independent increments, so that

$$\begin{aligned}\mathbf{P}(N_k(A)=0)|_{k=1} &= \prod_{j=1}^{n} \mathbf{P}(N_1(a_j,b_j]=0)\\ &= \prod_{j=1}^{n} \mathbf{P}(P(a_j,b_j]=0)\\ &= \mathbf{P}(P(A)=0),\end{aligned} \quad (8)$$

proving the required result. □

2.1. Computation and Approximation.

In this subsection, we shall present several methods by which the formulæ of Section 2 may be simplified and/or approximated so as to make them tractable for values of $k \geq 2$. In what follows, we shall suppress the limits on the various summations; often, they will also be replaced by somewhat larger but equivalent sets. We start by considering the Laplace transform L of *part* of (2), the key formula of Section 2:

$$e^{-(\alpha+\beta)t} L\left\{ \sum_{n=kx}^{\infty} \frac{(\alpha t)^n}{n!} \sum_y \binom{y+x}{x} \left(\frac{\beta}{\alpha}\right)^y \sum_j (-1)^j \binom{y+1}{j} \binom{n-kx-jk}{y} \right\}$$

$$=e^{-(\alpha+\beta)t}\sum_y \binom{y+x}{x}(\frac{\beta}{\alpha})^y \sum_j (-1)^j \binom{y+1}{j} \sum_{n=kx+jk+y}^{\infty} \frac{\alpha^n}{s^{n+1}} \binom{n-kx-jk}{y}$$

$$=e^{-(\alpha+\beta)t}\alpha^{kx}\sum_y \binom{y+x}{x}(\frac{\beta}{\alpha})^y \sum_j (-1)^j \binom{y+1}{j} \sum_{m=y}^{\infty} (\frac{\alpha}{s})^{m-y} \binom{m}{y} (\frac{\alpha}{s})^{jk+y} s^{-(kx+1)}$$

$$=e^{-(\alpha+\beta)t}\alpha^{kx}\sum_y \binom{y+x}{x}(\frac{\beta}{\alpha})^y \alpha^y s^{-(kx+1+y)}(1-\frac{\alpha}{s})^{-(y+1)} \sum_j (-1)^j \binom{y+1}{j} (\frac{\alpha}{s})^{jk}$$

$$=e^{-(\alpha+\beta)t}\alpha^{kx}\sum_y \binom{y+x}{x}(\frac{\beta}{s})^y s^{-(kx+1)}(1-\frac{\alpha}{s})^{-(y+1)}(1-(\frac{\alpha}{s})^k)^{y+1}$$

$$=e^{-(\alpha+\beta)t}\frac{\alpha^{kx}}{s^{kx+1}}\frac{(1-(\frac{\alpha}{s})^k)}{(1-\frac{\alpha}{s})}\sum_y \binom{y+x}{x}(\frac{\beta}{s})^y(\frac{1-(\frac{\alpha}{s})^k}{(1-\frac{\alpha}{s})})^y$$

$$=e^{-(\alpha+\beta)t}(\frac{\alpha}{s})^{kx}\frac{(1-(\frac{\alpha}{s})^k)}{(s-\alpha)}\frac{1}{(1-\frac{\beta(1-(\alpha/s)^k)}{(s-\alpha)})^{x+1}}, \tag{9}$$

where the summation over y has been taken, without loss of validity, to be over all of $\mathbf{Z}^+$; j has been summed from 0 to $y+1$ in the above interchange. It follows that $\mathbf{P}(N_{k,t}=x)$ can be reexpressed as

$$e^{-(\alpha+\beta)t}\alpha^{kx}L^{-1}\left\{\frac{1-(\alpha/s)^k}{s^{kx}(s-\alpha)}\frac{1}{(1-\frac{\beta(1-(\alpha/s)^k)}{(s-\alpha)})^{x+1}}\right\} \tag{10}$$

The above representation has several advantages. Note, in particular, that it assumes a particularly simple form when, for example, $x=0$ or $\alpha=\beta$. In general, the inverse Laplace transform in (10) can be found whenever the partial fraction decomposition there can be computed. For specific known values of α and β, this can be achieved, at least in principle, with the answer being expressible in terms of a finite combination of elementary functions. Approximate solutions can certainly be obtained in any case.

Our second method relies on the fact that the infinite sum in (2) can be approximated by a finite sum on using the Chernoff-Okamoto exponential inequalities for Poisson tails (see, for example, Dudley (1978)); we shall state these inequalities for ease of reference: Let $P(k,\lambda)$ and $Q(k,\lambda)$ represent, respectively, the Poisson lower and upper tail probabilities $\mathbf{P}(\text{Po}(\lambda) \le k)$ and $\mathbf{P}(\text{Po}(\lambda) \ge k)$, where $\text{Po}(\lambda)$ denotes the Poisson r.v. with mean λ. Then

$$P(k, \lambda) \le (\lambda/k)^k \exp(k - \lambda) \quad (k \le \lambda) \tag{11}$$

and

$$Q(k, \lambda) \le (\lambda/k)^k \exp(k - \lambda) \quad (k \ge \lambda) \tag{12}$$

By (12), $\mathbf{P}(T_t \ge N) \le ((\alpha+\beta)te/N)^N$ if N is larger than $(\alpha+\beta)t$; thus restricting the sum over n to the set $\{kx, kx+1, \ldots, N\}$, while computing the quantity $\mathbf{P}(N_{k,t} = x)$, leads to an error of at most $((\alpha + \beta)te/N)^N$. This is not entirely satisfactory, however, from the viewpoint of the statistical tests that will be considering in the next section, since we shall often need an error estimate on the upper tail $\mathbf{P}(N_{k,t} \ge y)$. This quantity may easily be bounded, however, on using the Markov inequality as follows: Since $\mathbf{E}(N_{k,t}) \le \mathbf{E}(T_t) = (\alpha + \beta)t < \infty$, $\mathbf{P}(N_{k,t} \ge y) = O(1/y)$; this last bound is admittedly crude, but it suffices, and can easily be improved. Now, suppose we wish to compute an upper tail probability such as $\mathbf{P}(N_{k,t} \ge x)$, which equals $\mathbf{P}(x \le N_{k,t} \le y - 1) + \mathbf{P}(N_{k,t} \ge y)$ (for some carefully chosen y) then the first component may be estimated on using the Chernoff-Okamoto bounds, whereas the second term can be tackled with Markov's inequality.

Finally, we will briefly consider the computation of the mean waiting time for the first A-run of length k; higher order moments are handled similarly. We have, by (3) and (6),

$$\mathbf{E}(E_k) = \frac{1}{\alpha+\beta} \sum_{n \ge 0} \mathbf{P}(L_t \le k - 1 | T_t = n) = \frac{1}{\alpha+\beta} \sum_{n \ge 0} \mathbf{P}(N_{k,t} = 0 | T_t = n) \tag{13}$$

The summands in (13) can easily be computed for each n (on using a conditional version of (3)). More importantly, $\mathbf{P}(N_{k,t} = 0 | T_t = n)$ is bounded below, for each $n \ge k$, by $\mathbf{P}(F_t \ge n - (k-1) | T_t = n)$, which equals

$$\sum_{j=0}^{k-1} \binom{n}{j} \left(\frac{\alpha}{\alpha+\beta}\right)^j \left(\frac{\beta}{\alpha+\beta}\right)^{n-j} \tag{14}$$

which is, in turn, larger than $k(\gamma/\alpha+\beta)^n$, where $\gamma = \min\{\alpha, \beta\}$. It follows that the tail of the series in (13) can be bounded below by a term of the form $\frac{A}{\alpha+\beta}(\frac{\gamma}{\alpha+\beta})^N$, which can be made arbitrarily small. We now consider an upper bound on the summands in (13). Observe that

$$\mathbf{P}(N_{k,t} = 0 | T_t = n) \leq (1 - (\frac{\alpha}{\alpha+\beta})^k)^{n/k}, \tag{15}$$

since if $N_{k,t} = 0$, there must necessarily have been no success runs of length k in any of the (approximately) $[n/k]$ adjacent (and non-overlapping) blocks of length k starting with the one corresponding to trials $\{1, 2, \ldots, k\}$. It follows as before that the tail of the series in (13) is bounded above as well by a term that is of order B^N; $B < 1$. These easy estimates provide us with a magnitude of the error incurred when approximating (13) by its finite analog.

2.2. Applications.

We shall present two basic applications, the second probably deeper than the first. Consider the following shock model: a device malfunctions due to the cumulative effect of k consecutive shocks, with each impinging Poissonian event independently causing such a shock to occur with probability p, $0 < p < 1$. If one has \$ X in one's annual repair budget, and if each breakdown costs \$ A, the models of Section 2 could be used to find the probability of exceeding this budget, or, conversely, to find a repair allocation (\$ Y, say) that would not be exceeded with probability $1-\lambda$. The expected time till failure of such a system could be found on using (13). The obvious question that arises, of course, is the following: what detrimental effects do l consecutive shocks cause, where $l \leq k-1$? None? Unfortunately, we are forced to answer 'yes' if our models are to be used meaningfully. In other words, the kind of situation that we must restrict ourselves to is one for which the system is 'self-healing', with a single 'success' being able to undo the effect of upto $k-1$ consecutive shocks. We might, for example, consider a device that is rented to customers who are either careful or negligent. A careful customer will service the device before use (change the oil,....) and consecutive rentals to k negligent customers will wreck the device. What we have outlined is clearly related to the so-called 'consecutive-k-out-of-n-failure' reliability systems; see Chryssaphinou and Papastavridis (1990) for an analysis of such a system in a random environment. The main difference between the two set-ups is that we consider a single device that must incur k consecutive shocks, whereas Chryssaphinou and Papastavridis (1990) deal with several linearly arranged components, each of which is independently subject to a shock with probability p; a single shock puts any component out of action, and subsequent events have a potential impact only on functioning components. The system fails when k consecutive components are knocked out. It is evident, then, that the two situations are, in a sense, opposites of each other. Our assumption of 'self-healing' is analogous to assuming, for example, that a consecutive k-out-of-n-failure pipeline system will be operative even if there are a very large number of segments of adjacent and malfunctioning components, provided that each segment is of length at most $k-1$. Such a (standard) assumption is no less unrealistic than ours.

We now address the problem of testing for the independence of the two Poisson processes $\{X_t : t \geq 0\}$ and $\{Y_t : t \geq 0\}$. We shall assume throughout that α and β are known. To begin, we recall that standard runs tests for the random arrangement of n symbols, n_1 of which are S (with the other n_2 symbols being F) are based either on the total number of runs (the Wald-Wolfowitz test) or on the length of the longest success run (Mosteller's test). Both criteria are discussed in Gibbons (1971). Both these tests are conditional on the values of n_1 and n_2 being known, but may easily be adapted to the situation where these quantities have not, as yet, been determined (e.g., the experiment may not have been conducted). In the latter case, it is essential that the value of p be known, so that the required conditioning can be performed. Other criteria for testing for randomness against various kinds of alternatives (due to Haldane, Smith, Barton, David, Mallows, O'Brien, Dyck etc.) are discussed in Agin and Godbole (1992). In the same article, a new test for randomness is proposed. It is based on the number of success runs of a fixed length k; since k is free to assume any value larger than 1, we obtain a versatile procedure which is a generalization of the 'length of the longest run' test (due to the fact that the longest run is of length at most $l-1$ iff there are no runs of length l). Furthermore, the new test was seen to be more powerful than the standard criteria for the alternatives of non-systematic and cyclical clustering.

If we suspect that the arrival orders of the A's and B's in our superimposed Poisson process are not random, then several of the abovementioned criteria can easily be adapted to serve as test statistics; in particular, a formula for the total number of runs in $[0,t]$ can be derived along the lines of Theorem 1 above, yielding, consequently, an analog of the Wald-Wolfowitz test. We do this in Section 3. Corollary 1 can be used to obtain a test analogous to Mosteller's. Either procedure can be used for testing for the independence of the two stochastic processes. Conditional versions of these tests will not pose any computational problems, and the approximations of Section 2.1 can be adapted for use with the unconditional versions. If we base an exact test on the statistic $N_{k,t}$, however, then we stand a better chance, as before, of detecting certain kinds of non-randomness. As an rather artificial example, we note that a repetitive arrival pattern such as $AAABAAAB\ldots$ would not lead to the rejection of H_0 using either of the standard tests (with $\alpha = 3$ and $\beta = 1$, say); on the other hand, the value of $N_{3,t}$ can be shown to be too large for most time intervals. Simulations, numerical details, exact distributions under alternative hypotheses, computations of power, and other verifications of the efficacy of our proposed procedures will be presented in a forthcoming article; for now, however, we note the following facts:

(a) The exact test based on $N_{k,t}$ is versatile enough to be used under a wide variety of situations; note, in particular, that we can run the processes until a prescribed number N of events have occurred (this yields synchronous observations and a conditional test) or for some fixed time T (asynchronous data is obtained), or combine these possibilities in some fairly arbitrary way (we might, for example, decide to observe the processes for time T, provided that $T + N$ exceeds a level Λ, and for time S otherwise). It is also possible to design sequential tests for the independence of the two processes.

(b) The test can be used for a variety of k's, and it is possible, in principle, to use whatever value of k that happens to be convenient (or necessary!) At the present time, work is in progress to lessen the arbitrariness that such a set-up entails; this could be achieved, in particular, if the joint distribution of the number of A-runs of lengths $(k_1, k_2, \ldots, k_r)$ could be found, at least for $r = 2$, and possibly in general, and if the corresponding test could be based on all these (very highly correlated) runs of varying lengths.

(c) A rather serious shortcoming of the method employed is that no attention is paid to the rate at which the two types of events occur in the interval $[0, t]$; in other words, our test is of little use in detecting the kind of non-randomness that manifests itself a time-dependent (as opposed to an order-dependent) way. This problem needs to be addressed in the future.

(d) If one type of occurrence is relatively rare, then no runs of a specified length may occur in a certain interval, and the test based on $N_{k,t}$ would yield a p-value of 1. One needs, therefore, to consider asymptotic distributions. We mention, in this context, the work of Godbole and Schaffner (1993), where the asymptotic Poissonness of the distribution of the number of runs of a certain length k was proved together with rates; both Bernoulli and Markov trials were considered. Asymptotic normality is proved under a different set of assumptions in Feller (1968). In the above rare event situation, our test may be based on waiting time statistics E_k and Γ_k as well.

(e) Questions of computation can be tackled as indicated in Section 2.1. In each case, the infinite series in question can be truncated at some level N, and error bounds obtained for the neglected terms. These can, moreover, be refined if necessary.

3. The Total Number of Runs of Superimposed Poisson Processes.

In this section we consider, as in Section 2, two independent and simultaneously observable Poisson processes $\{X_t\}$ and $\{Y_t\}$, $t \geq 0$, with parameters α and β respectively. The focus of our attention will, however, be the process $\{R_t : t \geq 0\}$, where R_t denotes the total number of runs of either process in the interval $[0, t]$; for example, the sequence $SSSFFSFSSFFF$ corresponds to $R_t = 6$. Letting T_t and F_t be as in Section 2, we first observe that the quantities $R_t = x$, $T_t = n$ and $F_t = y$ must satisfy $[x/2] \leq y \leq n - [x/2]$, so that we have

$$\begin{aligned}
\mathbf{P}(R_t = x) &= \sum_{n=x}^{\infty} \sum_{y=[x/2]}^{n-[x/2]} \mathbf{P}(R_t = x, T_t = n, F_t = y) \\
&= \sum_n \sum_y \mathbf{P}(T_t = n)\mathbf{P}(F_t = y | T_t = n)\mathbf{P}(R_t = x | T_t = n, F_t = y) \\
&= \sum_n \sum_y \frac{e^{-(\alpha+\beta)t}[(\alpha+\beta)t]^n}{n!} \binom{n}{y} \left(\frac{\beta}{\alpha+\beta}\right)^y \left(\frac{\alpha}{\alpha+\beta}\right)^{n-y} \phi(x, n, y) \quad (16)
\end{aligned}$$

where $\phi(x, n, y) = \mathbf{P}(R_t = x | T_t = n, F_t = y)$ is given, as in Gibbons (1971), by

$$\phi(x, n, y) = \frac{2\binom{y-1}{\frac{x}{2}-1}\binom{n-y-1}{\frac{x}{2}-1}}{\binom{n}{y}} \quad (x = 2, 4, \ldots) \tag{17}$$

and

$$\phi(x, n, y) = \frac{\binom{y-1}{(x-1)/2}\binom{n-y-1}{((x-1)/2)-1} + \binom{y-1}{((x-1)/2)-1}\binom{n-y-1}{(x-1)/2}}{\binom{n}{y}} \quad (x = 3, 5, \ldots) \tag{18}$$

It is easy to see, on the other hand, that

$$\mathbf{P}(R_t = 0) = e^{-(\alpha+\beta)t} \tag{19}$$

and

$$\begin{aligned}
\mathbf{P}(R_t = 1) &= \sum_{x=1}^{\infty} \frac{e^{-\alpha t}(\alpha t)^x}{x!} \frac{e^{-\beta t}(\beta t)^0}{0!} + \sum_{y=1}^{\infty} \frac{e^{-\beta t}(\beta t)^y}{y!} \frac{e^{-\alpha t}(\alpha t)^0}{0!} \\
&= e^{-(\alpha+\beta)t}(e^{\alpha t} + e^{\beta t} - 2) \quad (20)
\end{aligned}$$

Now it follows from (16), (17) and (18) that

$$\mathbf{P}(R_t = x) = 2e^{-(\alpha+\beta)t} \sum_{n=x}^{\infty} \sum_{y=[x/2]}^{n-[x/2]} \frac{(\alpha t)^n}{n!} (\frac{\beta}{\alpha})^y \binom{y-1}{(x/2)-1} \binom{n-y-1}{(x/2)-1} \qquad (x = 2, 4, \ldots) \quad (21)$$

and

$$\mathbf{P}(R_t = x) = e^{-(\alpha+\beta)t} \sum_{n=x}^{\infty} \sum_{y=[x/2]}^{n-[x/2]} \frac{(\alpha t)^n}{n!} (\frac{\beta}{\alpha})^y \bullet$$
$$\left[\binom{y-1}{(x-1)/2} \binom{n-y-1}{(x-3)/2} + \binom{y-1}{(x-3)/2} \binom{n-y-1}{(x-1)/2}\right] \quad (x = 3, 5, \ldots) \quad (22)$$

We have thus proved

Theorem 3. *Let R_t be the total number of runs of events associated with two independent Poisson processes $\{X_t\}$ and $\{Y_t\}$, $t \geq 0$, with parameters α and β respectively. Then the probability distribution of R_t is given by equations (19) through (22).* □

We next simplify the formulæ of Theorem 3 by using Laplace transform methods as in Section 2: Observe that for $x = 2, 4, \ldots$,

$$L\left[\sum_{n=x}^{\infty} \sum_{y=[x/2]}^{n-[x/2]} \frac{(\alpha t)^n}{n!} (\frac{\beta}{\alpha})^y \binom{y-1}{(x/2)-1} \binom{n-y-1}{(x/2)-1}\right]$$
$$= \sum_{n=x}^{\infty} \sum_{y=[x/2]}^{n-[x/2]} \frac{\alpha^n}{s^{n+1}} (\frac{\beta}{\alpha})^y \binom{y-1}{(x/2)-1} \binom{n-y-1}{(x/2)-1}$$
$$= \sum_{y=[x/2]}^{\infty} \binom{y-1}{(x/2)-1} (\frac{\beta}{\alpha})^y \sum_{n=y+(x/2)}^{\infty} \binom{n-y-1}{(x/2)-1} \frac{\alpha^n}{s^{n+1}}$$
$$= \sum_{y=[x/2]}^{\infty} \binom{y-1}{(x/2)-1} (\frac{\beta}{\alpha})^y (\frac{\alpha^{y+(x/2)}}{s^{y+1+(x/2)}}) \sum_{n=y+(x/2)}^{\infty} \binom{n-y-1}{(x/2)-1} (\frac{\alpha}{s})^{n-y-(x/2)}$$
$$= \sum_{y=[x/2]}^{\infty} \binom{y-1}{(x/2)-1} (\frac{\beta}{s})^y (\frac{\alpha^{x/2}}{s^{1+(x/2)}}) \sum_{m=(x/2)-1}^{\infty} \binom{m}{(x/2)-1} (\frac{\alpha}{s})^{m-(x/2)+1}$$

$$
\begin{aligned}
&= \frac{\alpha^{x/2}}{s^{(x/2)+1}} \sum_{y=x/2}^{\infty} \binom{y-1}{(x/2)-1} \left(\frac{\beta}{s}\right)^{y-(x/2)} \left(\frac{\beta}{s}\right)^{x/2} \frac{1}{(1-\frac{\alpha}{s})^{x/2}} \\
&= \frac{(\alpha\beta)^{x/2}}{s^{x+1}(1-\frac{\alpha}{s})^{x/2}} \sum_{k=(x/2)-1}^{\infty} \binom{k}{(x/2)-1} \left(\frac{\beta}{s}\right)^{k-(x/2)+1} \\
&= \frac{(\alpha\beta)^{x/2}}{s^{x+1}(1-\frac{\alpha}{s})^{x/2}(1-\frac{\beta}{s})^{x/2}} \\
&= \frac{(\alpha\beta)^{x/2}}{s(s-\alpha)^{x/2}(s-\beta)^{x/2}} \qquad (23)
\end{aligned}
$$

In a similar fashion, it may be verified that for $x = 3, 5, \ldots$,

$$
\begin{aligned}
&L\left[\sum_{n=x}^{\infty} \sum_{y=[x/2]}^{n-[x/2]} \frac{(\alpha t)^n}{n!} \left(\frac{\beta}{\alpha}\right)^y \left[\binom{y-1}{(x-1)/2}\binom{n-y-1}{(x-3)/2} + \binom{y-1}{(x-3)/2}\binom{n-y-1}{(x-1)/2}\right]\right] \\
&= (\alpha\beta)^{(x-1)/2} \frac{(\alpha+\beta)s - 2\alpha\beta}{s(s-\alpha)^{(x+1)/2}(s-\beta)^{(x+1)/2}} \qquad (24)
\end{aligned}
$$

leading to

Corollary 3. *The probability distribution of R_t can be expressed in terms of the inverse Laplace transforms of simple functions as follows:*

$$
\begin{aligned}
&\mathbf{P}(R_t = x) \\
&= e^{-(\alpha+\beta)t} \quad (x = 0) \\
&= e^{-(\alpha+\beta)t}(e^{\alpha t} + e^{\beta t} - 2) \quad (x = 1) \\
&= 2(\alpha\beta)^{x/2} e^{-(\alpha+\beta)t} L^{-1}\left[\frac{1}{s(s-\alpha)^{x/2}(s-\beta)^{x/2}}\right] \quad (x = 2, 4, \ldots) \\
&= (\alpha\beta)^{(x-1)/2} e^{-(\alpha+\beta)t} L^{-1}\left[\frac{(\alpha+\beta)s - 2\alpha\beta}{s(s-\alpha)^{(x+1)/2}(s-\beta)^{(x+1)/2}}\right] \quad (x = 3, 5, \ldots) \qquad (25)
\end{aligned}
$$

The expressions in Theorem 3 and Corollary 3 leave much to be desired from a computational point of view. Using standard formulæ for inverse Laplace transforms, (25) may be reexpressed, by considering the cases $\alpha = \beta$ and $\alpha \neq \beta$ separately, as follows: For $\alpha = \beta$,

$$\mathbf{P}(R_t = x)$$

$$=e^{-2\alpha t} \quad (x=0)$$

$$=2e^{-2\alpha t}\left\{(-1)^x + e^{\alpha t}\sum_{k=0}^{x-1}\frac{t^k}{k!(x-k-1)!}\frac{d^{x-k-1}}{ds^{x-k-1}}\left(\frac{\alpha^x}{s}\right)_{s=\alpha}\right\}, \quad (x \geq 1) \quad (26)$$

whereas the somewhat more complicated general case $\alpha \neq \beta$ has only been stated below for the even case:

$$\mathbf{P}(R_t = x) = 2e^{-(\alpha+\beta)t}\left\{1 + e^{\alpha t}\sum_{k=0}^{x/2}\frac{t^k}{k!((x/2)-k-1)!}\frac{d^{(x/2)-k-1}}{ds^{(x/2)-k-1}}\left(\frac{(\alpha\beta)^{x/2}}{s(s-\beta)^{x/2}}\right)_{s=\alpha}\right.$$

$$\left. + e^{\beta t}\sum_{k=0}^{x/2}\frac{t^k}{k!((x/2)-k-1)!}\frac{d^{(x/2)-k-1}}{ds^{(x/2)-k-1}}\left(\frac{(\alpha\beta)^{x/2}}{s(s-\alpha)^{x/2}}\right)_{s=\beta}\right\} \quad (27)$$

From now on, we just focus on the case $\alpha = \beta$:

Theorem 4. *Let R_t be the total number of runs, in the interval $[0, t]$, of particles associated with either of two independent Poisson processes $\{X_t\}$ and $\{Y_t\}$, $t \geq 0$, with the same intensity α. Then $\mathbf{P}(R_t = 0) = \exp\{-2\alpha t\}$, and for $x \geq 1$,*

$$\mathbf{P}(R_t = x) = 2e^{-\alpha t}(-1)^x \sum_{k=x}^{\infty}\frac{(-\alpha t)^k}{k!} \quad (28)$$

Proof. It can easily be verified that

$$\frac{d^{x-k-1}}{ds^{x-k-1}}\frac{\alpha^x}{s} = \frac{\alpha^x(-1)^{x-k-1}(x-k-1)!}{s^{x-k}} \quad (29)$$

so that (26) yields

$$\begin{aligned}\mathbf{P}(R_t = x) =& 2e^{-2\alpha t}\left[(-1)^x + e^{\alpha t}\sum_{k=0}^{x-1}\frac{t^k}{k!}\frac{\alpha^x(-1)^{x-k-1}}{\alpha^{x-k}}\right] \\ =& 2e^{-2\alpha t}(-1)^x\left[1 - e^{\alpha t}\sum_{k=0}^{x-1}\frac{(-\alpha t)^k}{k!}\right] \\ =& 2e^{-\alpha t}(-1)^x\sum_{k=x}^{\infty}\frac{(-\alpha t)^k}{k!}\end{aligned} \quad (30)$$

This concludes the proof of Theorem 4; it may readily be verified that

$$\lim_{x\to\infty} \mathbf{P}(R_t = x) = 0 \tag{31}$$

and that $\sum_{x=0}^{\infty} \mathbf{P}(R_t = x) = 1$. □

Lemma 1.

$$\begin{aligned}(a)\ &\sum_{x=1}^{k}(-1)^x x\\ &=\frac{k}{2} \quad (k = 2, 4, \ldots)\\ &=-\frac{k+1}{2} \quad (k = 1, 3, \ldots)\end{aligned} \tag{32}$$

$$(b)\ \sum_{x=1}^{k}(-1)^x x^2 = \frac{(-1)^k k(k+1)}{2} \tag{33}$$

$$(c)\ \sum_{k=1}^{\infty}\frac{k(k+1)(\alpha t)^k}{k!} = e^{\alpha t}[(\alpha t)^2 + 2\alpha t] \tag{34}$$

Proof. The proof of (32) is elementary, while (33) may be exhibited to be true using induction. Finally,

$$\begin{aligned}\sum_{k=1}^{\infty}\frac{k(k+1)(\alpha t)^k}{k!} &=\sum_{k=1}^{\infty}\frac{k(\alpha t)^k}{(k-1)!} + \sum_{k=1}^{\infty}\frac{(\alpha t)^k}{(k-1)!}\\ &=\sum_{k=1}^{\infty}\frac{(k-1)(\alpha t)^k}{(k-1)!} + 2\sum_{k=1}^{\infty}\frac{(\alpha t)^k}{(k-1)!}\\ &=(\alpha t)^2\sum_{k=2}^{\infty}\frac{(\alpha t)^{k-2}}{(k-2)!} + 2\alpha t\sum_{k=1}^{\infty}\frac{(\alpha t)^{k-1}}{(k-1)!}\\ &=(\alpha t)^2 e^{\alpha t} + 2\alpha t e^{\alpha t}\end{aligned} \tag{35}$$

proving (34). □

Theorem 5.

$$(a)\ \mathbf{E}(R_t) =(\alpha t + \frac{1}{2}) - \frac{1}{2}e^{-2\alpha t} \tag{36}$$

$$(b)\ \mathbf{V}(R_t) =(\alpha t - \frac{1}{4}) + (\alpha t + \frac{1}{2})e^{-2\alpha t} - \frac{1}{4}e^{-4\alpha t} \tag{37}$$

$$(c)\ M_t(u) =e^{-2\alpha t} - \frac{2e^{u-\alpha t}}{1+e^u}(e^{-\alpha t} - e^{\alpha t e^u}) \tag{38}$$

where $M_t(u)$ represents the (moment) generating function of R_t.

Proof. (36) and (37) may be proved from first principles on using (28) and Lemma 1; we prefer, however, to deduce expressions for $\mathbf{E}(R_t)$ and $\mathbf{V}(R_t)$ from (38), which we turn to next: By (28),

$$\begin{aligned}
M_t(u) =& e^{-2\alpha t} + \sum_{x=1}^{\infty} e^{ux}(2e^{-\alpha t}(-1)^x \sum_{k=x}^{\infty} \frac{(-\alpha t)^k}{k!}) \\
=& e^{-2\alpha t} + 2e^{-\alpha t} \sum_{k=1}^{\infty} \frac{(-\alpha t)^k}{k!} \sum_{x=1}^{k} (-e^u)^x \\
=& e^{-2\alpha t} + 2e^{-\alpha t} \sum_{k=1}^{\infty} \frac{(-\alpha t)^k}{k!} \frac{(-e^u) - (-e^u)^{k+1}}{1+e^u} \\
=& e^{-2\alpha t} - \frac{2e^{-\alpha t}e^u}{1+e^u} [\sum_{k=1}^{\infty} \frac{(-\alpha t)^k}{k!} - \sum_{k=1}^{\infty} \frac{(\alpha t e^u)^k}{k!}] \\
=& e^{-2\alpha t} - \frac{2e^{-\alpha t}e^u}{1+e^u} [(e^{-\alpha t} - 1) - (e^{\alpha t e^u} - 1)]; \qquad (39)
\end{aligned}$$

this proves (38). Tedious but straightforward differentiation now yields the required expressions (36) and (37) for $\mathbf{E}(R_t)$ and $\mathbf{V}(R_t)$ respectively. □

Note that $\mathbf{E}(R_t) \sim \alpha t$ and that $\mathbf{V}(R_t) \sim \alpha t$ as $t \to \infty$; with this in mind, we next prove the following central limit theorem for R_t:

Theorem 6. *Let R_t be the total number of runs, in the interval $[0, t]$, of particles associated with either of two independent Poisson processes $\{X_t\}$ and $\{Y_t\}$, $t \geq 0$, with the same intensity α. Then*

$$\lim_{t\to\infty} \mathbf{P}(\frac{R_t - \alpha t}{\sqrt{\alpha t}} \leq x) = \frac{1}{\sqrt{2\pi}} \int_{-\infty}^{x} e^{\frac{-t^2}{2}} dt \qquad (40)$$

Proof. Denoting the moment generating function of $(R_t - \alpha t)/\sqrt{\alpha t}$ by $N_t(u)$, we have

$$N_t(u) = e^{-u\sqrt{\alpha t}} M_t(\frac{u}{\sqrt{\alpha t}})$$

$$=e^{-u\sqrt{\alpha t}}\left[e^{-2\alpha t}-\frac{2e^{-\alpha t}e^{\frac{u}{\sqrt{\alpha t}}}}{1+e^{\frac{u}{\sqrt{\alpha t}}}}(e^{-\alpha t}-e^{\alpha te^{\frac{u}{\sqrt{\alpha t}}}})\right]$$

$$=e^{-(u\sqrt{\alpha t}+2\alpha t)}-\frac{2e^{-(u\sqrt{\alpha t}+2\alpha t-\frac{u}{\sqrt{\alpha t}})}}{1+e^{\frac{u}{\sqrt{\alpha t}}}}+\frac{2}{1+e^{\frac{u}{\sqrt{\alpha t}}}}e^{-(u\sqrt{\alpha t}+\alpha t-\frac{u}{\sqrt{\alpha t}}-\alpha te^{\frac{u}{\sqrt{\alpha t}}})}$$

$$=M_1+M_2+\frac{2}{1+e^{\frac{u}{\sqrt{\alpha t}}}}M_3 \quad \text{say} \tag{41}$$

where M_1 and M_2 may be seen to tend to zero as $t\to\infty$ and

$$\lim_{t\to\infty}\log M_3=\lim_{t\to\infty}\left[-(u\sqrt{\alpha t}+\alpha t-\frac{u}{\sqrt{\alpha t}}-\alpha te^{\frac{u}{\sqrt{\alpha t}}})\right]$$
$$=\lim_{x\to 0}(-\frac{1}{x^2}-\frac{u}{x}+ux+\frac{e^{ux}}{x^2})$$
$$=\frac{u^2}{2} \tag{42}$$

on using L'Hôpital's rule. In other words, $N_t(u)\to e^{\frac{u^2}{2}}$ as $t\to\infty$, proving Theorem 6. □

Standard arguments (see, e.g., Zhang (1990)) now lead to the following

Corollary 4. *With the same notation as in Theorem 6, and with* $\mathbf{E}(R_t)$ *and* $\mathbf{V}(R_t)$ *given by (36) and (37) respectively,*

$$\lim_{t\to\infty}\mathbf{P}(\frac{R_t-\mathbf{E}(R_t)}{\sqrt{\mathbf{V}(R_t)}}\le x)=\frac{1}{\sqrt{2\pi}}\int_{-\infty}^{x}e^{\frac{-t^2}{2}}dt \tag{43}$$

Our next goal is to prove exponential probability inequalities for the tail of the distribution of R_t; we are currently investigating the question of how these may be used, in conjunction with a maximal inequality, to prove a law of the iterated logarithm for R_t. We first establish some preliminary results:

Lemma 2. $\mathbf{P}(R_t\ge y)=1$ *for* $y=0$ *and*

$$\mathbf{P}(R_t\ge y)=2e^{-\alpha t}\sum_{k\ge y,k\text{ odd}}\frac{(\alpha t)^k}{k!}\quad(k=1,3,\ldots)$$
$$=2e^{-\alpha t}\sum_{k\ge y,k\text{ even}}\frac{(\alpha t)^k}{k!}\quad(k=2,4,\ldots) \tag{44}$$

Proof. By (28),

$$\mathbf{P}(R_t \geq y) = \sum_{x=y}^{\infty} 2e^{-\alpha t}(-1)^x \sum_{k=x}^{\infty} \frac{(-\alpha t)^k}{k!}$$

$$= 2e^{-\alpha t} \sum_{k=y}^{\infty} \frac{(-\alpha t)^k}{k!} \sum_{x=y}^{k} (-1)^x \tag{45}$$

The proof of Lemma 2 is completed on observing that $\sum_{x=y}^{k}(-1)^x$ equals 1, -1 or 0 according as both y and k are even; both y and k are odd; or one of the two quantities is even (and the other odd). □

Lemma 3. *Let R_t be the total number of runs, in the interval $[0,t]$, of particles associated with either of two independent Poisson processes $\{X_t\}$ and $\{Y_t\}$, $t \geq 0$, with the same intensity α. Then*

$$\mathbf{P}(R_t \geq y) \leq 2\mathbf{P}(\text{Po}(\alpha t) \geq y), \quad (y \geq 0) \tag{46}$$

and

$$\mathbf{P}(R_t \geq y) \geq \mathbf{P}(\text{Po}(\alpha t) \geq y), \quad (y \geq \alpha t) \tag{47}$$

where Po(λ) *denotes the Poisson r.v. with parameter* λ.

Proof. (46) follows directly from Lemma 3, since

$$\sum_{k \geq y; k \text{ odd or } k \text{ even}} \frac{e^{-\alpha t}(\alpha t)^k}{k!} \leq \mathbf{P}(\text{Po}(\alpha t) \geq y) \tag{48}$$

Now, observe that

$$\frac{(\alpha t)^k}{k!} \geq \frac{(\alpha t)^{k+1}}{(k+1)!} \quad (k \geq \alpha t) \tag{49}$$

so that if $y \geq \alpha t$ is odd, we have

$$\sum_{k \geq y; k \text{ odd}} 2\frac{e^{-\alpha t}(\alpha t)^k}{k!} = 2e^{-\alpha t} \sum_{j=0}^{\infty} \frac{(\alpha t)^{y+2j}}{(y+2j)!}$$

$$\geq e^{-\alpha t} \sum_{j=0}^{\infty} [\frac{(\alpha t)^{y+2j}}{(y+2j)!} + \frac{(\alpha t)^{y+2j+1}}{(y+2j+1)!}]$$

$$= \mathbf{P}(\text{Po}(\alpha t) \geq y); \tag{50}$$

this proves (47) for odd y's; the even case is treated similarly. □

Theorem 7. *Let R_t be the total number of runs, in the interval $[0,t]$, of particles associated with either of two independent Poisson processes $\{X_t\}$ and $\{Y_t\}$, $t \geq 0$, with the same intensity α. Then*

(a) For any $\epsilon > 0$,

$$\mathbf{P}\left(\frac{R_t - \alpha t}{\sqrt{\alpha t}} > \epsilon\right) \leq 2e^{-\frac{\epsilon^2}{2}+\frac{\epsilon^3}{2\sqrt{\alpha t}}} \tag{51}$$

(b) For any $\gamma > 0$, there exists $\epsilon(\gamma)$ such that $\epsilon > \epsilon(\gamma)$ implies that

$$\mathbf{P}\left(\frac{R_t - \alpha t}{\sqrt{\alpha t}} > \epsilon\right) \geq e^{-(1+\gamma)\frac{\epsilon^2}{2}} \tag{52}$$

Proof. By Lemma 3, (11), (12) and the fact that $\log(1+x) \geq x - \frac{x^2}{2}$,

$$\begin{aligned}\mathbf{P}(R_t > \alpha t + \epsilon\sqrt{\alpha t}) \leq & 2\mathbf{P}(\mathrm{Po}(\alpha t) > \alpha t + \epsilon\sqrt{\alpha t}) \\ \leq & 2\left(\frac{\alpha t}{\alpha t + \epsilon\sqrt{\alpha t}}\right)^{\alpha t + \epsilon\sqrt{\alpha t}} e^{\epsilon\sqrt{\alpha t}} \\ = & 2e^{-(\alpha t + \epsilon\sqrt{\alpha t})\log(1+\frac{\epsilon}{\sqrt{\alpha t}})} e^{\epsilon\sqrt{\alpha t}} \\ \leq & 2e^{-(\alpha t+\epsilon\sqrt{\alpha t})(\frac{\epsilon}{\sqrt{\alpha t}} - \frac{\epsilon^2}{2\alpha t})} e^{\epsilon\sqrt{\alpha t}}\end{aligned} \tag{53}$$

which yields (51). Now, by (47), Lemma 3, the Bohman inequality

$$\mathbf{P}(\mathrm{Po}(\lambda) > k) \geq \mathbf{P}(\mathrm{N}(0,1) > \frac{k-\lambda}{\sqrt{\lambda}}) \ (k \geq \lambda); \tag{54}$$

[see, e.g., Shorack and Wellner (1986)] and the Mill's ratio inequality [which can also be found in Shorack and Wellner (1986)], we have

$$\begin{aligned}\mathbf{P}(R_t > \alpha t + \epsilon\sqrt{\alpha t}) \geq & \mathbf{P}(\mathrm{N}(0,1) > \epsilon) \\ \geq & \left(\frac{1}{\epsilon} - \frac{1}{\epsilon^3}\right)\frac{1}{\sqrt{2\pi}} e^{-\frac{\epsilon^2}{2}}.\end{aligned} \tag{55}$$

We need, therefore, to show that for any $\gamma > 0$, there exists $\epsilon(\gamma)$ such that $\epsilon > \epsilon(\gamma)$ implies $\frac{1}{\sqrt{2\pi}}(\frac{1}{\epsilon} - \frac{1}{\epsilon^3}) \geq e^{-\frac{\gamma\epsilon^2}{2}}$, or, equivalently, that for any $\gamma > 0$, $\lim_{\epsilon\to\infty} e^{\frac{\gamma\epsilon^2}{2}} \frac{1}{\sqrt{2\pi}}$ $(\frac{1}{\epsilon} - \frac{1}{\epsilon^3}) \geq 1$. The last inequality is evidently true. □

References

Agin, M. A. and A. P. Godbole (1992). A new exact test for randomness, in *Computing Science and Statistics: Proceedings of the 22nd Symposium on the Interface*, C. Page and R. LePage, eds., Springer Verlag, New York.

Chryssaphinou, O. and S. Papastavridis (1990). Reliability of a consecutive-k-out-of-n system in random environment, *J. Appl. Probab.* **27**, 452-458.

Dudley, R. M. (1978). Central limit theorems for empirical measures, *Ann. Probab.* **6**, 899-929.

Feller, W. (1968). *An Introduction to Probability Theory and its Applications, Vol. 1, 3rd Edition.* John Wiley and Sons, Inc., New York.

Gibbons, J. D. (1971). *Nonparametric Statistical Inference.* McGraw-Hill, New York.

Godbole, A. P. (1990a). Specific formulæ for some success run distributions, *Statist. Probab. Lett.* **10**, 119-124.

Godbole, A. P. (1990b). On hypergeometric and related distributions of order k, *Commun. Statist.* **19 A**, 1291-1301.

Godbole, A. P. and A. A. Schaffner (1993). Improved Poisson approximations for word patterns, *Adv. Appl. Probab.* **25**, 334-347.

Karlin, S. and H. M. Taylor (1975). *A First Course in Stochastic Processes, 2nd Edition.* Academic, New York.

Karr, A. F. (1986). *Point Processes and their Statistical Inference.* Marcel Dekker, New York.

Panaretos, J. and E. Xekalaki (1986). On some distributions arising from certain generalized sampling schemes, *Commun. Statist.* **15 A**, 873-891.

Philippou, A. N., C. Georghiou and G.N. Philippou (1983). A generalized geometric distribution and some of its properties, *Statist. Probab. Lett.* **1**, 171-175.

Philippou, A. N. and F. Makri (1986). Successes, runs and longest runs, *Statist. Probab. Lett.* **4**, 211-215.

Shorack, G. and J. Wellner (1986). *Empirical Processes with Applications to Statistics.* John Wiley and Sons, Inc., New York.

Zhang, H. (1990). Exact distributions and limit theorems for runs of superimposed Poisson processes. M.S. Thesis, Michigan Technological University, Houghton.

CIRCULAR OVERLAPPING SUCCESS RUNS

M.V. Koutras, G.K. Papadopoulos and S.G. Papastavridis
University of Athens

Let $M_{n,k}^c$ be the number of success runs of length k in a sequence of n Bernoulli trials arranged on a circle (so that the first trial is assumed to be next to the nth). An exact formula is given for the probability $P(M_{n,k}^c = x)$, $x \geq 0$. Employing the Chen-Stein method, a total variation upper bound is derived for the rate of convergence of $M_{n,k}^c$ to the Poisson distribution. Finally, several limit theorems are presented, exploring the asymptotic distribution of $M_{n,k}^c$ for large $(n \to \infty)$ sequences of Bernoulli trials.

CIRCULAR SUCCESS RUNS, OVERLAPPING RUNS, CHEN-STEIN METHOD, MARKOV CHAINS, POISSON CONVERGENCE, BINOMIAL DISTRIBUTION OF ORDER K.

1. Introduction

The study of the distribution of success runs in a sequence of Bernoulli trials goes back at least a couple of centuries; the origin of run-related problems can be attributed to De Moivre (1756). Leaving aside old history and concentrating on more recent publications, models pertaining to success runs have been studied in the following contexts: non-parametric statistical inference (see Gibbons (1971), Agin and Godbole (1992) and references therein), quality control (see Wolfowitz (1943), Prairie et al (1962)), combinatorial analysis (see David and Barton (1962)) and reliability theory (see Chao et al (1993) or Papastavridis and Koutras (1993)).

Let $X_1, X_2, \ldots, X_n$ denote the outcomes (0 or 1) of n independent Bernoulli trials with success probabilities $p_i = P(X_i = 1)$ and failure probabilities $q_i = 1-p_i = P(X_i = 0), i = 1, 2, \ldots, n$. If k is a positive integer $(k \leq n)$, *a success run of length k* is usually defined as an uninterrupted sequence of k 1's. Perhaps the most popular way of enumerating success runs of length k is through the *non-overlapping* count

A. P. Godbole and S. G. Papastavridis (eds.), Runs and Patterns in Probability, 287–305.

proposed by Feller (1968), i.e., the search for the next run starts anew after the completion of the previous run. The distribution of the respective random variable has been named *Binomial distribution of order k* and was extensively studied by Aki (1985), Philippou and Makri (1986), Hirano (1986), Aki and Hirano (1988), Godbole (1990), (1990a), Fu (1993) etc.

Though Feller's way of counting is suitable for deriving theoretical results, it depends on the specific statistical problem which way of counting should be adopted. For example, as demonstrated by Goldstein (1990), in DNA sequence matching, the statistic related to the number of success runs of length *at least k*, can be used for deciding whether two DNA sequences are similar or not. Recently, Ling (1988) introduced an alternative procedure of enumerating the number of success runs in a sequence of Bernoulli trials. By his way of counting, in each sequence of more than k successes, every surplus (after the kth) success gives a contribution of 1 to the random variable $M_{n,k}$ which enumerates the runs. For example, the sequence 011100111101101 results in $M_{15,2} = 6, M_{15,3} = 3, M_{15,4} = 1$. The success runs, when enumerated by Ling's counting, have been named *overlapping success runs* and the respective distribution *Type II Binomial distribution of order k.* This type of distribution was studied by Hirano et al (1991), Godbole (1991, 1992) and Chryssaphinou et al (1993).

If the linear arrangement of the n Bernoulli trials is turned to a circular one so that the outcome of the first Bernoulli trial is considered next to the nth trial's outcome, a new class of distribution is created. The resulting success runs are called *circular (non-overlapping or overlapping) success runs,* and their distributions *circular Binomial distributions of order k: (type I and II* respectively*).* The non-overlapping case has been recently explored by Makri and Philippou (1993), Alevizos et al (1992), Koutras et al (1993). For the overlapping iid case ($p_i = p, i = 1, 2, \ldots, n$), Makri and Philippou (1993) provided an exact formula involving multinomial coefficients that are summed up over an index set determined by a diophantine equation. Chryssaphinou et al (1993), working with generating functions, derived simple recursive formulae for both iid and non-iid cases.

In the present article, we study the distribution of the number $M_{n,k}^c$ of overlapping success runs of length k in a circular sequence of non-iid Bernoulli trials. In Section 2, the probability function of $M_{n,k}^c$ is expressed through a single summation involving matrix products of proper Markov-chain transition matrices. In the development of our result, use was made of the Markov approach proposed by Fu and Koutras (1993) for the study of the respective linear model. In Section 3, the Chen-Stein method is employed to obtain a total variation upper bound for the rate of convergence of $M_{n,k}^c$ to a Poisson distribution. The remainder of Section 3 is devoted to the examination of the asymptotic behaviour of $M_{n,k}^c$ for the case of an infinite ($n \to \infty$) sequence of Bernoulli trials.

2. Overlapping circular success runs: the Markov chain approach

Let $X_1, X_2, \ldots, X_n$ be independent Bernoulli random variables with success probabilities

$$p_i = P(X_i = 1) = 1 - P(X_i = 0) = 1 - q_i \quad i = 1, 2, \ldots, n.$$

Following Ling (1988), we introduce the random variables

$$Y_i = \prod_{j=i}^{i+k-1} X_j, \; i = 1, 2, \ldots, n. \tag{2.1}$$

where it is conventional to set $X_{n+j} = X_j$, $j = 1, 2, \ldots, k-1$. Then, the sums

$$M_{n,k} = \sum_{i=1}^{n-k+1} Y_i, \; M_{n,k}^c = \sum_{i=1}^{n} Y_i \tag{2.2}$$

enumerate the number of overlapping success runs of length k in the linear and circular sequence $X_1, X_2, \ldots, X_n$ respectively.

The probability mass function of $M_{n,k}$ in the *iid case* $p_i = p$ $i = 1, 2, \ldots, n$ was expressed by Ling (1988) as a double sum of multinomial terms requiring the solution of a set of simultaneous diophantine equations. Besides, recurrence relations were given for both probability function (pf) and probability generating function (pgf). Hirano et al (1991) solved Ling's recurrence relation for the pgf and derived alternative expressions for the pf and pgf of $M_{n,k}$. The non-iid case was treated by Chryssaphinou et al (1993) who provided recurrence relations and Fu and Koutras (1993) who gave an exact formula based on the matrix product of proper transition matrices.

Regarding the random variable $M_{n,k}^c$, two different approaches have been used so far. Makri and Philippou (1993), working with combinatorial arguments were led to a double summation formula analogous to the linear iid case and a formula expressing $P(M_{n,k}^c = x)$ through the pf of $M_{n,k}$. On the other hand, Chryssaphinou et al (1993), employing generating function techniques, provided a set of simple and computationally manageable recurrence relations. Moreover, in the non-iid case, a double summation formula was given expressing $P(M_{n,k}^c = x)$ in terms of the pf's of $M_{i,j}$.

In Theorem 2.1 below, we derive a single summation formula for $P(M_{n,k}^c = x)$ valid for both iid and non-iid cases. Our proof takes advantage of the Markov-chain approach developed by Fu and Koutras (1993), which makes it feasible to effectively keep track of the precise final state of *linear* subsequences of $X_1, X_2, \ldots, X_n$.

In order to make our exposition self contained, we recall first some relevant definitions and necessary results from Fu and Koutras (1993). Consider the Markov chain $\{W_0(t) : t = 0, 1, \ldots, n\}$ defined on the finite state space $\Omega = \{(x, i) : x = 0, 1, \ldots, N-1 \text{ and } i = -1, 0, \ldots, k-1\} \cup \{(N, -1)\} - \{(0, -1)\}, N = n - k + 1$, as follows: for any sequence of outcomes of length t, say $SFS \ldots SFSS \ldots S$ (where there are m S's in the last group); let m be the number of trailing successes counting backwards and x be the number of overlapping success runs of length k up to the t-th trial. We define $W_0(t) = (x, m)$ if $m \leq k-1$ and $W_0(t) = (x, -1)$ if $m \geq k$.

Let $A_t = (\alpha_{ij})$ be the $k \times k$ matrix with

$$\alpha_{ij} = \begin{cases} q_t & \text{if } j = 1, \quad i = 1, 2, \ldots, k \\ p_t & \text{if } j = i + 1, \quad i = 1, 2, \ldots, k - 1 \\ 0 & \text{otherwise} \end{cases}$$

and let $\alpha_1 = (1, 0, \ldots, 0), \alpha_k = (0, 0, \ldots, 1)$ be unit (row) vectors of the space R^k. Fu and Koutras (1993) proved that

$$P(M_{n,k} = x) = P(W_0(n) = (x, j) : j \leq k - 1 | W_0(0) = (0, 0)) = e_1 (\prod_{t=1}^{n} \Lambda_t) U_i' \quad (2.3)$$

where $e_j = (0, \ldots, 1, \ldots, 0), j = 1, 2, \ldots, N(k + 1)$ are the unit (row) vectors of $R^{N(k+1)}$,

$$U_0' = \sum_{j=1}^{k} e_j, \quad U_N' = e_{N(k+1)}$$

$$U_x' = \sum_{j=0}^{k} e_{x(k+1)+j}, \quad 1 \leq x \leq N - 1$$

and

$$\Lambda_t = \begin{vmatrix} A_t & p_t\alpha_k' & & & & & & \\ & & q_t\alpha_1 & p_t & & & & \\ & & A_t & p_t\alpha_k' & & & & \\ & & & \cdot & \cdot \; \cdot & & & \\ & & & & \cdot \; \cdot & \cdot & & \\ & & & & & & q_t\alpha_1 & p_t \\ & & & & & & A_t & p_t\alpha_k' \\ & & & & & & & 1 \end{vmatrix} \quad (2.4)$$

is a $N(k + 1) \times N(k + 1)$ bidiagonal blocked matrix. Before proceeding to the statement of our result, the following additional notation is required. If

$$I \subseteq \{1, 2, \ldots, N(k + 1)\}$$

we define

$$U(I) = \sum_{j \in I} e_j$$

and for each $x(0 \leq x \leq N)$, we introduce the auxiliary index sets

$$I_i^{(1)}(x) = \begin{cases} \{x(k+1)+j : \ 1 \leq j \leq k-i\} \cup \{(x-i)(k+1)\} & (i = 0, 1, \ldots, k-1) \\ \emptyset & (i \geq k) \end{cases}$$

$$I_i^{(2)}(x) = \begin{cases} \emptyset & \text{if } i = 0 \\ \{(x-j)(k+1)+k-i+j : \max(1, i-k+1) \leq j \leq i\} & \text{if } i \geq 1. \end{cases}$$

Finally let

$$I_i(x) = (I_i^{(1)}(x) \cup I_i^{(2)}(x)) \cap \{1, 2, \ldots, N(k+1)\}, \ i \geq 0. \tag{2.5}$$

Theorem 2.1.

The probability function of the number $M_{n,k}^c$ of circular overlapping success runs of length $k \geq 2$ in $n(n \geq k)$ Bernoulli trials is given by

$$P(M_{n,k}^c = x) = \begin{cases} \prod_{j=1}^{n} p_j & \text{if } x = n \\ \sum_{i=1}^{n} q_i \left(\prod_{j \neq i} p_j \right) & \text{if } x = n-k \\ \sum_{i=0}^{x+k-1} a_i(x) \beta_i & \text{if } 0 \leq x < n-k \end{cases} \tag{2.6}$$

where

$$\beta_i = q_{i+1} \prod_{j=0}^{i} p_j, \quad i = 0, 1, \ldots, n-2$$

$$a_i(x) = e_1 \left(\prod_{t=i+2}^{n} \Lambda_t \right) U'(I_i(x)).$$

(Convention: $p_0 = P(X_0 = 1) = 1$).

Proof. For $x = n$ and $x = n - k$, the result follows immediately. Moreover, it is evident that $P(M_{n,k}^c = x) = 0$ for all $n - k < x \leq n - 1$. Assume next that $0 \leq x < n - k$. Since

$$\{M^c_{n,k} = x\} \subseteq \bigcup_{i=0}^{x+k-1} \{X_{i+1} = 0, X_j = 1 \;\; for \;\; j = 0, 1, \ldots, i\}$$

a direct application of the total probability theorem yields

$$P(M^c_{n,k} = x) = \sum_{i=0}^{x+k-1} P(M^c_{n,k} = x|(1 - X_{i+1}) \prod_{j=0}^{i} X_j = 1) P((1 - X_{i+1}) \prod_{j=0}^{i} X_j = 1), \tag{2.7}$$

and due to the independence of $X_1, X_2, \ldots, X_n$,

$$P((1-X_{i+1}) \prod_{j=0}^{i} X_j = 1) = E((1-X_{i+1}) \prod_{j=0}^{i} X_j) = \beta_i \quad i = 0, 1, \ldots, x+k-1. \tag{2.8}$$

Regarding the conditional probabilities which appear in the RHS of (2.7), we observe the following:

i) If $i = 0$, the event $\{M^c_{n,k} = x|X_1 = 0\}$ implies that x overlapping success runs should be formed in the *linear* sequence $X_1, X_2, \ldots, X_n$. Hence, making use of a shifted version of (2.3), we obtain

$$P(M^c_{n,k} = x|X_1 = 0) = P(W_2(n) = (x, r), r \leq k - 1|W_0(1) = (0, 0))$$

which yields

$$P(M^c_{n,k} = x|X_1 = 0) = e_1(\prod_{t=2}^{n} \Lambda_t) U'(I_0(x)) = a_0(x). \tag{2.9}$$

ii) If $1 \leq i < k$, the event $\{M^c_{n,k} = x|(1 - X_{i+1}) \prod_{j=0}^{i} X_j = 1\}$ induces that the random variable $W_{i+1}(n)$ describing the *linear* sequence $X_{i+2}, \ldots, X_n$ should take the value $(x - i, -1)$ or (x, j) with $0 \leq j \leq k - i - 1$ or $(x - (i + j - k + 1), j)$ with $0 \leq j \leq k - 1$ and $j + i \geq k$. An appropriate modification of (2.3) yields now

$$P(M_{n,k}^c = x | (1 - X_{i+1}) \prod_{j=0}^{i} X_j = 1) = P(W_{i+2}(n) = (x - i, -1) | W_0(i+1) = (0,0))$$

$$+ \sum_{j=0}^{k-i-1} P(W_{i+2}(n) = (x, j) | W_0(i+1) = (0,0))$$

$$+ \sum_{j=\max(0,i-k)}^{k-1} P(W_{i+2}(n) = (x + k - i - j - 1, j) | W_0(i+1) = (0,0))$$

and it can be easily verified that

$$P(M_{n,k}^c = x | (1 - X_{i+1}) \prod_{j=0}^{i} x_j = 1) = e_1(\prod_{t=i+2}^{n} \Lambda_t) U'(I_i(x)) = a_i(x). \tag{2.10}$$

iii) For $i \geq k$, working in a similar fashion as in (ii), we derive

$$P(M_{n,k}^c = x | (1 - X_{i+1}) \prod_{j=0}^{i} X_j = 1) = P(W_{i+2}(n) = (x - i, -1) | W_0(i+1) = (0,0))$$

$$+ \sum_{j=0}^{k-1} P(W_{i+2}(n) = (x + k - i - j - 1, j) | W_0(i+1) = (0,0))$$

which, on using definition (2.5), simplifies to

$$P(M_{n,k}^c = x | (1 - X_{i+1}) \prod_{j=0}^{i} x_j = 1) = e_1(\prod_{t=i+2}^{n} \Lambda_t) U'(I_i(x)) = a_i(x). \tag{2.11}$$

The proof of Theorem 2.1 is now easily completed by combining (2.7)-(2.11).

As an illustration of how Theorem 2.1 could be applied, consider the case $n = 5,\ k = 2,\ p_i = \frac{1}{i+10},\ 1 \leq i \leq 5$. It is evident that $N = 4$, and for the computation of $P(M_{5,2}^c = x),\ 0 \leq x < 3$, we are going to need the first rows of the matrix products

$\prod_{t=i+2}^{5} \Lambda_t$, $0 \leq i \leq x+1$, where Λ_t are 12×12 matrices. The basic steps of the required calculations are summarized in the following tables:

First row, in two displays, of $\prod_{t=i+2}^{5} \Lambda_t$

$i=0$	.922650	.061508	.004396	.010256	.000397	.000336	.000427	0
$i=1$	.928205	.061905	.004396	.005128	0	.000336	0	0
$i=2$	.933333	.061905	.004762	0	0	0	0	0
$i=3$	.933333	.066667	0	0	0	0	0	0

$i=0$	0	.000031	0	0
$i=1$	0	0	0	0
$i=2$	0	0	0	0
$i=3$	0	0	0	0

Index sets $I_i(x)$

$x \setminus i$	0	1	2	3
0	{1,2}	{1}	-	-
1	{3,4,5}	{4,2}	{1}	-
2	{6,7,8}	{7,5,3}	{4,2}	{1}

	$i=0$	$i=1$	$i=2$	$i=3$	$\sum_{i=0}^{x+1} a_i(x)\beta_i$
β_i	.909091	.083333	.006993	.000541	-
$a_i(0)$	.984158	.928205	-	-	.9720
$a_i(1)$	.015049	.067033	.933333	-	.0258
$a_i(2)$	.000763	.004396	.061905	.933333	.0020

	$P(M^c_{5,2} = x)$	$xP(M^c_{5,2} = x)$
$x = 0$	.9720	.0000
$x = 1$	.0258	.0258
$x = 2$	.0020	.0040
$x = 3$	.0002	.0006
$x = 4$	.0000	.0000
$x = 5$	.0000	.0000
	1.0000	$E(M^c_{5,2}) = .0304$

It goes without saying that for large values of n and k, the derivation of the pf of $M^c_{n,k}$ through Theorem 2.1 is extremely laborious to be done by hand. Nevertheless, the development of a computer program for general n and k is a relatively easy task, and the computing time required is quite reasonable.

In the iid case $p_i = p,\ i = 1, 2, \ldots, n$, we have

Corollary 2.1. *The probability function of the number $M^c_{n,k}$ of circular overlapping success runs of length $k \geq 2$ in $n (n \geq k)$ iid Bernoulli trials is given by*

$$P(M^c_{n,k} = x) = \begin{cases} p^n & \text{if } x = n \\ nqp^{n-1} & \text{if } x = n-k \\ q\left\{a_0(x) + \sum_{i=1}^{x+k-1} a_i(x)p^i\right\} & \text{if } 0 \leq x < n-k \end{cases}$$

where $a_i(x)$ are the sums of the elements of the first row of Λ^{n-i-1} whose coordinates belong to the index set $I_i(x)$ ($\Lambda = \Lambda_t \quad t = 1, 2, \ldots, n$).

Proof. Immediate.

It is worth mentioning that for $x = 0$, the computation of $P(M^c_{n,k} = x)$ could be alternatively performed by the following:

Corollary 2.2. *The probability of zero occurrences of circular overlapping success runs of length $k \geq 2$ in $n (n \geq k)$ Bernoulli trials is given by*

$$P(M^c_{n,k} = 0) = \sum_{i=0}^{k-1} a_i(x)\beta_i$$

where $a_i(x)$ *is the sum of the elements of the first row of* $\prod_{t=i+2}^{n} A_t$, *whose coordinates belong to the index set* $I_i(0)$.

Proof. It is an immediate consequence of (2.6) for $x = 0$ if we take into account that, due to the special structure of the blocked matrix Λ_t, the upper leftmost $k \times k$ block of the matrix product $\prod_{t=i+2}^{n} \Lambda_t$ equals $\prod_{t=i+2}^{n} A_t$.

As a final remark, we mention that the approach used in this article can be easily modified to cover the case of a two-state Markov-chain (instead of Bernoulli trials). A *linear* model of this kind was introduced and studied recently by Hirano and Aki (1993).

3. Bounds and Limits

In this section, we provide approximations and bounds for the probability funciton of the random variable $M^c_{n,k}$ through the Chen-Stein method (see Barbour et al (1992)). Furthermore, using these bounds, we study the limiting behaviour of the random variable $M^c_{n,k}$ for large (i.e. $n \to \infty$) circular sequences of Bernoulli variables.

Let $X_1, X_2, \ldots, X_n$ be a sequence of n independent Bernoulli variables with success probabilities $p_i = P(X_i = 1) = 1 - q_i \;\; i = 1, 2, \ldots, n$. We introduce the following notations:

$$p_{i+n} = p_i, \; i = 1, 2, \ldots, k-1$$

$$\gamma_j = \prod_{i=j}^{j+k-1} p_i, \; j = 1, 2, \ldots, n$$

$$\gamma = \sup_{1 \le j \le n} \gamma_j, \; p = \sup_{1 \le j \le n} p_j, \; \lambda = \sum_{j=1}^{n} \gamma_j > 0.$$

In the next theorem, the distribution of the random variable $M^c_{n,k}$ is approximated by a Poisson distribution with mean λ; an estimation of the error of the approximation is also provided.

Theorem 3.1. *For the number $M_{n,k}^c$ of the overlapping occurences of circular success runs of length k in a sequence of n independent Bernoulli trials, we have*

$$\left|P(M_{n,k}^c = x) - e^{-\lambda}\frac{\lambda^x}{x!}\right| \leq (1-e^{-\lambda})((2k-1)\gamma + 2p\frac{1-p^{k-1}}{1-p}) \tag{3.1}$$

for all $x \geq 0$.

Proof. It is clear that with Y_j denoting the indicator of the event that a run starts at the jth position,

$$E(M_{n,k}^c) = \sum_{j=1}^{n} E(Y_j) = \sum_{j=1}^{n} \gamma_j = \lambda.$$

Let $d(W_1, W_2)$ denote the total variation distance between two discrete random variables W_1 and W_2, namely

$$d(W_1, W_2) = \sup_{A \subseteq Z+} |P(W_1 \in A) - P(W_2 \in A)|.$$

For every $\alpha \in \{1, 2, \ldots, n\}$, we define the following "neighbourhood of dependence" of α:

$B_\alpha = \{\beta : 1 \leq \beta \leq n$ *and* Y_α *and* Y_β *contain at least one common variable* $X_i\}$.

In order to establish a Poisson approximation through the well known Chen-Stein method, the following three quantities should be evaluated or bounded from above (see Arratia et al (1989)).

$$b_1 = \sum_{\alpha=1}^{n} E(Y_\alpha) \sum_{\beta \in B_\alpha} E(Y_\beta)$$

$$b_2 = \sum_{\alpha=1}^{n} \sum_{\beta \in B_\alpha - \{\alpha\}} E(Y_\alpha Y_\beta)$$

$$b_3 = \sum_{\alpha=1}^{n} E|E\{Y_\alpha - p_\alpha | \sigma(Y_\beta : \beta \notin B_\alpha)\}|.$$

Due to the definition of B_α, for every $\beta \notin B_\alpha$, the random variables Y_β are independent of Y_α, and therefore

$$b_3 = 0. \tag{3.2}$$

In addition, on making use of the obvious inequality

$$\sum_{\beta \in B_\alpha} E(Y_\alpha) \leq |B_\alpha| \sup_{1 \leq j \leq n} \gamma_j$$

we may write

$$b_1 \leq |B_\alpha| \sup_{1 \leq j \leq n} \gamma_j \sum_{\alpha=1}^{n} E(Y_\alpha) = (2k-1)\gamma\lambda. \tag{3.3}$$

Finally, we observe that

$$\begin{aligned} \sum_{\beta \in B_\alpha \backslash \{\alpha\}} E(Y_\alpha Y_\beta) =& E(Y_\alpha) \sum_{\beta \in B_\alpha \backslash \{\alpha\}} E\Big(\prod_{j=\beta, j \notin \{\alpha,\ldots,\alpha+k-1\}}^{\beta+k-1} X_j \Big) \\ \leq& 2E(Y_\alpha) \sum_{i=1}^{k-1} p^i \end{aligned}$$

which yields immediately

$$b_2 \leq 2\lambda \sum_{i=1}^{k-1} p^i = 2\lambda p \frac{1-p^{k-1}}{1-p}. \tag{3.4}$$

Inequality (3.1) is now effortlessly deduced if we apply the basic inequality of the Chen-Stein method described in Theorem 1 of Arratia et al (1989), namely

$$\sup_{x \geq 0} \left| P(M_{n,k}^c = x) - e^{-\lambda} \frac{\lambda^x}{x!} \right| \leq (b_1 + b_2) \frac{(1 - e^{-\lambda})}{\lambda} + b_3$$

and take into account (3.2)-(3.4).

As indicated by the extensive numerical computational experimentation we have done, the approximating upper and lower bounds given by (3.1), namely

$$LB = e^{-\lambda}\frac{\lambda^x}{x!} - (1-e^{-\lambda})((2k-1)\gamma + 2p\frac{1-p^{k-1}}{1-p})$$

$$UB = e^{-\lambda}\frac{\lambda^x}{x!} + (1-e^{-\lambda})((2k-1)\gamma + 2p\frac{1-p^{k-1}}{1-p})$$

are quite close to the exact values $P(M^c_{n,k} = x)$ (especially for Bernoulli sequences where the occurrence of success runs of length at least k is rare).

The next corollary provides a simplified (but less sharp) upper bound for the LHS of (3.1), which will prove very useful for constructing limit theorems for large circular Bernoulli trials.

Corollary 3.1.

$$\left|P(M^c_{n,k} = x) - e^{-\lambda}\frac{\lambda^x}{x!}\right| \leq (2k-1)\gamma + (2k-2)p \leq 2k(\gamma + p) \qquad (3.5)$$

for all $x \geq 0$.

Proof. Immediate.

The following corollary, referring to the iid case, is directly derived from Theorem 3.1.

Corollary 3.2. *If* $p_i = p$ *when* $i = 1, 2, \ldots, n$ *then*

$$\left|P(M^c_{n,k} = x) - e^{-np^k}\frac{(np^k)^x}{x!}\right| \leq (1-e^{-np^k})\left((2k-1)p^k + 2p\frac{1-p^{k-1}}{1-p}\right)$$

for all $x \geq 0$.

We now continue our investigation with the study of the limiting distribution of $M^c_{n,k}$ as $n \to \infty$.

Assume that $X_1, X_2, \ldots, X_n$ is an infinite sequence of independent Bernoulli variables with success probabilities $p_{in} = P(X_i = 1) = 1 - q_{in}$, $i = 1, 2, \ldots, n$ and

M^c_{n,k_n} is the number of circular overlapping success runs of length k_n. Introducing again the notations

$$p_{i+n,n} = p_{in} \quad i = 1, 2, \ldots, k-1$$

$$\gamma_{jn} = \prod_{i=j}^{j+k-1} p_{in} \quad j = 1, 2, \ldots, n$$

and

$$\lambda(n) = \sum_{j=1}^{n} \gamma_{jn}$$

we have the next theorems.

Theorem 3.2. *If*

$$a. \quad \lim_{n\to\infty} \sum_{j=1}^{n} \gamma_{jn} = \mu > 0$$

$$b. \quad \lim_{n\to\infty} \left[k_n \sup_{1\le j\le n} p_{jn} \right] = 0$$

then

$$\lim_{n\to\infty} P(M^c_{n,k_n} = x) = e^{-\mu} \frac{\mu^x}{x!} \tag{3.6}$$

for all $x \geq 0$.

Proof. Since

$$\sup_{1\le j\le n} p_{jn} \le k_n \sup_{1\le j\le n} p_{jn}$$

assumption (b) yields

$$\lim_{n\to\infty} \sup_{1\le j\le n} p_{jn} = 0. \tag{3.7}$$

Moreover,

$$k_n \sup_{1\le j\le n} \gamma_{jn} \le k_n \left(\sup_{1\le j\le n} p_{jn} \right)^{k_n}$$

and in view of (3.7), we conclude

$$\lim_{n\to\infty}\left[k_n \sup_{1\le j\le n} \gamma_{jn}\right] = 0. \tag{3.8}$$

Relation (3.8) and condition (b) guarantee that the rightmost term of inequality (3.5) tends to 0 as $n \to \infty$. The desired result follows now easily, if we apply Corollary (3.1) and take into account assumption (a).

In the iid case, Theorem 3.2 yields the next.

Corollary 3.3. *Let $p_{in} = p(n)$ and $i = 1, 2, \ldots, n$ (iid case). If*

$$a. \quad \lim_{n\to\infty}\left[np(n)^{k_n}\right] = \mu > 0$$

$$b. \quad \lim_{n\to\infty}\left[k_n p(n)\right] = 0$$

then

$$\lim_{n\to\infty} P(M^c_{n,k} = x) = e^{-\mu}\frac{\mu^x}{x!}$$

for all $x \ge 0$.

Note that in the special case, $k_n = k$ (fixed), $n = 1, 2, \ldots$, the first assumption of Corollary 3.3 is sufficient for establishing the Poisson convergence of $M^c_{n,k}$ (condition (b) is easily checked to be true if (a) is valid).

References

Agin, M.A. and Godbole, A.P. (1992). A new exact runs test for randomness. *Computing Science and Statistics* (C. Page and R. LePage eds.), Proceedings of the 22nd Symposium on the Interface, New York, Springer Verlag, 281-285.

Aki, S. (1985). Discrete distributions of order k on a binary sequence, *Ann. Inst. Statist. Math.* **37**, Part A, 205-224.

Aki, S. and Hirano, K. (1988). Some characteristics of the binomial distribution of order k and related distributions, *Statistical Theory and Data Analysis II*, (ed. K. Matusita), 211-222. North-Holland.

Alevizos, P.D., Papastavridis, S.G. and Sypsas P. (1992). Reliability of cyclic m-consecutive-k-out-of-n: F systems, *Reliability Quality Control and Risk Assessment*, Proc. IASTED Conference (to appear).

Arratia, R., Goldstein, L. and Gordon, L. (1989). Two moments suffice for Poisson approximations: the Chen-Stein method, *Ann. Prob.*, **17**, 9-25.

Barbour, A.D., Holst, L. and Janson, S. (1992). *Poisson Approximation*, Oxford University Press.

Chao, M.T., Fu. J.C. and Koutras, M.V. (1993). A survey of the reliability studies of consecutive-k-out-of-n: F systems and its related systems, *Preprint*.

Chryssaphinou, O., Papastavridis, S. and Tsapelas, T. (1993). On the number of overlapping success runs in a sequence of independent Bernoulli trials, *Applications of Fibonacci Numbers*, Vol **5** (*to appear*).

David, F.N. and Barton, D.E. (1962). *Combinatorial Chance*, Griffins, London.

Feller, W. (1968). *An Introduction to Probability Theory and Its Applications*, Vol. I, 3rd ed. Wiley, New York.

Fu, J.C. (1993). Poisson convergence in reliability and coin tossing, *Statistica Sinica* (to appear).

Fu, J.C. and Koutras, M.V. (1993). Distribution theory of runs: A Markov chain approach, *Preprint*.

Gibbons, J.D. (1971). *Nonparametric Statistical Inference,* McGraw-Hill, New York.

Goldstein, L. (1990). Poisson approximation in DNA sequence matching, *Comm. in Stat. Theory and Methods,* **19(11)**, 4167-4179.

Godbole, A.P. (1990a). Specific formulae for some success run distributions, *Stat. Prob. Lett.*, **10**, 119-124.

Godbole, A.P. (1990b). Degenerate and Poisson convergence criteria for success runs, *Stat. Prob. Lett.*, **10**, 247-255.

Godbole, A.P. (1991). Poisson approximations for runs and patterns of rare events, *Adv. Appl. Prob.*, **23**, 851-865.

Godbole, A.P. (1992). The exact and asymptotic distribution of overlapping success runs, *Comm. in Stat.*, **A21**, 953-967.

Hirano, K. (1986). Some properties of the distributions of order k, *Fibonacci Numbers and Applications.* (Philippou, A.N., Horadam, A.F. and Bergum, G.E. eds), D. Reidel, 43-53.

Hirano, K. and Aki, S. (1993). On the number of occurrences of success runs of length k in a two state Markov chain. *Statistica Sinica,* (to appear).

Hirano, K., Aki, S., Kashiwagi, N. and Kuboki, H. (1991). On Ling's binomial and negative binomial distributions of order k, *Stat. Prob. Lett.*, **11**, 503-509.

Ling, K.D. (1988). On binomial distributions of order k, *Stat. Prob. Lett.*, **6**, 247-250.

Koutras, M.V., Papadopoulos, G.K. and Papastavridis, S.G. (1993). Runs on a circle, *Preprint.*

Makri, F.S. and Philippou, A.N. (1993). Binomial distributions of order k on the circle. In this volume.

De Moivre, A. (1756). *The Doctrine of Chance,* Chelsea Publishing Co., 3rd Edition.

Philippou, A.N. and Makri, F.S. (1986). Successes, runs and longest runs, *Stat. Prob. Lett.,* **4**, 101-105.

Papastavridis, S.G. (1990). m-consecutive-k-out-of-n: F system, *IEEE Trans. Reliability,* **R-39**, 386-388.

Papastavridis, S.G. and Koutras, M.V. (1993). Consecutive-k-out-of-n systems, *New trends in system reliability evaluation,* (K.B. Misra Ed.)

Prairie, R.P., Zimmer, W.J. and Brookhouse, J.K. (1962). Some acceptance sampling plans based on the theory of runs, *Technometrics,* **4**, 177-185.

Wolfowitz, J. (1943). On the theory of runs with some applications to quality control, *Ann. Math. Stat.,* **14**, 280-288.

ASYMPTOTIC JOINT DISTRIBUTION OF COVER TIMES

Tamás F. Móri

Eötvös Loránd University

Considering a set of patterns of length n over a fixed finite alphabet, we consider the corresponding cover time, defined as the time until each pattern appears at least once in an infinite sequence of independent, uniformly distributed random letters. We are interested in the joint distribution of cover times associated with a collection of subsets and after suitable normalization we determine the limit distribution as $n \to \infty$. Generalizations in terms of point processes are also presented.

PATTERN; RUN; WAITING TIME; COVER TIME; BONFERRONI INEQUALITIES; PROBABILITY GENERATING FUNCTION; POISSON APPROXIMATION; POISSON PROCESS; MARSHALL-OLKIN DISTRIBUTION

1. Introduction.

Let $\mathcal{X}$ be a finite alphabet of size d ($d \geq 2$). Finite strings over $\mathcal{X}$ (i.e., elements of $\mathcal{X}^n$, $n \geq 1$) will be called *patterns* or *words*. Consider an infinite sequence $X_1, X_2, \ldots$ of i.i.d. random letters distributed uniformly in $\mathcal{X}$. For any pattern $A \in \mathcal{X}^n$ define

$$T(A) = \inf\{N \geq n : (X_{N-n+1}X_{N-n+2}\ldots X_N) \equiv A\},$$

the *waiting time* till A is resulted by n consecutive random letters.

For every $n = 1, 2, \ldots$ let H_n be a subset of $\mathcal{X}^n$ and $W(H_n)$ the number of random letters needed till each pattern of H_n appears at least once, i.e.,

$$W(H_n) = \max\{T(A) : A \in H_n\}.$$

This is *the maximum waiting time* or *cover time* associated with H_n.

A. P. Godbole and S. G. Papastavridis (eds.), Runs and Patterns in Probability, 307–327.

Several papers have been devoted to the study of cover times of Markov chains and random walks on graphs. In most of these studies, the considered Markov chains are supposed rapidly mixing; in case of random walks on graphs this property is usually achieved by considering only strongly symmetric graphs. General results can be found in a sequence of papers published in *J. Theor. Probab.* **2** (1989) (references [2], [3], [10], [23]; see also [4], [20]). Roughly speaking, the less is assumed on the structure of the graph, the weaker, though quite general, theorems can be stated.

In the particular cases of highly symmetric graphs, it often turns out that the distributions of the cover times are approximately of Gumbel type, see [1], [6], [12], [13]. In our case, the sequence of random letters can be converted into a symmetric, nearest neighbor random walk on the de Bruijn digraph $B(d,n)$ in a natural way: $B(d,n)$ is defined with the vertex set $\mathcal{X}^n$ and edge set $\mathcal{X}^{n+1}$ so that arc $x_1x_2\dots x_{n+1}$ goes from vertex $x_1\dots x_n$ to vertex $x_2\dots x_{n+1}$, thus the states of the random walk are given by the overlapping blocks of n consecutive random letters. In spite of the numerous advantageous properties these de Bruijn graphs possess, they do not seem, at first sight, to satisfy those strong regularity conditions which might allow general methods to be applied simply for them. However, classical methods can still be fruitful, where a general setup fails to yield sufficiently sharp results. Several theorems on the weak and a.s. asymptotic properties of maximum waiting times have been published, see [7], [15], [16], [17], [18], [19]. In [17] it is shown that the limit distribution of the suitably normalized maximum waiting times is of Gumbel type, as n and $|H_n|$ tend to infinity. The speed of convergence is estimated in [16], where the following result is proved.

Let us normalize the maximum waiting time by introducing

$$Y(H_n) = d^{-n}W(H_n) - \log|H_n|.$$

Let $F(y) = \exp\{-e^{-y}\}$, the standard Gumbel distribution function.

Proposition 1.1. [16] *Set* $q = \frac{1}{6}\frac{d-1}{d+1}$. *Then*

$$\sup_y |P(Y(H_n) \le y) - F(y)| = o(|H_n|^{-q})$$

uniformly in H_n, *as* $|H_n| \to \infty$.

Thus, in the most interesting case of $H_n = \mathcal{X}^n$, the convergence is exponentially fast, which makes it possible to derive strong results on the a.s. behaviour of these waiting times [16]. This univariate result has been extended to the joint distribution of several maximum waiting times in [19]. It has turned out that under some additional hypotheses of technical character, maximum waiting times tend to be independent as $n \to \infty$. More precisely, the following theorem holds.

Suppose $|H_n| \to \infty$ in such a way that

$$\log|H_n| \le \frac{5}{4}\log|H_{n+1}| \qquad \text{for every } n. \tag{1.1}$$

This condition is obviously met when $|H_n|$ is increasing for $n \ge 1$.

Proposition 1.2. [19] *Let* $n_1 < n_2 < \cdots < n_m$ *be arbitrary positive integers, and let* $y = (y_1, \ldots, y_m)$ *run over* $\mathbf{R}^m$. *Under hypothesis* (1.1) *we have*

$$\sup_y |P(Y(H_{n_1}) \le y_1, \ldots, Y(H_{n_m}) \le y_m) - \prod_{i=1}^m F(y_i)| \le \text{const.} \sum_{i=1}^m |H_{n_i}|^{-6q^2},$$

with an absolute constant.

The aim of the present paper is to study the joint distribution of maximum waiting times associated with a fixed number of subsets of $\mathcal{X}^n$. It will be shown that cover times of disjoint subsets are approximately independent if n is large enough. We first extend the univariate limit result of [16] to a Poisson approximation theorem for the number of patterns not seen before a given moment (Section 2). Then, in Section 3, we prove the weak convergence of suitably defined point processes to independent standard Poisson processes. Finally, the joint distribution of the normalized maximum waiting times will be studied in Section 4. First, the joint distribution of cover times of pairwise disjoint subsets will be approximated with a product probability measure, then the general case of overlapping subsets will be treated by reducing it to the disjoint case. In Section 2 we will often refer to the method of [16]. Parts of Section 3 will resemble the argumentation of [19].

2. Bonferroni inequalities and Poisson approximation

Let $B_1, B_2, \ldots, B_n$ be arbitrary events in an arbitrary probability space. Let $I(.)$ denote the indicator function of the event in the brackets. Let N denote the number of events that occur, thus $N = \sum_{i=1}^{n} I(B_i)$. Introduce $g_N(z) = E(z^N)$, $0 \le z \le 1$, the probability generating function of N.

Bonferroni inequalities, that is, inequalities of the form

$$P(N = k) \le (\ge) \sum_{r=0}^{n} c_r S_r \quad \text{or} \quad P(N \ge k) \le (\ge) \sum_{r=0}^{n} c_r S_r,$$

where $S_0 = 1$ and for $r \ge 1$ $S_r = \sum_{1 \le i_1 < \cdots < i_r \le n} P(B_{i_1} \cap \cdots \cap B_{i_r})$, are often used for approximating the distribution of the random variable N. For example, one can easily apply them to prove that the distribution of N converges to the Poisson distribution with parameter $\lambda > 0$ as $n \to \infty$ and the events $A_i = A_{i,n}$ vary in such a way that $S_r \to \frac{\lambda^r}{r!}$. For a great variety of Bonferroni type inequalities with various methods of construction and proof see the relevant contributions in [8].

The following simple lemma shows how Bonferroni type or general linear estimators for the probability that none of the events occur can be converted into estimators of the probability generating function of N.

Lemma 2.1. Let $c(H), H \subset \{1, 2, \ldots, n\}$ *be real constants for which*

$$P(N = 0) \le (\ge) \sum_{H} c(H) P(\bigcap_{i \in H} B_i) \tag{2.1}$$

holds, whatever the events B_i, $1 \le i \le n$ *be. Then*

$$g_N(z) \le (\ge) \sum_{H} c(H) P(\bigcap_{i \in H} B_i)(1 - z)^{|H|}, \quad 0 \le z \le 1. \tag{2.2}$$

Proof. After suitably extending the probability space if necessary let us introduce random events $C_1, \ldots, C_n$, independently of each other and of the events $B_1, \ldots, B_n$, each with probability $1-z$. Let us apply (2.1) to the events $B_i' = B_i \cap C_i$. With the notation $N' = \sum_{i=1}^{n} I(B_i')$ we have

$$P(N'=0) = E(z^N) = g_N(z),$$

and

$$P(\bigcap_{i\in H} B_i') = P(\bigcap_{i\in H} B_i)(1-z)^{|H|},$$

thus we obtain just (2.2). □

Classical Bonferroni inequalities sometimes fail to be sufficiently sharp because the sums S_r are sensitive to strongly irregular behaviour of relatively few r-tuples of events. This disadvantage can be removed by using the graph-sieve of Rényi [21]. Let us introduce some notations before formulating it.

Let $B_1, \ldots, B_n$ be arbitrary events and G a simple graph on

$$V(G) = \{1, 2, \ldots, n\}.$$

Let $\mathcal{H}'$ denote the collection of independent subsets of G (i.e., spanning no edge), and $\mathcal{H}''$ the collection of those subsets which span exactly one edge. Furthermore, let

$$S_0' = 1, \quad S_r' = \sum_{\substack{H\in\mathcal{H}' \\ |H|=r}} P(\bigcap_{i\in H} B_i), \quad S_r'' = \sum_{\substack{H\in\mathcal{H}'' \\ |H|=r}} P(\bigcap_{i\in H} B_i), \qquad r \geq 1.$$

Then

$$\sum_{r=0}^{2m+1} (-1)^r S_r' - \sum_{r=1}^{m} S_{2r+1}'' \leq P(N=0) \leq \sum_{r=0}^{2m} (-1)^r S_r' + \sum_{r=1}^{m} S_{2r}'',$$

consequently, for every m, $1 \leq m \leq n$, we have

$$|P(N=0) - \sum_{r=0}^{m-1} (-1)^r S_r'| \leq S_m' + \sum_{r=2}^{m} S_r''. \tag{2.3}$$

This formula was applied in [17] and [16] to obtain the limit distribution of maximum waiting times. It helped to leave subsets with strongly overlapping patterns out of consideration. Now, combining the estimations derived in [16] with Lemma 2.1 above, we can prove asymptotic Poissonity for the number of patterns not seen before a certain moment.

As in the Introduction, let $H_n \subset \mathcal{X}^n$ for $n \geq 1$, $y \in \mathbf{R}$ fixed, $F(y) = \exp\{-e^{-y}\}$, and define

$$N = N_n = |\{A \in H_n : T(A) > d^n(\log |H_n| + y)\}|. \tag{2.4}$$

Theorem 2.1. *Set* $q = \frac{1}{6}\frac{d-1}{d+1}$. *Then*

$$|g_N(z) - \exp\{-e^{-y}(1-z)\}| \leq \text{ const. } |H_n|^{-q} \tag{2.5}$$

uniformly in $z \in [0,1]$, $H_n \subset \mathcal{X}^n$ *and* y *satisfying*

$$|H_n|^{-q} \leq F(y) \leq 1 - |H_n|^{-q}. \tag{2.6}$$

Proof. The proof will be analogous to that of Lemma 8 in [16]. Considering the events $B(A) = \{T(A) > d^n(\log |H_n|+y)\}$, $A \in H_n$, we are going to apply Lemma 2.2 to them. Let N denote the number of patterns appearing after time $d^n(\log |H_n|+y)$, i.e., the number of $B(A)$'s that occur. Let us define a simple graph on vertex set H_n in the following way. Two vertices are to be connected if they overlap to a certain extent. Overlap between patterns can be measured by the so-called *leading number* or *correlation* (see e.g., [11]). According to that, for arbitrary patterns $A = (a_1a_2 \ldots a_n)$ and $A' = (a'_1a'_2 \ldots a'_n)$ the *leading number of* A *over* A' is defined as

$$A * A' = \sum_{i=1}^{n} \varepsilon_i d^i,$$

where $\varepsilon_i = 1$ if $(a_{n-i+1} \ldots a_n) \equiv (a'_1 \ldots a'_i)$, otherwise $\varepsilon_i = 0$. This quantity is extremely useful when dealing with waiting times, for instance, the expected number of letters drawn till the first pattern of H_n is observed can be expressed as a function of these leading numbers. Particularly, $E(T(A)) = A{*}A$. With $q' = \min\{\frac{1}{6}, \frac{1}{5}\frac{d-1}{d+1}\} > q$, $\delta = |H_n|^{-2q'}$ let A, $A' \in H_n$ span an edge if any of the following inequalities hold:

$$A * A > (a+\delta^2)d^n, \quad A' * A' > (1+\delta^2)d^n, \quad A * A' > 2\delta d^n, \quad A' * A > 2\delta d^n. \tag{2.7}$$

We refer to the following results of [16]. Suppose $|H_n|$ is large enough, greater than d^{1000}, say. With the notation $m = \lceil \log |H_n| \rceil$ we then have

$$|S_1' - e^{-y}| \le |H_n|^{-q}, \tag{2.8}$$

$$|S_r' - \frac{1}{r!}e^{-ry}| \le \frac{1}{r!}e^{-ry}|H_n|^{-2q}, \quad 2 \le r \le m, \tag{2.9}$$

$$S_2'' \le (\frac{1}{2!}e^{-2y}F(y) + 1)|H_n|^{-q}, \tag{2.10}$$

$$S_r'' \le \frac{1}{r!}e^{-ry}F(y)|H_n|^{-q}, \quad 2 < r \le m. \tag{2.11}$$

We are going to apply Lemma 2.1 to inequality (2.3), then substitute (2.8)–(2.11) into the estimation thus obtained. Let us start from the following simple decomposition.

$$\begin{aligned}
&|g_N(z) - \exp\{-e^{-y}(1-z)\}| \\
\le&\Big|g_N(z) - \sum_{r=0}^{m-1}(-1)^r S_r'(1-z)^r\Big| \\
&+\Big|\sum_{r=0}^{m-1}(-1)^r S_r'(1-z)^r - \sum_{r=0}^{m-1}(-1)^r\frac{1}{r!}e^{-ry}(1-z)^r\Big| \\
&+\Big|\sum_{r=0}^{m-1}(-1)^r\frac{1}{r!}e^{-ry}(1-z)^r - \exp\{-e^{-y}(1-z)\}\Big|.
\end{aligned} \tag{2.12}$$

By Lemma 2.1, the first term in the right hand side of (2.12) is bounded by

$$\begin{aligned}
&S_m'(1-z)^m + \sum_{r=2}^{m} S_r''(1-z)^r \\
\le&|S_m' - \frac{1}{m!}e^{-my}| + \sum_{r=2}^{m} S_r'' + \frac{1}{m!}e^{-my}.
\end{aligned} \tag{2.13}$$

The second term in the right hand side of (2.12) and the first term in (2.13) together are not greater than

$$\begin{aligned}
&\sum_{r=1}^{m}\left|S'_r - \frac{1}{r!}e^{-ry}\right| \\
\leq &|H_n|^{-q} + |H_n|^{-2q}\sum_{r=2}^{m}\frac{1}{r!}e^{-ry} \\
\leq &|H_n|^{-q} + |H_n|^{-2q}F(y)^{-1} \\
\leq &2|H_n|^{-q}.
\end{aligned} \tag{2.14}$$

Here we have applied (2.8), (2.9) and finally (2.6).

The second term in (2.13) can be estimated by (2.9) and (2.10):

$$\sum_{r=2}^{m} S''_r \leq |H_n|^{-q}F(y)\sum_{r=2}^{m}\frac{1}{r!}e^{-ry} + |H_n|^{-q} \leq 2|H_n|^{-q}. \tag{2.15}$$

Finally, the last term in (2.12) is bounded by $\frac{1}{m!}e^{-my}$, which, on the other hand, is less than $\frac{e^m}{m^m}(qm)^m < e^{-qm}$, thus from (2.12)–(2.15) it follows that

$$|g_N(z) - \exp\{-e^{-y}(1-z)\}| \leq 4|H_n|^{-q} + \frac{2}{m!}e^{-my} \leq 6|H_n|^{-q}$$

for y satisfying (2.4). □

Lemma 2.2 above provides an alternative proof to Theorem 2 of [17]. In [17] there is only a hint of a proof which would utilize the generalization of the graph-sieve due to Galambos.

Corollary 2.1. *The limit distribution as $|H_n| \to \infty$ of the random variable N defined in (2.4) is Poissonian with parameter e^{-y}.*

Remark 2.1. Applying Lemma 3.1 of [19] to N one obtains that

$$E(N) = |H_n|e^{-x} + \mathrm{O}\Big(\log|H_n|(1+x)\exp\{-\frac{d-1}{d}x\}\Big),$$

$$\mathrm{Var}(N) \leq E(N) + \mathrm{O}\Big(|H_n|\log|H_n|(1+x)\exp\{-\frac{d-1}{d}2x\}\Big)$$

where $x = \log|H_n| + y$, thus

$$E(N) = e^{-y} + O(|H_n|^{\frac{-(d-1)}{d}} \log^3|H_n|) = e^{-y} + O(|H_n|^{-5q}), \qquad (2.16)$$

and

$$\mathrm{Var}(N) \le e^{-y} + O(|H_n|^{-5q}) + O(|H_n|^{\frac{-(2d-3)}{d}} \log^4|H_n|) = e^{-y} + O(|H_n|^{-5q}) \quad (2.17)$$

which agrees well with Theorem 2.1 (in fact, (2.17) is valid with equality, too).

Remark 2.2. Theorem 2.1 provides an estimate, though crude, to the rate of convergence of individual probabilities to the corresponding term of the Poisson distribution. Namely, it follows that for every fixed nonnegative integer k

$$\left|P(N=k) - \frac{1}{k!}e^{-ky}e^{-e^{-y}}\right| \le \text{const. } 2^{k+1}|H_n|^{-q2^{-k}}, \qquad (2.18)$$

where the constant is the same as in (2.5). This assertion can be proved by induction on k. For $k = 0$, (2.18) follows from (2.5) by substituting $z = 0$. Suppose we have proved (2.18) for $k < r$. Let $z = |H_n|^{-q2^{-r}}$, then $|H_n|^{-q2^{-k}} z^k \le z^{r+1}$. Hence

$$\begin{aligned}\left|\sum_{k=0}^{r-1} P(N=k)z^k - \sum_{k=0}^{r-1} \frac{1}{k!}(e^{-y}z)^k e^{-e^{-y}}\right| &\le \text{const. } \sum_{k=0}^{r-1} 2^{k+1}|H_n|^{-q2^{-k}} z^k \\ &\le \text{const. } (2^{r+1}-2)z^{r+1}\end{aligned}$$

by the induction hypothesis, and

$$\left|\sum_{k=r+1}^{\infty} P(N=k)z^k - \sum_{k=r+1}^{\infty} \frac{1}{k!}(e^{-y}z)^k e^{-e^{-y}}\right| \le z^{r+1}.$$

Combining these with (2.5) we obtain that

$$\left|P(N=r)z^r - \frac{1}{r!}(e^{-y}z)^r e^{-e^{-y}}\right| \le \text{const. } (2^{r+1}-1)z^{r+1} + z^{2^r} \le \text{const. } (2z)^{r+1},$$

which is tantamount to (2.18) for $k = r$.

3. Convergence to Poisson processes

In this section we are going to deal with the joint distribution of maximum (and other) waiting times associated with disjoint subsets of patterns. Before formulating the problem let us start with a lemma interesting on its own.

Let $H_n \subset \mathcal{X}^n$ and $y \in \mathbf{R}$ satisfying (2.6). Let us consider the (random) subset of H_n containing all patterns not seen before $d^n(\log|H_n| + y)$ (its size was denoted by N in Corollary 2.1).

Lemma 3.1. *The probability that (2.7) does not hold for some* $A, A' \in H_n$ *not seen before* $d^n(\log|H_n| + y)$ *is less than* const. $|H_n|^{-q}$, *where the constant is universal.*

Proof. The probability in question is clearly less than

$$\sum_{\substack{A,A' \in H_n \\ A,A' \text{violate } (2.7)}} P(B(A) \cap B(A')) = S_2'' \leq (\frac{1}{2!}e^{-2y}F(y) + 1)|H_n|^{-q} < 2|H_n|^{-q}$$

by (2.10), if $|H_n|$ is large enough. □

Now we are in position to formulate and prove our main result. Let us introduce a few notations.

Suppose for every $n \geq 1$ we are given k pairwise disjoint subsets $H_1, H_2, \ldots, H_k$ of $\mathcal{X}^n$ (for the sake of readability dependence on n will not be indicated in the sequel). Let $\mathcal{B}$ denote the family of Borel subsets of $\mathbf{R}_+$. Define

$$\xi_i(B) = |\{A \in H_i : T(A) \in B\}|, \ B \in \mathcal{B}, \ 1 \leq i \leq k.$$

This is the point process (random measure) defined by the waiting times for patterns of H_i. The corresponding arrival times are

$$W_i(\ell) = \min\{t > 0 : \xi_i((t, +\infty)) < \ell\}, \ 1 \leq \ell \leq |H_i|.$$

Thus $W_i(1)$ is the time of covering H_i (the time when the last pattern arrives), $W_i(2)$ is the time of arrival of the last but one pattern, etc.; obviously, $W_i(1) > W_i(2) > \cdots$. In earlier papers the linear normalization

$$Y_i(\ell) = d^{-n}W_i(\ell) - \log|H_i|$$

was used; this time we would rather turn to another transform of $W_i(\ell)$ so as to obtain nicer limits. Let

$$f_i(t) = |H_i| \exp\{-d^{-n}t\}, \quad t \geq 0,$$

$$Z_i(\ell) = f_i(W_i(\ell)) = \exp\{-Y_i(\ell)\},$$

and the corresponding random point measure

$$\zeta_i(B) = |\{\ell \geq 1 : Z_i(\ell) \in B\}|, \quad B \in \mathcal{B}.$$

Clearly, $\zeta_i = \xi_i \circ \mathrm{Inv}\ f_i$.

Theorem 3.1. *Suppose $n \to \infty$ and $H_1, H_2, \ldots, H_k$ vary in such a way that*

$$\lim_{n\to\infty} |H_i| = \infty, \quad 1 \leq i \leq k,$$

and

$$|H_i| = o(|H_j| \log |H_j|) \quad 1 \leq i,\, j \leq k. \tag{3.1}$$

Then the point processes $(\zeta_1, \zeta_2, \ldots, \zeta_k)$ converge weakly to i.i.d. homogeneous Poisson processes with unit intensity.

Remark 3.1. In [19] independence of cover times associated with different pattern lengths was proved. The proof was based on the observation that those cover times were getting more and more separated. This cannot occur when condition (3.1) is fulfilled, hence for the proof of Theorem 3.1 new ideas are needed. When the quantities $\log |H_i|$, $1 \leq i \leq k$, differ in the order of magnitude, the method of [19] can be applied to derive the same conclusion as in Theorem 3.1.

Proof of Theorem 3.1. Though Theorem 3.1 does not say anything about the rate of convergence, in the proof we also try to give estimations at every step if possible, for later references. According to Proposition 3.22 of [22] it is sufficient to check that (3.2) and (3.3) below hold for every $B_1, B_2, \ldots, B_k \in \mathcal{T}$, where $\mathcal{T}$ is a basis of relatively compact open sets such that $\mathcal{T}$ is closed under finite unions and intersections and the limit process (as random point measure) attaches 0 to ∂B with probability 1 for every $B \in \mathcal{T}$. Our choice for $\mathcal{T}$ is the ring of finite unions of bounded intervals which are of positive distance from 0. Thus, the two assertions to be checked are

$$\lim_{n\to\infty} P(\zeta_i(B_i) = 0,\ 1 \le i \le k) = \exp\{-\sum_{i=1}^{k} \lambda(B_i)\}, \tag{3.2}$$

$$\lim_{n\to\infty} E(\zeta_i(B_i)) = \lambda(B_i), \quad 1 \le i \le k, \tag{3.3}$$

where λ stands for the Lebesgue measure on $\mathbf{R}_+$.

Thus, let $B_i \in \mathcal{T}$, $a_i = \inf\ B_i$, $b_i = \sup\ B_i$, $1 \le i \le k$, these are the only quantities not depending on n. Introduce $C_i = \text{Inv}\ f_i(B_i) \in \mathcal{T}$. Rewriting (3.2) and (3.3) we have to prove that

$$\lim_{n\to\infty} P(\xi_i(C_i) = 0,\ 1 \le i \le k) = \exp\{-\sum_{i=1}^{k} \lambda(B_i)\}, \tag{3.4}$$

and

$$\lim_{n\to\infty} E(\xi_i(C_i)) = \lambda(B_i), \quad 1 \le i \le k. \tag{3.5}$$

Let $H_0 = \cup_{i=1}^{k} H_i$ and define $z = \log|H_0| - \log\log|H_0|^q + R$ with the remainder term R satisfying

$$\lim_{n\to\infty} R = \infty, \quad R = o(\log\log|H_0|). \tag{3.6}$$

Let $\mathcal{F} = \sigma\{X_t : t \le d^n z\}$ and $\mathcal{H}_i = \{A \in H_i : T(A) > d^n z\}$, $N_i = |\mathcal{H}_i|$, $0 \le i \le k$, these are $\mathcal{F}$-measurable. We first deal with the conditional probability $P(\xi_i(C_i) = 0,\ 1 \le i \le k|\mathcal{F})$. It can be approximated by the (conditional) probability that the first occurrence after time $d^n z + n$ of every $A \in \mathcal{H}_i$ does not fall in C_i. From (3.1) and (3.6) it follows that

$$\inf\ C_i = \text{Inv}\ f_i(b_i) = d^n(\log|H_i| - \log\ b_i) > d^n z + n$$

if n is large enough. Thus,

$$P(\xi_i(C_i) = 0,\ 1 \le i \le k|\mathcal{F}) \ge \varphi(\mathcal{H}_1, \mathcal{H}_2, \ldots, \mathcal{H}_k), \tag{3.7}$$

where for $M_1 \subset H_1, \ldots, M_k \subset H_k$

$$\varphi(M_1, \ldots, M_k) = P(T(A) \notin C_i - d^n z - n,\ A \in M_i,\ 1 \le i \le k).$$

On the other hand, the error of (3.6) is clearly less than the conditional (given $\mathcal{F}$) probability that any element of H_0 is first observed in the interval $(d^n z,\ d^n z+n)$, and this can be estimated by the expected number of elements of H_0 appearing in that interval. Thus (2.16) can be applied. Since $F(z - \log|H_0|) \ge |H_0|^{-q}$, (2.6) is satisfied, hence we have

$$\begin{aligned}
&E(P(\xi_i(C_i) = 0,\ 1 \le i \le k|\mathcal{F}) - \varphi(\mathcal{H}_1, \mathcal{H}_2, \ldots, \mathcal{H}_k)) \\
\le&\exp\{-(z - \log|H_0|)\} - \exp\{-(z - \log|H_0|) - nd^{-n}\} + O(|H_0|^{-q}) \\
\le&(1 - \exp\{-nd^{-n}\}) \log|H_0|^q + O(|H_0|^{-q}) \\
=&O(|H_0|^{-q}).
\end{aligned} \tag{3.8}$$

It was first proved in [14] that waiting times for different patterns can be quite well approximated by independent, exponentially distributed random variables with common expectation d^n provided that the overlap between the patterns is negligible. More precisely, suppose that any two patterns of a collection $A_1, \ldots, A_r \in H_0$ satisfy (2.7) with $\delta = |H_0|^{-2q}$. Let $b = 2nd^{-n}$, suppose $rb < \frac{1}{5}$. Then by Theorem 1 of [14] we have

$$\begin{aligned}
\exp\{-r^2 b - (1 + rb)\sum_{j=1}^{r} t_j\} &\le P(T(A_j) > (A_j * A_j)t_j,\ 1 \le j \le r) \\
&\le \exp\{r^2 b - (1 - 2\delta)\sum_{j=1}^{r} t_j\}.
\end{aligned}$$

We are going to apply this to $\mathcal{H}_0$ (conditionally w.r.t. N_0), on the event where $\mathcal{H}_0$ satisfies (2.7) (this event has probability tending to 1 by Lemma 3.1). Thus, in our case, N_0 will play the role of r. By (2.16) and (2.17) we clearly have $E(N_0) \to \infty$ and $\mathrm{Var}(N_0) = \mathrm{O}(E(N_0))$, hence $N_0 \sim E(N_0) = \mathrm{o}(\log|H_0|)$ and $N_0^2 b = \mathrm{o}(|H_0|^{-2q})$ in probability. Furthermore, we want to apply the above estimation only with t_j belonging to the boundary of one of the sets

$$D_i = d^{-n}C_i - z - nd^{-n} = -\log B_i + \log|H_i| - z - nd^{-n}, \quad 1 \le i \le k.$$

That is, we shall always have

$$t_j \le \mathrm{O}(1) + \log|H_i| - z \sim \log\log|H_0|$$

for every $j \le r$ and this holds uniformly. Hence $\sum_{j=1}^{N_0} t_j = \mathrm{o}(\log|H_0|)$ in probability.

If $A_1, \ldots, A_r$ satisfy (2.7), and both r and $\sum_{j=1}^{r} t_j$ are $\mathrm{o}(\log|H_0|)$, then

$$\begin{aligned} P(T(A_j) > d^n t_j,\ 1 \le j \le r) &\le \exp\{r^2 b - (1-2\delta)\sum_{j=1}^{r} t_j\} \\ &= \exp\{-\sum_{j=1}^{r} t_j\}(1 + \mathrm{O}(r^2 b + \delta \sum_{j=1}^{r} t_j)) \\ &= \exp\{-\sum_{j=1}^{r} t_j\} + \mathrm{O}(|H_0|^{-2q}), \end{aligned}$$

and

$$\begin{aligned} P(T(A_j) > d^n t_j,\ 1 \le j \le r) &\ge \exp\{-r^2 b - (1+rb)(1+\delta^2)\sum_{j=1}^{r} t_j\} \\ &= \exp\{-\sum_{j=1}^{r} t_j\}(1 + \mathrm{O}(r^2 b + (rb + \delta^2)\sum_{j=1}^{r} t_j)) \\ &= \exp\{-\sum_{j=1}^{r} t_j\} + \mathrm{O}(|H_0|^{-2q}), \end{aligned}$$

that is,

$$\Big|P(T(A_j) > d^n t_j,\ 1 \le j \le r) - \exp\{-\sum_{j=1}^{r} t_j\}\Big| \le \mathrm{const.}\, |H_0|^{-2q}. \tag{3.9}$$

Let us apply this estimation to compute $\varphi(\mathcal{H}_1, \mathcal{H}_2, \ldots, \mathcal{H}_k)$ on the event where $\mathcal{H}_0$ satisfies (2.7). Suppose every B_i is built of at most $r-1$ disjoint intervals, then the same is true for $D_i = d^{-n}C_i - z - nd^{-n}$, and $\mathbf{R}_+ \setminus D_i$ consists of at most r intervals, one of which is infinite. Thus we have to compute the probability that a vector of N_0 waiting times falls into a product set D composed from finite unions of disjoint (finite or infinite) intervals. It is easy to see that D is the union of at most r^{N_0} N_0-dimensional rectangles, finite or infinite. The probability of each rectangle can be expressed as an alternating sum of terms associated with each of its vertices. Each term is of the form $P(T(A_j) > d^n t_j,\ j \in M)$, where $M \subset \mathcal{H}_0$ and for some indices j the sign $>$ may be replaced with $\geq$. There are at most 2^{N_0} of those terms. In this way the probability of D is obtained as a sum of at most $(2r)^{N_0}$ terms, each of them fulfills the conditions of (3.9). Consequently, if we approximate the probability of D by replacing the waiting times with i.i.d. exponentially distributed random variables, the error we made is $\mathrm{O}((2r)^{N_0}|H_0|^{-2q}) = \mathrm{O}(|H_0|^{-q})$ (here the constant involved in the O notation depends on the sets B_i through the bound r).

Thus, with probability tending to 1,

$$\varphi(\mathcal{H}_1, \mathcal{H}_2, \ldots, \mathcal{H}_k) = \prod_{i=1}^{k}(1-p_i)^{N_i} + \mathrm{O}(|H_0|^{-q}),$$

where

$$p_i = \int_{D_i} e^{-t}dt = \lambda(\exp\{-D_i\}) = \exp\{z + nd^{-n} - \log|H_i|\}\lambda(B_i).$$

By (2.16) and (2.17), $N_i \sim E(N_i) \sim |H_i|e^{-z}$, hence $N_i p_i \to \lambda(B_i)$ in probability for every i, $1 \leq i \leq k$. Consequently,

$$\varphi(\mathcal{H}_1, \mathcal{H}_2, \ldots, \mathcal{H}_k) \to \exp\{-\sum_{i=1}^{k}\lambda(B_i)\}$$

in probability, which proves (3.4) in virtue of (3.8).

Let us turn to the proof of (3.5). Let the interval (t_i, u_i) be a component of B_i. Denote Inv $f_i(t_i)$ and Inv $f_i(u_i)$ by t_i' and u_i', resp., then (u_i', t_i') is a component of

C_i. From (2.16) it follows that

$$\begin{aligned} E(\xi_i((u_i', t_i'))) &= E(\xi_i((u_i', \infty))) - E(\xi_i([t_i', \infty))) \\ &= \exp\{-d^{-n}u_i' + \log|H_i|\} - \exp\{-d^{-n}t_i' + \log|H_i|\} + O(|H_i|^{-q}) \\ &= f_i(u_i') - f_i(t_i') + O(|H_i|^{-q}) \\ &= t_i - u_i + O(|H_i|^{-q}), \end{aligned}$$

hence (3.5). The proof is complete. □

Remark 3.2. In the proof we did not use the asymptotic Poissonity of N_i, $0 \le i \le k$, only (2.16) and (2.17) which come from [19]. Therefore we obtained another (though fairly indirect) proof of Theorem 2 of [17].

4. Joint distribution of maximum waiting times

From Theorem 3.1 it is easy to read off the limiting joint distribution of maximum waiting times associated with pairwise disjoint sets of patterns.

Let $H_1, H_2, \ldots, H_k$ be pairwise disjoint subsets of $\mathcal{X}^n$ for every $n \ge 1$, and $Y_i = d^{-n}W(H_i) - \log|H_i|$, $1 \le i \le k$, the corresponding maximum waiting times normalized linearly.

Theorem 4.1. *Suppose the conditions of Theorem 3.1 are fulfilled. Then*

$$\lim_{n\to\infty} P(Y_1 \le y_1, \ldots, Y_k \le y_k) = \prod_{i=1}^{k} F(y_i)$$

for arbitrary real numbers $y_1, y_2, \ldots y_k$.

Remark 4.1. Looking through the proof of (3.4) one can derive bounds for the rate of convergence in Theorem 4.1. Indeed, $\{Y_i \le y_i\} = \{\zeta_i((0, e^{-y_i}]) = 0\}$, thus $r = 1$ and we obtain that

$$\varphi(\mathcal{H}_1, \mathcal{H}_2, \ldots, \mathcal{H}_k) - \prod_{i=1}^{k} F(y_i) = \exp\{\sum_{i=1}^{k}(N_i \log(1 - p_i) + \exp\{-y_i\})\} + R',$$

where $E(R') \leq \text{const.}|H_0|^{-q}$. Now,

$$\begin{aligned} &N_i \log(1 - p_i) + \exp\{-y_i\} \\ =&N_i(p_i + \log(1 - p_i)) + (E(N_i) - N_i)p_i + (\exp\{-y_i\} - E(N_i)p_i) \\ =&I_1 + I_2 + I_3. \end{aligned}$$

Here $I_1 = O(p_i)$, $I_2 = O(p_i \text{Var}^{\frac{1}{2}}(N_i)) = O(\sqrt{p_i})$ and $I_3 = O(|H_0|^{-q})$. Since

$$p_i = \exp\{-y_i + z + nd^{-n} - \log|H_i|\} = (\log|H_0|)^{-1+o(1)},$$

the rate we could get is extremely slow compared with that stated in Proposition 1.1 and is very far from being sharp. This is why it has not been included in Theorem 4.1.

Next we turn to the general case of not necessarily disjoint subsets $H_1, \ldots, H_k \subset \mathcal{X}^n$. Define $H_0 = \cup_{i=1}^k H_i$ and for $I \subset \{1, 2, \ldots, k\}$, $I \neq \emptyset$, let $H_I = (\cap_{i \in I} H_i) \cap (\cap_{i \notin I} H_i^c)$. This time we apply the same transformation to each coordinate variable: let $f_0(t) = |H_0| \exp\{-d^{-n}t\}$ and $Z_i = f_0(W(H_i))$, $1 \leq i \leq k$.

Theorem 4.2. *Suppose $n \to \infty$ and $H_1, H_2, \ldots, H_k$ vary in such a way that*

$$\lim_{n \to \infty} |H_i| = \infty, \quad 1 \leq i \leq k,$$

and

$$\lim_{n \to \infty} \frac{|H_I|}{|H_0|} = \lambda_I$$

exists for every $I \neq \emptyset$. Then

$$\lim_{n \to \infty} P(Z_1 > y_1, \ldots, Z_k > y_k) = \exp\{-\sum_{I \neq \emptyset} \lambda_I \max_{i \in I} y_i\}$$

for arbitrary real numbers $y_1, y_2, \ldots, y_k$, that is, the limit distribution of the random vector $(Z_1, Z_2, \ldots, Z_k)$ is multivariate exponential of Marshall-Olkin type.

Proof. (For derivation and properties of the Marshall-Olkin distribution see [9], p.270.) Suppose first that every λ_I is positive. Then

$$\begin{aligned}
&P(Z_i > y_i,\ 1 \le i \le k)\\
=&P(d^{-n}W(H_i) - \log|H_0| < -\log y_i,\ 1 \le i \le k)\\
=&P(d^{-n}W(H_I) - \log|H_0| < -\log y_i,\ I \ne \emptyset,\ i \in I)\\
=&P\Big(d^{-n}W(H_I) - \log|H_0| < -\log\big(\max_{i\in I} y_i\big),\ I \ne \emptyset\Big)\\
=&P\left(d^{-n}W(H_I) - \log|H_I| < -\log\left(\frac{|H_I|}{|H_0|}\right) - \log\big(\max_{i\in I}\ y_i\big),\ I \ne \emptyset\right).
\end{aligned}$$

Since the subsets H_I are pairwise disjoint and $\lim_{n\to\infty} \frac{|H_I|}{|H_J|}$ is finite and positive for every $I \ne J$, we can apply Theorem 4.1 to obtain the limit distribution of the variables $Y_I = d^{-n}W(H_I) - \log|H_I|$. It follows that

$$\lim_{n\to\infty} P(Y_I < -\log\left(\frac{|H_I|}{|H_0|}\right) - \log\big(\max_{i\in I} y_i\big),\ I \ne \emptyset)$$

$$= \prod_{I\ne\emptyset} F\left(-\log\left(\frac{|H_I|}{|H_0|}\right) - \log\big(\max_{i\in I} y_i\big)\right)$$

which is equal to $\exp\{-\sum_{I\ne\emptyset} \lambda_I \max_{i\in I} y_i\}$, as stated.

If some of the limits λ_I happen to be 0, we can majorize $P(Z_i > y_i,\ 1 \le i \le k)$ by decreasing our sets H_i. Thus, let $\mathcal{I} = \{I : \lambda_I = 0\}$ and $H_i' = H_i \backslash (\cup_{I\in\mathcal{I}} H_I)$, $1 \le i \le k$. Clearly, $\frac{|H_0'|}{|H_0|} \to 1$, $H_I' = \emptyset$ for $I \in \mathcal{I}$, and $W(H_i') \le W(H_i)$, thus

$$\begin{aligned}
&P(Z_i > y_i, 1 \le i \le k)\\
\le&P(|H_0|\exp\{-d^{-n}W(H_i')\} > y_i, 1 \le i \le k)\\
=&P\left(d^{-n}W(H_I) - \log|H_I| < -\log\left(\frac{|H_I|}{|H_0|}\right) - \log\big(\max_{i\in I}\ y_i\big), I \notin \mathcal{I}\right)\\
\to&\exp\{-\sum_{I\notin\mathcal{I}} \lambda_I \max_{i\in I}\ y_i\}\\
=&\exp\{-\sum_{I\ne\emptyset} \lambda_I \max_{i\in I}\ y_i\}.
\end{aligned}$$

On the other hand, enlarging of the sets H_i leads to lower estimations if H_0 does not change. Therefore define $H_i' = H_i \cup (\cup_{I\in\mathcal{I}} H_I)$. In this way $H_0' = H_0$ and for every $I \neq \emptyset$, $\lim_{n\to\infty} \frac{|H_I'|}{|H_0|} = \lambda_I$, but $H_I' = \emptyset$ for $I \in \mathcal{I}$, $I \neq \{1,2,\ldots,k\}$. If $\{1,2,\ldots,k\} \notin \mathcal{I}$, we can apply Theorem 4.1 for the new sets and we obtain that

$$\begin{aligned}
&P(Z_i > y_i,\ 1 \le i \le k)\\
\ge&P(Z_i' > y_i,\ 1 \le i \le k)\\
=&P\left(d^{-n}W(H_i') - \log|H_I'| < -\log\left(\frac{|H_I'|}{|H_0|}\right) - \log\left(\max_{i\in I} y_i\right),\ I \notin \mathcal{I}\right)\\
\to&\exp\{-\sum_{I\notin\mathcal{I}} \lambda_I \max_{i\in I} y_i\}\\
=&\exp\{-\sum_{I\neq\emptyset} \lambda_I \max_{i\in I} y_i\}.
\end{aligned}$$

If $\{1,2,\ldots,k\}$ would belong to $\mathcal{I}$, that is, $\lim_{n\to\infty} \frac{|\cap_{i=1}^k H_i|}{|\cup_{i=1}^k H_i|} = 0$, we have to modify the above construction by fixing a $J \in \mathcal{I}$, then picking $H_\varepsilon \subset H_J$, a small subset for which $\lim_{n\to\infty} \frac{|H_\varepsilon|}{|H_0|} = \varepsilon > 0$, and including H_ε in the union defining H_i'. In this way, the limit $\lim_{n\to\infty} \frac{|H_I'|}{|H_0|} = \lambda_I'$ exists for every $I \neq \emptyset$, and

$$\lambda_I' = \begin{cases} \lambda_I + \varepsilon & \text{when } I = \{1,2,\ldots,k\},\\ \lambda_I - \varepsilon & \text{when } I = J,\\ \lambda_I & \text{otherwise;} \end{cases}$$

moreover, $H_I' = \emptyset$ still holds for $I \in \mathcal{I}$, $I \neq \{1,2,\ldots,k\}$. In the same way as above, one gets the lower bound

$$\liminf_{n\to\infty} P(Z_i > y_i,\ 1 \le i \le k) \ge \exp\{-\sum_{I\neq\emptyset} \lambda_I' \max_{i\in I} y_i\}.$$

Since $\varepsilon > 0$ can be arbitrarily small, the proof is complete. □

Remark 4.2. Similarly to the proof of Theorem 4.2, it can be derived from Theorem 3.1 that in the case of overlapping subsets $H_1, H_2, \ldots, H_k$ the k-dimensional point process $(\xi_1 \circ \mathrm{Inv}\ f_0, \ldots, \xi_k \circ \mathrm{Inv}\ f_0)$ converges weakly to a general k-variate Marshall-

Olkin process $(\eta_1, \dots, \eta_k)$ defined in the following way: for $I \notin \mathcal{I}$ let η_I be independent homogeneous Poisson processes with intensity λ_I, and $\eta_i = \sum_{I: i \in I} \eta_I$, $1 \le i \le k$ (see [5]).

Acknowledgement

This work was supported by the Hungarian National Foundation for Scientific Research, Grant No 1905.

References

[1] ALDOUS, D. J. (1983) On the time taken by random walks on finite groups to visit every state. *Z. Wahrsch. verw. Geb.* **62**, 361-393.

[2] ALDOUS, D. J. (1989) An introduction of covering problems for random walks on graphs. *J. Theor. Probab.* **2**, 87-89.

[3] ALDOUS, D. J. (1989) Lower bounds for covering times for reversible Markov chains and random walks on graphs. *J. Theor. Probab.* **2**, 91-100.

[4] ALDOUS, D. J. (1991) Threshold limits for cover times. *J. Theor. Probab.* **4**, 197-211.

[5] COX, D. R. AND LEWIS, P. A. W. (1972) Multivariate point processes. In: *Proc. Sixth Berkeley Symp. Math. Statist. Prob., Vol. 3.* Univ. California Press, Los Angeles, 401-408.

[6] DEVROYE, L. AND SHIBI, A. (1990) Random walks on highly symmetric graphs. *J. Theor. Probab.* **3**, 497-514.

[7] ERDŐS, P. AND CHEN, R. W. (1988) Random walks on $\mathbf{Z}_2^n$. *J. Multivar. Anal.* **25**, 111-118.

[8] GALAMBOS, J. AND KÁTAI, I., eds. (1992) *Probability Theory and Applications — Essays to the Memory of József Mogyoródi.* Kluwer, Dordrecht.

[9] JOHNSON, N. L. AND KOTZ, S. (1972) *Distributions in Statistics: Continuous Multivariate Distributions.* Wiley, New York.

[10] KAHN, J. D., LINIAL, N., NISAN, N. AND SAKS, M. E. (1989) On the cover time of random walks on graphs. *J. Theor. Probab.* **2**, 121-128.

[11] LI, S. R. (1980) A martingale approach to the study of occurrence of sequence patterns in repeated experiments. *Ann. Probab.* **8**, 1171-1176.

[12] MATTHEWS, P. (1988) Covering problems for Markov chains. *Ann. Probab.* **16**, 1215-1228.

[13] MATTHEWS, P. (1989) Some sample path properties of a random walk on the cube. *J. Theor. Probab.* **2**, 129-146.

[14] MÓRI, T. F. (1985) Large deviation results for waiting times in repeated experiments. *Acta Math. Hungar.* **45**, 213-221.

[15] MÓRI, T. F. (1987) On the expectation of the maximum waiting time. *Annales Univ. Sci. Budapest., Sect. Comput.* **7**, 111-115.

[16] MÓRI, T. F. (1990) More on the waiting time till each of some given patterns occurs as a run. *Can. J. Math.* **42**, 915-932.

[17] MÓRI, T. F. (1991) On the waiting time till each of some given patterns occurs as a run. *Probab. Th. Rel. Fields* **87**, 313-323.

[18] MÓRI, T. F. (1992) Random walks on de Bruijn graphs (in Russian) *Teor. Veroyatn. Primen.* **37**, 194-197.

[19] MÓRI, T. F. (1992) Maximum waiting times are asymptotically independent. *Combinatorics, Probability & Computing,* to appear.

[20] PALACIOS, J. L. (1992) Expected cover times of random walks on symmetric graphs. *J. Theor. Probab.* **5**, 597-600.

[21] RÉNYI, A. (1961) A general method for proving theorems in probability theory and some of its applications. In: *Selected Papers of A. Rényi, Vol. 2,* Akadémiai Kiadó, Budapest, 1976, 393-398.

[22] RESNICK, S. I. (1987) *Extreme Values, Regular Variation, and Point Processes.* Springer, New York.

[23] ZUCKERMAN, D. (1989) Covering times of random walks on bounded degree trees and other graphs. *J. Theor. Probab.* **2**, 147-157.

Index

A

B

H

I

N

O

P

Other *Mathematics and Its Applications* titles of interest:

A.V. Ivanov and N.N. Leonenko: *Statistical Analysis of Random Fields.* 1989, 256 pp. ISBN 90-277-2800-3

V. Paulauskas and A. Rackauskas: *Approximation Theory in the Central Limit Theorem. Exact Results in Banach Spaces.* 1989, 176 pp. ISBN 90-277-2825-9

R.Sh. Liptser and A.N. Shiryayev: *Theory of Martingales.* 1989, 808 pp. ISBN 0-7923-0395-4

S.M. Ermakov, V.V. Nekrutkin and A.S. Sipin: *Random Processes for Classical Equations of Mathematical Physics.* 1989, 304 pp. ISBN 0-7923-0036-X

G. Constantin and I. Istratescu: *Elements of Probabilistic Analysis and Applications.* 1989, 488 pp. ISBN 90-277-2838-0

S. Albeverio, Ph. Blanchard and D. Testard (eds.): *Stochastics, Algebra and Analysis in Classical and Quantum Dynamics.* 1990, 264 pp. ISBN 0-7923-0637-6

Ya.I. Belopolskaya and Yu.L. Dalecky: *Stochastic Equations and Differential Geometry.* 1990, 288 pp. ISBN 90-277-2807-0

A.V. Gheorghe: *Decision Processes in Dynamic Probabilistic Systems.* 1990, 372 pp. ISBN 0-7923-0544-2

V.L. Girko: *Theory of Random Determinants.* 1990, 702 pp. ISBN 0-7923-0233-8

S. Albeverio, PH. Blanchard and L. Streit: *Stochastic Processes and their Applications in Mathematics and Physics.* 1990, 416 pp. ISBN 0-9023-0894-8

B.L. Rozovskii: *Stochastic Evolution Systems. Linear Theory and Applications to Non-linear Filtering.* 1990, 330 pp. ISBN 0-7923-0037-8

A.D. Wentzell: *Limit Theorems on Large Deviations for Markov Stochastic Process.* 1990, 192 pp. ISBN 0-7923-0143-9

K. Sobczyk: *Stochastic Differential Equations. Applications in Physics, Engineering and Mechanics.* 1991, 410 pp. ISBN 0-7923-0339-3

G. Dallaglio, S. Kotz and G. Salinetti: *Distributions with Given Marginals.* 1991, 300 pp. ISBN 0-7923-1156-6

A.V. Skorohod: *Random Processes with Independent Increments.* 1991, 280 pp. ISBN 0-7923-0340-7

L. Saulis and V.A. Statulevicius: *Limit Theorems for Large Deviations.* 1991, 232 pp. ISBN 0-7923-1475-1

A.N. Shiryaev (ed.): *Selected Works of A.N. Kolmogorov, Vol. 2: Probability Theory and Mathematical Statistics.* 1992, 598 pp. ISBN 90-277-2795-X

Yu.I. Neimark and P.S. Landa: *Stochastic and Chaotic Oscillations.* 1992, 502 pp. ISBN 0-7923-1530-8

Other *Mathematics and Its Applications* titles of interest:

Y. Sakamoto: *Categorical Data Analysis by AIC.* 1992, 260 pp.
ISBN 0-7923-1429-8

Lin Zhengyan and Lu Zhuarong: *Strong Limit Theorems.* 1992, 200 pp.
ISBN 0-7923-1798-0

J. Galambos and I. Katai (eds.): *Probability Theory and Applications.* 1992, 350 pp. ISBN 0-7923-1922-2

N. Bellomo, Z. Brzezniak and L.M. de Socio: *Nonlinear Stochastic Evolution Problems in Applied Sciences.* 1992, 220 pp. ISBN 0-7923-2042-5

A.K. Gupta and T. Varga: *Elliptically Contoured Models in Statistics.* 1993, 328 pp. ISBN 0-7923-2115-4

B.E. Brodsky and B.S. Darkhovsky: *Nonparametric Methods in Change-Point Problems.* 1993, 210 pp. ISBN 0-7923-2122-7

V.G. Voinov and M.S. Nikulin: *Unbiased Estimators and Their Applications. Volume 1: Univariate Case.* 1993, 522 pp. ISBN 0-7923-2382-3

V.S. Koroljuk and Yu.V. Borovskich: *Theory of* U-*Statistics.* 1993, 552 pp.
ISBN 0-7923-2608-3

A.P. Godbole and S.G. Papastavridis (eds.): *Runs and Patterns in Probability: Selected Papers.* 1994, 358 pp. ISBN 0-7923-2834-5

MIX
Papier aus verantwortungsvollen Quellen
Paper from responsible sources
FSC® C105338
FSC
www.fsc.org

If you have any concerns about our products,
you can contact us on
ProductSafety@springernature.com

In case Publisher is established outside the EU,
the EU authorized representative is:
Springer Nature Customer Service Center GmbH
Europaplatz 3, 69115 Heidelberg, Germany

Printed by Libri Plureos GmbH
in Hamburg, Germany